Computer Graphics: Systems and Applications

H. Hagen H. Müller G. M. Nielson (Eds.)

Focus on Scientific Visualization

With 231 Figures

Springer-Verlag
Berlin Heidelberg New York
London Paris Tokyo
Hong Kong Barcelona
Budapest

Editors

Hans Hagen
Fachbereich Informatik
Universität Kaiserslautern
Postfach 3049
W-6750 Kaiserslautern, FRG

Heinrich Müller
Fachbereich Informatik
Lehrstuhl VII
Universität Dortmund
Postfach 500500
W-4600 Dortmund 50, FRG

Gregory M. Nielson
Computer Science Department
Arizona State University
Tempe, AZ 85287, USA

ISBN 3-540-54940-4 Springer-Verlag Berlin Heidelberg New York
ISBN 0-387-54940-4 Springer-Verlag New York Berlin Heidelberg

Library of Congress Cataloging-in-Publication Data.
Focus on scientific visualization/H. Hagen, H. Müller, G. M. Nielson.
p. cm. – (Computer graphics, systems and applications) Includes bibliographical references and index.
ISBN 0-387-54940-4 (U. S.)
1. Science-Methodology. 2. Visualization-Technique. I. Hagen, Hans, 1953- . II. Müller, Heinrich. III. Nielson, Gregory. IV. Series: Symbolic computation. Computer graphics-Systems and applications. Q175.F63 1993 502.85'66–dc20 92-27324 CIP

Printed in Hong Kong

Cover Design: H. Lopka, Ilvesheim
33/3140 - 5 4 3 2 1 0 - Printed on acid-free paper

Preface

One of the important themes being nurtured under the aegis of *Scientific Visualization* is the utilization of the broad bandwidth of the human sensory system in steering and interpreting complex processes and simulations involving voluminous data sets across diverse scientific disciplines. Since vision dominates our sensory input, strong efforts have been made to bring the power of mathematical abstraction and modeling to our eyes through the mediation of computer graphics. This interplay between various application areas and their specific problem solving visualization techniques has been emphasized in this book.

The intent of the book is to present the state of the art in visualization techniques both as an overview for the inquiring scientist, and as a solid platform from which developers may extend existing techniques or devise new ones to meet the specific needs of their problems. The book contains some chapters dedicated to surveys and tutorials of specific topics as well as some original work not previously published, but in all cases the emphasis has been on presenting the extensive detail necessary for others to reconstruct the techniques and algorithms. Far from detracting from the value of the book, we feel that a certain redundancy between some chapters will usefully serve the reader in giving the measure of the significance of some techniques and in promoting the completeness of each chapter as a standalone document.

One of our secondary goals in crafting this volume has been to provide a vehicle for the teaching of state of the art techniques in scientific visualization. The "red threads" weaving within and between chapters may help to organize and prepare advanced courses or seminars on special interest areas of visualization. The exhaustive bibliographies appended to many of the chapters point out where to get further information to complete the course material. An index has also been provided to conveniently locate related information. On the other hand, the book is not intended to suggest or limit the extent of the entire curriculum of scientific visualization. Such an encylopedic task would certainly be very difficult to achieve given the rapid development of this field and the inherent heterogeneity of the application areas and fundamental techniques.

The first three chapters of the book cover application areas that have strong needs for scientific visualization techniques due to the enormity of the data sets and simulations that comprise typical problems. Specifically, fluid flow visualization, volume visualization in medicine, and environmental protection problems are discussed. Each chapter explains how the chosen visualization techniques meet the requirements of the specific application.

The second set of chapters has been divided into four sections that explain the fundamentals of scientific visualization. A section on *Data Structuring and Data Administration* details data structures used in visualization algorithms and the implications of visualization for a scientific data base system. *Data Modeling* is a section of four chapters covering volume synthesis principles, surface interpolation, volumetric and surface on surface generation, and curve and surface interrogation. The largest section consists of six chapters on *Rendering.* The specific topics include sorting for polyhedron compositing, mixing volume and surface rendering, the volume priority z-buffer, Fourier volume rendering, a shading algorithm for radiosity rendering, and X-ray tracing. The final section is devoted to a chapter on auditory representation of scientific data.

Initial planning for this work was made during the first Dagstuhl seminar on scientific visualization which took place during August 26-30 of 1991. The IBFI Schloss Dagstuhl was founded in 1990 and is located in southwest Germany between Saarbrücken and Trier. It offers the opportunity of one week meetings bringing together the most significant worldwide researchers on topics of importance in computer science. The Dagstuhl seminar on scientific visualization was attended by some forty participants from the United States, Germany, the Netherlands, France, Russia, and Austria. The participants were invited to contribute to this volume and a list of these authors and their addresses can be found immediately after the table of contents. The editors wish to express special thanks to them for their impressive efforts.

The contributions were collected by electronic mail and FTP by the Institute of Compter Science at the University of Freiburg, Germany. The disparate documents were converted into the uniform style of this volume using the LaTeX formatting system, an imposing task but one that we feel has been well rewarded by the result. Particular thanks go to Jörg Winckler for his steady support in all questions concerning LaTeX and for developing the present style of this book based on previous work by Achim Stößer. Detlef Ruprecht, Michael Stark, Keith Voegele and Hansjörg Scherberger also provided instrumental support of the editing process. Last but not least we thank the Springer-Verlag for making this book possible, and for the hints improving the form of the manuscript.

We hope that the work of the contributing authors and our support staff have combined into a book that will serve the needs of our potential readers in the visualization community.

Freiburg, Germany
August 1992

Hans Hagen
Heinrich Müller
Gregory M. Nielson

Contents

Contributors

R. Daniel Bergeron
Department of Computer Science, Kingsbury Hall
University of New Hampshire, Durham, NH 03824, USA
rdb@cs.unh.edu

Michael Bomans
Institut für Mathematik und Datenverarbeitung in der Medizin
Universität Hamburg, Martinistraße 52, 2000 Hamburg 20, Germany

Ralf Denzer
Fachbereich Informatik, Universität Kaiserslautern
Postfach 3049, 6750 Kaiserslautern, Germany

Martin Frühauf
Fraunhofer-Institut für Graphische Datenverarbeitung
Wilhelminenstr. 7, 6100 Darmstadt, Germany
fruehauf@agd.fhg.de

Michel Grave
ONERA DMI/CC
B.P. 72, F-92322 Chatillon Cedex, France

Georges G. Grinstein
Graphics Research Laboratory, Department of Computer Science,
University of Massachusetts at Lowell, Lowell, MA 01854, USA
grinstein@cs.ulowell.edu

Ernst Gschwind
Hewlett-Packard GmbH
Herrenberger Str. 130, 7030 Böblingen, Germany

Hans Hagen
Fachbereich Informatik, Universität Kaiserslautern
Postfach 3049, 6750 Kaiserslautern, Germany
hagen@informatik.uni-kl.de

Stefanie Hahmann
Fachbereich Informatik, Universität Kaiserslautern
Postfach 3049, 6750 Kaiserslautern, Germany
hahmann@informatik.uni-kl.de

Karl-Heinz Höhne
Institut für Mathematik und Datenverarbeitung in der Medizin
Universität Hamburg, Martinistraße 52, 2000 Hamburg 20, Germany
hoehne@imdm.uke.uni-hamburg.dbp.de

Philip Jacob
Fachbereich Informatik, Universität Kaiserslautern
Postfach 3049, 6750 Kaiserslautern, Germany
jacob@informatik.uni-kl.de

Arie Kaufman
Computer Science Department, SUNY at Stonybrook
Stonybrook, NY 11794-4400, USA
ari@sbcs.sunysb.edu

Fred Kitson
Hewlett Packard Labs, 1501 Page Mill Road, Building 3U
P.O. Box 10490, Palo Alto, CA 94303-0969, USA
kitson@hplabs.hp.com

Arnold Klingert
Institut für Betriebs- und Dialogsysteme, Universität Karlsruhe
Postfach 6980, 7500 Karlsruhe, Germany
ajk@ira.uka.de

Ulrich Lang
Rechenzentrum der Universität Stuttgart
Allmandring 30, 7000 Stuttgart 80, Germany
ulang@rus.uni-stuttgart.de

Rolf van Lengen
Fachbereich Informatik, Universität Kaiserslautern
Postfach 3049, 6750 Kaiserslautern, Germany
lengen@informatik.uni-kl.de

Tom Malzbender
Hewlett Packard Labs, 1501 Page Mill Road
P.O. Box 10490, Palo Alto, CA 94303-0969, USA
malzbend@hplabs.hp.com

Nelson Max
L-301, Lawrence Livermore National Laboratory
P.O. Box 808, Livermore, CA 94550, USA
max2@llnl.gov

Loren D. Meeker
Department of Mathematics, Kingsbury Hall
University of New Hampshire, Durham, NH 03824, USA

Heinrich Müller
Institut für Informatik, Universität Freiburg
Rheinstr. 10, 7800 Freiburg, Germany
mueller@informatik.uni-freiburg.de

Yasuo Nakajima
Nissan Motor C. Ltd.
560-2, Okatsukoku, Atsugi-City, Kanagawa, 243-01, Japan

Gregory M. Nielson
Computer Science Department, Arizona State University
Tempe, AZ 85287, USA
nielson@enuxva.eas.asu.edu

Michael Pommert
Institut für Mathematik und Datenverarbeitung in der Medizin
Universität Hamburg, Martinistraße 52, 2000 Hamburg 20, Germany
pommert@imdm.uke.uni-hamburg.dbp.de

Frits Post
Delft University of Technology, Department of Technical Informatics
P.O. Box 356, 2600 AJ Delft, Netherlands
frits@duticg.tudelft.nl

Martin Riemer
Institut für Mathematik und Datenverarbeitung in der Medizin
Universität Hamburg, Martinistraße 52, 2000 Hamburg 20, Germany

Alfred Schmitt
Institut für Betriebs- und Dialogsysteme, Universität Karlsruhe
Postfach 6980, 7500 Karlsruhe, Germany
aschmitt@ira.uka.de

Thomas Schreiber
Fachbereich Informatik, Universität Kaiserslautern
Postfach 3049, 6750 Kaiserslautern, Germany
schreib@informatik.uni-kl.de

Stuart Smith
Institute for Visualization and Perception Research,
Computer Science Department,
University of Massachussetts Lowell, Lowell, MA, 01854, USA
stu@cs.ulowell.edu

Ted M. Sparr
Department of Computer Science, Kingsbury Hall
University of New Hampshire, Durham, NH 03824, USA

Achim Stößer
Institut für Betriebs- und Dialogsysteme, Universität Karlsruhe
Postfach 6980, 7500 Karlsruhe, Germany
stoesser@ira.uka.de

Ulf Tiede
Institut für Mathematik und Datenverarbeitung in der Medizin
Universität Hamburg, Martinistraße 52, 2000 Hamburg 20, Germany

Theo van Walsum
Delft University of Technology, Department of Technical Informatics
P.O. Box 356, 2600 AJ Delft, Netherlands

Burkard Wördenweber
Hella KG, Postfach 2840, 4780 Lippstadt, Germany

Fluid Flow Visualization

Frits H. Post, Theo van Walsum

This chapter presents an overview of techniques for visualization of fluid flow data. As a starting point, a brief introduction to experimental flow visualization is given. The rest of the chapter concentrates on computer graphics flow visualization. A pipeline model of the flow visualization process is used as a basis for presentation. Conceptually, this process centres around visualization mapping, or the translation of physical flow parameters to visual representations. Starting from a set of standard mappings partly based on equivalents from experimental visualization, a number of data preparation techniques is described, to prepare the flow data for visualization. Next, a number of perceptual effects and rendering techniques are described, and some problems in visual presentation are discussed. The chapter ends with some concluding remarks and suggestions for future development.

1 Introduction

For centuries, fluid flow researchers have been studying fluid flows in various ways, and today fluid flow is still an important field of research. The areas in which fluid flow plays a role are numerous. Gaseous flows are studied for the development of cars, aircraft and spacecrafts, and also for the design of machines such as turbines and combustion engines. Liquid flow research is necessary for naval applications, such as ship design, and is widely used in civil engineering projects such as harbour design and coastal protection. In chemistry, knowledge of fluid flow in reactor tanks is important; in medicine, the flow in blood vessels is studied. Numerous other examples could be mentioned. In all kinds of fluid flow research, visualization is an key issue.

1.1 Purposes and Problems of Flow Visualization

Flow visualization probably exists as long as fluid flow research itself. Until recently, experimental flow visualization, as described in Sect. 2, has been the main visualization aid in fluid flow research. Experimental flow visualization techniques are applied for several reasons:

- to get an impression of fluid flow around a scale model of a real object, without any calculations;

- as a source of inspiration for the development of new and better theories of fluid flow;
- to verify a new theory or model.

Though used extensively, these methods suffer from some problems. A fluid flow is often affected by the experimental technique, and not all fluid flow phenomena or relevant parameters can be visualized with experimental techniques. Also, the construction of small scale physical models, and experimental equipment such as wind tunnels are expensive, and experiments are time consuming.

Recently a new type of visualization has emerged: *computer-aided visualization.* The increase of computational power has led to an increasing use of computers for numerical simulations. In the area of fluid dynamics, computers are extensively used to calculate velocity fields and other flow quantities, using numerical techniques to solve the governing Navier-Stokes equations. This has led to the emergence of Computational Fluid Dynamics (CFD) as a new field of research and practice.

To analyse the results of the complex calculations, computer visualization techniques are necessary. Humans are capable of comprehending much more information when it is presented visually, rather than numerically. By using the computer not only for calculating the numerical data, but also for visualizing these data in an understandable way, the benefits of the increasing computational power are much greater.

The visualization of fluid flow simulation data may have several different purposes. One purpose is the verification of theoretical models in fundamental research. When a flow phenomenon is described by a model, this flow model should be compared with the "real" fluid flow. The accuracy of the model can be verified by calculation and visualization of a flow with the model, and comparison of the results with experimental results. If the numerical results and the experimental flow are visualized in the same way, a qualitative verification by visual inspection can be very effective. Research in numerical methods for solving the flow equations can be supported by visualizing the solutions found, but also by visualization of intermediate results during the iterative solution process.

Another purpose of fluid flow visualization is the analysis and evaluation of a design. For the design of a car, an aircraft, a harbour, or any other object that is functionally related with fluid flow, calculation and visualization of the fluid flow phenomena can be a powerful tool in design optimization and evaluation. In this type of applied research, communication of flow analysis results to others, including non-specialists, is important in the decision making process.

In practice, often both experimental and computer-aided visualization will be applied. Fluid flow visualization using computer graphics will be inspired by experimental visualization. Following the development of 3D flow solution techniques, there is especially an urgent need for visualization of 3D flow patterns. This presents many interesting but still unsolved problems to computer graphics research. Flow data are different in many respects from the objects and surfaces traditionally displayed by 3D computer graphics. New techniques are emerging for generating informative images of flow patterns; also, techniques are being developed to transform the flow visualization problem to display of traditional graphics primitives.

1.2 Overview

The main purpose of this chapter is to give an introduction to fluid flow visualization with computer graphics. It was primarily written for the computer graphics students interested in scientific visualization, and therefore some knowledge of basic computer graphics concepts and techniques is assumed. No knowledge of fluid dynamics is assumed; readers interested in the principles of fluid dynamics theory are referred to text books on fluid dynamics (such as Batchelor [1]). This chapter will also be useful for CFD specialists who want to know more about techniques available to visualize their data. An attempt has been made to provide a connection of the visualization process with the CFD simulation process. For basic graphics techniques they should refer to text books on computer graphics (such as Foley et al. [13]).

Before we turn to computer-aided visualization flow techniques, we will briefly discuss some concepts and methods from experimental flow visualization (Sect. 2). In Sect. 3, a pipeline model of the flow visualization process is introduced, and two important stages of the pipeline are described: visualization mapping, and data preparation and analysis techniques. Section 4 describes computer graphics presentation techniques, based on a set of standard forms of flow visualization described in Sect. 3. Some additional topics in rendering and animation will also be discussed. Finally, in Sect. 5, some conclusions are drawn and directions for research are indicated.

2 Experimental Flow Visualization

Experimental flow visualization has a long history, and today a wide variety of techniques to visualize fluid flows is known. Merzkirch [41] and Yang [65] give an overview of the fundamental techniques for flow visualization and describe a number of applications of these techniques. Van Dyke [58] shows many excellent pictures of experimental fluid flow experiments.

Following Merzkirch [41], three basic types of visualization can be distinguished: adding foreign material (2.1), optical techniques (2.2), and adding heat and energy (2.3).

2.1 Addition of Foreign Material

Time lines, streak lines and path lines play an important role in experimental fluid flow visualization techniques:

- *Time lines* are lines that, once released in the fluid, are moved and transformed by the fluid flow. The motion and formation of the line, which is often released perpendicular to the flow, shows the fluid flow. In practice, time lines often consist of a row of small particles, such as hydrogen bubbles.

- A *streak line* arises when dye is injected in the flow from a fixed position. Injecting the dye for a period of time gives a line of dye in the fluid, from which the fluid flow can be seen.

- A *path line* is the path of a particle in the fluid. Imagine a light-emitting particle in the flow. A path line is obtained when a photographic plate is exposed for several seconds.

We will see in Sect. 3 and 4 that time lines, streak lines, and path lines also play an important role in computer graphics flow visualization. In steady flows, streak lines and path lines coincide. In that case, streak lines and path lines are identical to *stream lines*, lines that are everywhere tangent to the velocity field.

Time lines, streak lines or path lines can be visualized by injecting foreign material in the fluid, which in general is transparent. As already mentioned, streak lines can be created by injecting dye in the (liquid) fluid. An example of an application of flow visualization using dye is shown in Fig. A of the color illustrations of this chapter at the end of the book. In case of a gas flow, smoke can be injected in the fluid. Though conceptually simple, these techniques require very careful application. The injection itself can disturb the flow. The density and temperature of the injected material, if not equal to that of the flow at the injection spot, can also influence the flow field.

Path lines are generated by adding small particles to the fluid. An example of suitable particles for liquid fluid is magnesium powder; in gaseous fluids, oil droplets can be used. The velocity can be measured by photographing the motion of the particles with a known exposure time (see Fig. 1).

Electrolytic and photochemical techniques can be used to generate time lines. Dye injection techniques are difficult to apply, because the use of several dye injectors would disturb the flow significantly. Electrolytic techniques only need a small and thin electrode in the fluid. A well known technique is the hydrogen bubble technique. This technique is based on the electrolysis of water. When electrodes are inserted in the fluid and a voltage is applied to the electrodes, hydrogen bubbles are formed at the cathode and oxygen bubbles are formed at the anode. By isolating some parts of the electrodes and by varying the voltage level, all kinds of combinations of time and streak lines can be obtained.

Photochemical techniques can be applied if the fluid should not be disturbed at all. These techniques also produce a visible tracer (dye) in the fluid by focusing a bundle of light onto a point in the fluid, or by using a laser to produce a dye along a line in the fluid. The dye is produced by a photochemical reaction.

Special techniques exist to study the flow field on a surface. One way to visualize flow fields in the neighbourhood of a surface is to fix *tufts* (small threads)

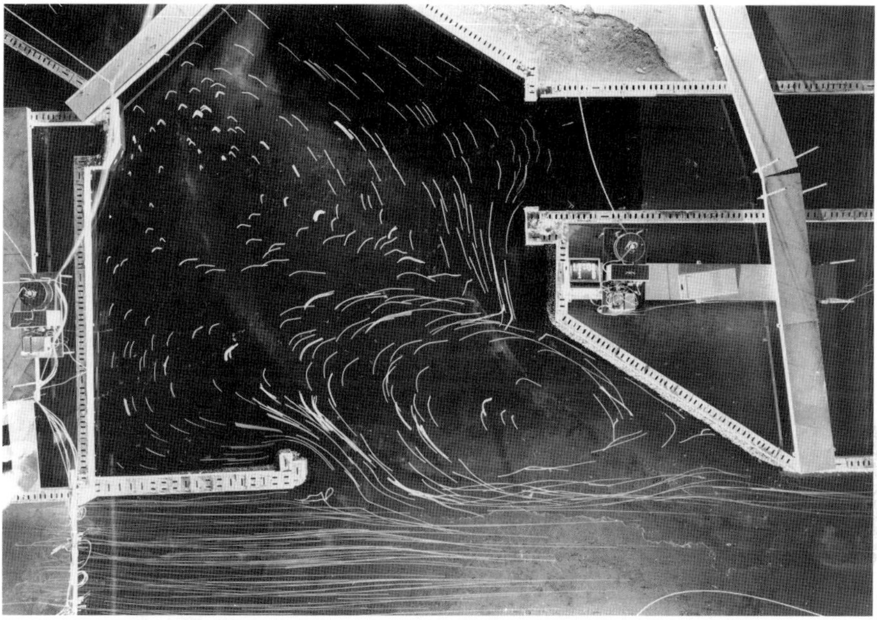

Fig. 1. Flow in a harbour model, visualized with particles (Courtesy Delft Hydraulics).

at several points on the surface. The orientation of the threads indicates the direction of the velocity. Another way is to coat the surface with a viscous material such as oil. The fluid flow will produce patterns in the viscous material, see Fig. 2. These patterns contain information on both the direction and the magnitude of the fluid velocity.

2.2 Optical Techniques

Optical flow visualization techniques are based on the effect that a change in density causes a change in the light refraction index. These techniques can only be applied when the density of the fluid is not constant.

There are three basic optical techniques: shadow techniques, schlieren techniques and interferometry. The simplest technique of these is the shadow technique. Shadowgraphs are created by passing a parallel light beam through a moving fluid (Fig. 3). The density variations will cause some of the light rays to be refracted. If the light beams are focused on a photographic plate, the refraction of some of the light beams results in dark and light areas, see Fig. 4.

The schlieren technique resembles the shadowgraph technique. It also uses a parallel light beam, but now two diaphragms are used. One diaphragm (inserted before the flow region) selects the light beams which are passing through the flow

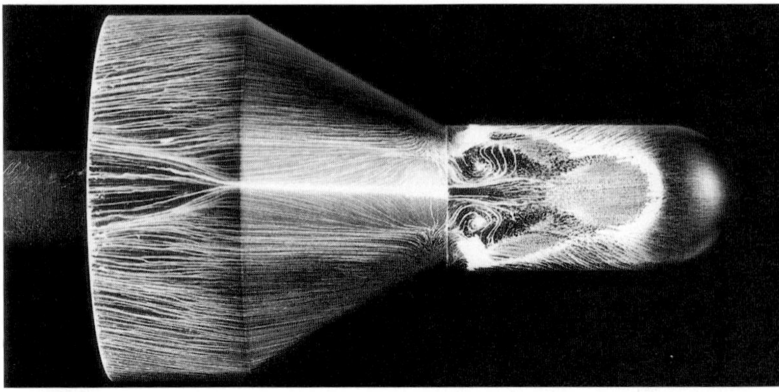

Fig. 2. Oil flow pattern visualizes the skin friction distribution of an axisymmetric model at incidence in a supersonic flow (Courtesy High-Speed Laboratory, Department of Aerospace Engineering, Delft University of Technology).

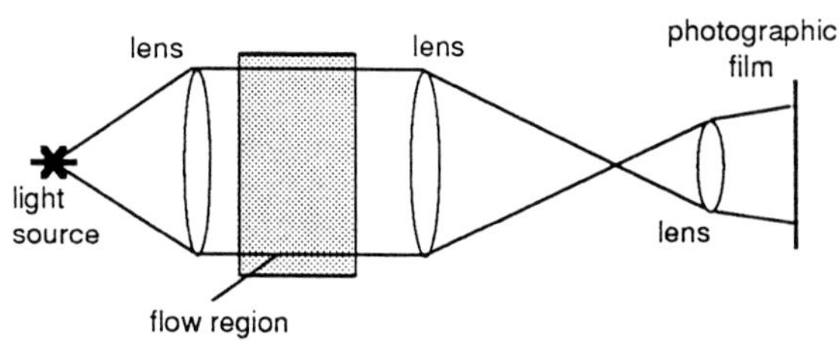

Fig. 3. Schematic arrangement of a shadowgraph system.

region. A second diaphragm (after the flow region) stops some of the refracted light beams. This technique is used to visualize density gradients.

Interferometry is based on the fact that a change in density not only results in a refraction of the light, but also in a phase shift. In an interferometer parallel light is split into two beams. One of the beams enters the flow field, the other beam does not enter the flow field. When both beams are merged and projected on the same photographic plate, interference occurs when the phase of one of the beams is shifted by a change of density in the fluid flow.

2.3 Addition of Heat and Energy

Instead of adding material to a fluid flow, it is possible to add energy to the fluid to visualize its flow. Especially in cases were the techniques of 2.1 and 2.2 cannot be applied, for example with low density fluid flows, these techniques are useful.

Heat can be added to a fluid to artificially change the density. If the pressure is constant, adding heat to a portion of the fluid results in a lower density. The

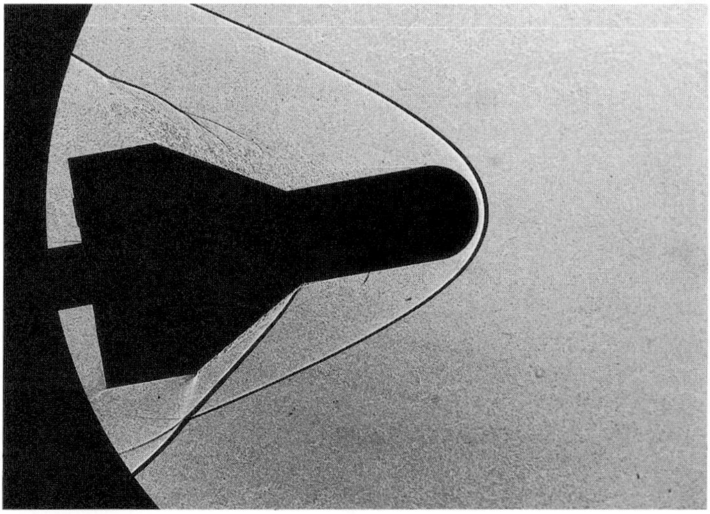

Fig. 4. Shadowgraph visualizes the density distribution in the flow around an axisymmetric model in a supersonic flow (Courtesy High-Speed Laboratory, Department of Aerospace Engineering, Delft University of Technology).

fluid can then be visualized with techniques as described in 2.2. Visualizing velocity profiles in gaseous flows is possible with the spark tracing technique. Two electrodes produce a spark discharge. This discharge ionizes the path of the spark. The light-emitting ionized fluid elements can be traced through the fluid.

A technique which can be used even for very low density fluid flows is electron-beam flow visualization. A beam of electrons traverses the gaseous fluid. When electrons collide with gas molecules, these gas molecules will be excited and emit radiation. The intensity of the radiation is approximately proportional with the density of the fluid. By moving the electron beam, the entire flow area can be scanned.

3 Computer Graphics Flow Visualization

Experimental flow visualization is a starting point for flow visualization using computer graphics. The process of computer visualization is described in general, and applied to CFD (3.1). The heart of the process is the translation of physical to visual variables. Fluid mechanics theory and practice help to identify a set of "standard" forms of visualization (3.2). To prepare the flow data to be cast in visual form, several types of operations may have to be performed on the data (3.3). A special type of operation, analysis of the topology of a fluid flow data set, is described separately (3.4).

3.1 The Flow Visualization Process

Scientific visualization with computer-generated images can be generally conceived as a three-stage pipeline process (Haber and McNabb [21]). We will use an extended version of this process model here (see Fig. 5), and discuss its application to flow data visualization.

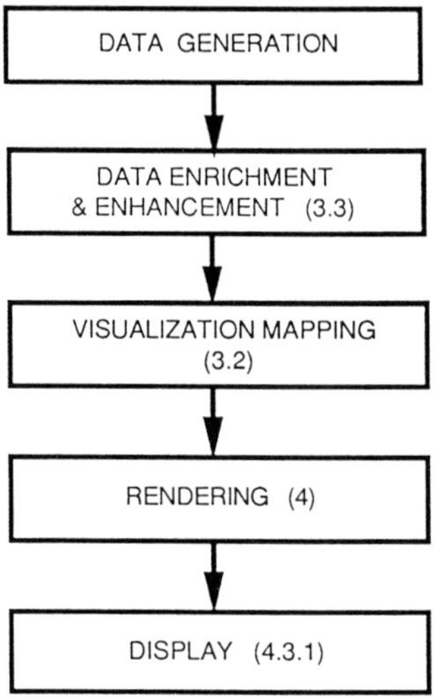

Fig. 5. A pipeline model of the visualization process.

- *Data generation*: production of numerical data by measurement or numerical simulations. Flow data can be based on flow measurements, or can be derived from analysis of images obtained with experimental visualization techniques as described in Sect. 2, using image processing (Yang [65]). Numerical flow simulations often produce velocity fields, sometimes combined with scalar data such as pressure, temperature, or density. In the rest of this paper, we will be mainly concerned with data generated by CFD simulations.

- *Data enrichment and enhancement*: modification or selection of the data, to reduce the amount or improve the information content of the data. Examples are domain transformations, sectioning, thinning, interpolation, sampling, and noise filtering. We will discuss some of these operations in 3.3.

- *Visualization mapping*: translation of the physical data to suitable visual primitives and attributes. This is the central part of the process; the conceptual mapping involves the "design" of a visualization: to determine what we want to see, and how to visualize it. Abstract physical quantities are cast into a visual domain of shapes, light, colour, and other optical properties. Some "standard" conceptual visualization mappings of flow data will be discussed in 3.2.

The actual mapping is carried out by computing derived quantities from the data suitable for direct visualization. For flow visualization, an example of this is the computation of particle paths from a velocity field. As this type of operations is closely related to the operations of the enrichment and enhancement stage, we will discuss these together in 3.3 under the heading Data Preparation.

- *Rendering*: transformation of the mapped data into displayable images. Typical operations here are viewing transformations, lighting calculations, hidden surface removal, scan conversion, and filtering (anti-aliasing and motion blur). Rendering techniques for flow visualization will be discussed in Sect. 4.

- *Display*: showing the rendered images on a screen. A display can be direct output from the rendering process, or be simply achieved by loading a pixel file into a frame buffer; it may also involve other operations such as image file format translation, data (de)compression, and colour map manipulations. For animation, a series of precomputed rendered images may be loaded into main memory of a workstation, and displayed using a simple playback program.

3.2 Flow Visualization Mappings

The style of visualization using numerically generated data is suggested by the formalism underlying the numerical simulations. The two different analytical formulations of flow: Eulerian and Lagrangian, can be used to distinguish two classes of visualization styles:

- *Eulerian*: physical quantities are specified in fixed locations in a 3D field. Visualization tends to produce static images of a whole study area. A typical Eulerian visualization is an arrow plot, showing flow direction arrows at all grid points.

- *Lagrangian*: physical quantities are linked to small particles moving with the flow through an area, and are given as a function of starting position and time. Visualization often leads to dynamic images (animations) of moving particles, showing only local information in the areas where particles move.

Other styles may be called *indirect* visualization of flow fields. A velocity field can be characterized by deriving certain scalar quantities, which can be visualized using volume rendering techniques (see 4.2.6). Examples are the magnitude of velocity or vorticity, or helicity (Buning [7]). Also, instead of individual particles, concentration of particles per volume unit can be computed, and visualized as a scalar concentration field (Hin et al. [32]). A second type of indirect visualization shows the effect of the flow on the geometry of certain objects, eg. deformation of spheres (Geiben and Rumpf [17]).

There is a number of visual representations, that can be conceived as "standard" visualization mappings in the sense of the process model described in 3.1, some of which are directly based on experimental flow visualization: velocity arrows, stream lines, stream surfaces (ribbons), streak lines, path lines, time lines (surfaces), and contours. They can be visualized using obvious graphical primitives such as points, lines, and surfaces. In Sect. 4, we will also discuss some more advanced presentation techniques.

Arrows are icons for vectors, and the most obvious way to visualize a velocity field with computer graphics is an arrow plot. Usually, arrows are tied to grid points, and indicate both direction and speed at these points. In this way, a complete view of a vector field can be given; this is a clear example of an Eulerian style visualization. Used in this way, arrows can give reasonable results in 2D, but there are numerous problems in the 3D case. We will discuss some of these problems in 4.2. A special type of arrow is the *stream vector*, where the tail of the arrow follows a stream line.

Stream lines, *streak lines*, *path lines* (or *particle traces*), and *time lines* are directly derived from experimental visualization (see 2.1). They can be generated and rendered as 2D or 3D curves (see Fig. 6, 7, 8). In 3.3, some techniques used for computation of these curves will be discussed; we will return to rendering in 4.2 and 4.3.

In 2D, stream lines characterize a general directional flow pattern very well (Fig. 6). Also, they give an indirect cue to velocity magnitude, because the distance between stream lines is inversely proportional to velocity.

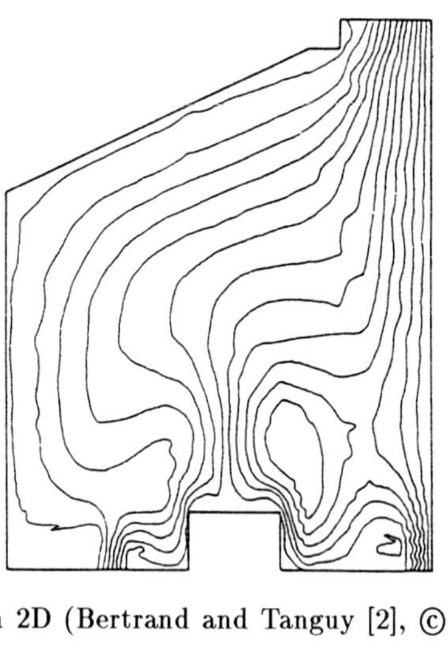

Fig. 6. Stream lines in 2D (Bertrand and Tanguy [2], ©1988, John Wiley & Sons, Ltd. By permission).

In 3D, stream lines can also be used, but additional depth cues are needed to locate the curves in space. One possibility is to generate a *stream surface*, interpolating a set of adjacent stream lines (Hultquist [33]). A surface interpolating only two adjacent stream lines is called a *stream ribbon* (Buning [7], Hultquist [33]).

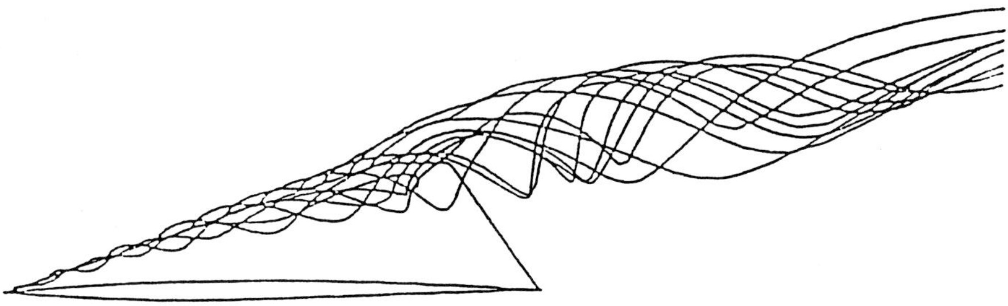

Fig. 7. Particle traces (Strid et al. [53], ©1989, von Karman Institute for Fuid Dynamics. By permission).

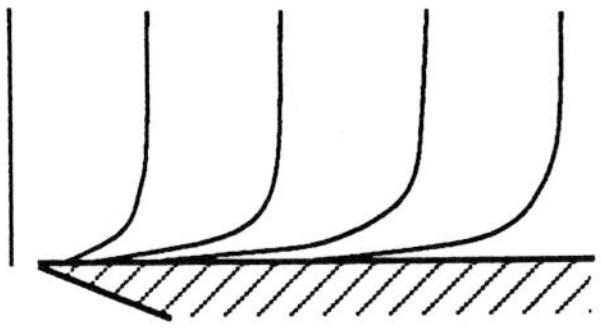

Fig. 8. Time lines.

Streak lines and path lines (see Fig. 7) can be visualized as static curves, and this can be interpreted either as a time interval in a steady flow, or as a single instant in an unsteady flow. Animation can show the curves dynamically as they change over time. Particles can also be visualized as moving dots or objects, or as a moving texture (Van Wijk [60, 62]; see 4.2.4 and 4.3.5).

Time lines drawn for a series of instants clearly show the distribution of velocity (Fig. 8). When a 2D array of particles is released at one instant in a 3D flow field, they can define a *time surface* at each instant.

Contours are curves (in 2D) or surfaces (in 3D) where a given scalar physical variable has a constant value. Contours, also called *iso-curves* and *iso-surfaces*, do not have counterparts in experimental visualization. Their applicability is limited to scalar data, such as pressure, temperature, or velocity magnitude, or scalar quantities derived from a vector field. An example of the latter is the display of pressure contours on the surface of an airplane fuselage (see Fig. 9).

In 3D, contour surfaces have the advantage of visualization with a full range of visual depth cues, but a problem is the display of multiple contours in one picture. Dynamic probing of a scalar field by interactively varying the contour value, and showing a changing contour surface is a good solution. This is hard to perform in real time, but this may be simulated by showing sets of precomputed contour surfaces; other forms of pre-processing, such as span filtering (Gallagher [15]) can also be used to speed up this process.

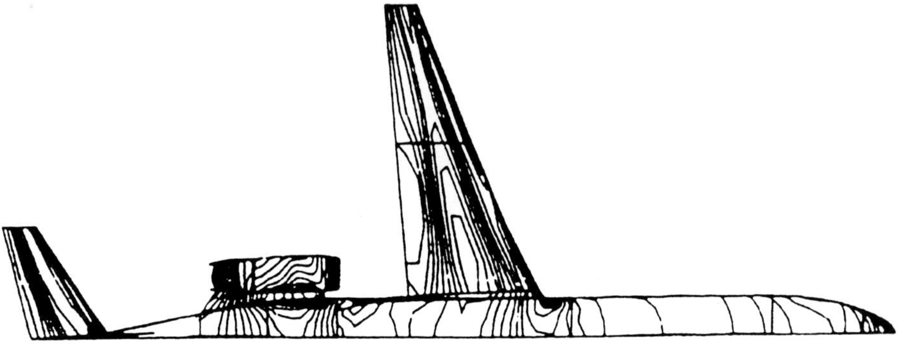

Fig. 9. Pressure contours on a surface.

3.3 Data Preparation

Visualization usually starts with "raw data" resulting from the data generation process, and often these data are not suitable for direct visualization in the given form. Data preparation consists of a variety of operations, including data enrichment and enhancement, and computation of derived quantities for a certain visualization mapping.

We will give some examples of operations performed at this stage. Some operations are applicable to any type of data, and some are based on the presence of a grid, in which data values are given for the node positions.

3.3.1 Filtering

Measured data will always contain noise and outlyers, or peak values, which may disturb visualization. The data can be viewed as samples from a continuous signal, and the image as a reconstruction of that signal. In terms of signal processing, the source signal may contain too many high frequency components, caused by measurement noise and peaks. Filtering can be applied to remove these spurious high frequencies.

3.3.2 Data Selection

To reduce the amount of data to be visualized, and to concentrate on the most interesting parts or features of the data, data selection techniques must be applied. The simplest selection is a global thinning by sub-sampling, or by aggregation and averaging of the whole data field. Also, a part may be cut out by clipping the data against a given volume. More sophisticated selection can be done by calculating some "interestingness" index for each grid cell, and only visualizing cells with a high value of this index. Measures for this may be local extreme values of a quantity, or large gradients, such as sudden changes in velocity. The computed indices can be treated as a scalar field, and volume rendering may be applied for visualization. Thresholding, or more advanced image processing

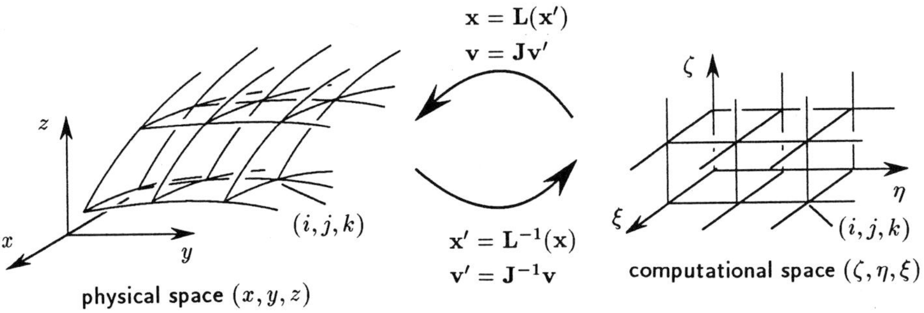

Fig. 10. Transition between physical space P and computational space C.

techniques can be used to decompose the data in meaningful parts. A last group of techniques is the extraction of specific flow features or patterns, such as flow field topology or vortex cores (Helman and Hesselink [27]; Globus et al. [20]); see Sect. 3.4.

3.3.3 Domain Transformations

The process of numerical simulation and visualization of fluid flow is typically performed in three different domains:

- the *physical domain* P: here, the equations of motion are defined; the domain is discretized, often into a curvilinear, boundary-conforming grid, fitting the surface of objects; the flow variables (velocity, density, pressure, etc.) are computed in the grid points of P.
- the *computational domain* C: the equations of motion are transformed to this domain; it is discretized to suit the needs of numerical computation, often into a uniform rectangular grid, and thus deformed with respect to P.
- the *graphical domain* G: often also discretized, to suit the needs of graphics processing. There is no generally accepted representation of G. As visualization often directly refers to physical reality, the shape of objects in G is usually the same as in P. Often a regular or hierarchical, rectangular discretization is used. G is populated with geometric primitives and attributes, such as shapes and colours, which must be ultimately expressed in pixels.

Transitions between these domains must be performed, often in both directions (see Fig. 10). The first step is discretization or grid generation, as a part of the preprocessing for numerical computations. Grids can be of several types: structured or unstructured, rectilinear or curvilinear, and combinations of these (Speray and Kennon [51]). Even more complex situations may occur when grids are *staggered*, which means that data values (such as velocity vector components) are computed in different faces of each grid cell. We will restrict the discussion here to structured grids, with regular hexahedral topology; the geometry of the grids can be curvilinear, resulting in cells with a warped brick shape, or orthogonal, resulting in cells with a cubical shape (Cartesian grid) or a rectangular brick shape.

The discretization in P for a flow simulation often is the definition of a structured, curvilinear grid with cell indices (i, j, k), such that the grid point nearest to the origin of the grid has coordinates (i, j, k). A general point of P is denoted as $\mathbf{x}_p(x, y, z)$. Velocity vectors $\mathbf{v}_p(i, j, k) = (u, v, w)$ are computed at each grid point.

Computational space C is usually discretized as a regular orthogonal Cartesian grid with the same cell indices (i, j, k), and points $\mathbf{x}_c(\xi, \eta, \zeta)$. Velocities at the grid points of C are $\mathbf{v}_c(i, j, k) = (u', v', w')$. Generally, a single global transformation between P and C is not known, but for each neighbourhood of a grid point (i, j, k) in P a local transformation $\mathbf{L}$ can be determined. Transformations for other points in P must then be derived by interpolation between grid values.

$\mathbf{L}$ specifies the local transformation of a grid point (i, j, k) from C to P as $\mathbf{x}_p = \mathbf{L}(\mathbf{x}_c)$; similarly, $\mathbf{L}^{-1}$ is used to transform a point from P to C (see Fig. 10). The Jacobian matrix $\mathbf{J}$ of $\mathbf{L}$, defined analytically as $\mathbf{J} = \partial\mathbf{L}/\partial\mathbf{x}_c$, can be used to transform a vector quantity from C to P. For instance, $\mathbf{v}_p = \mathbf{J} \cdot \mathbf{v}_c$. Again, the inverse $\mathbf{J}^{-1}$ is used to transform a vector from P to C.

In general, the mappings are only known at discrete points. As a consequence, the Jacobian must be approximated by finite differences. For a grid point (i, j, k) of C, the columns of $\mathbf{J}$ may, for example, be approximated by:

$$\mathbf{J}\mathbf{e}_1 = x_{i+1,j,k} - x_{i,j,k} \tag{3.1}$$

$$\mathbf{J}\mathbf{e}_2 = x_{i,j+1,k} - x_{i,j,k} \tag{3.2}$$

$$\mathbf{J}\mathbf{e}_3 = x_{i,j,k+1} - x_{i,j,k} \tag{3.3}$$

where $\mathbf{x}_{i,j,k}$ are the coordinates of grid point (i, j, k) in C, and $\mathbf{e}_i$ the unit vector in the direction of x_i. Another possibility would be to use central differences: $\mathbf{J}\mathbf{e}_1 = (\mathbf{x}_{i+1,j,k} - \mathbf{x}_{i-1,j,k})/2$; other types of differences can also be used.

As visualization often directly refers to physical reality, G must be undeformed with respect to P. Also, a new discretization is desirable in G, to support the operations in the rendering stage. The transition to another grid usually involves a resampling of the data field, and this has several disadvantages. Especially the transition from a boundary-conforming, locally refined curvilinear grid in P to a uniform orthogonal grid in G may lead to a severe waste of storage space, or to loss of information, depending on the resolution of the regular grid. In areas where the resolution of the P grid is higher than the G grid, data may be lost, while in low resolution areas of P, oversampling will lead to many identical data points in G. A partial solution is the use in G of a hierarchical grid type of which the resolution can vary locally, such as the octree (Samet [47]).

Often, the boundary conformance will be lost, so that object geometry must be represented separately in G. Another important point is the degradation of

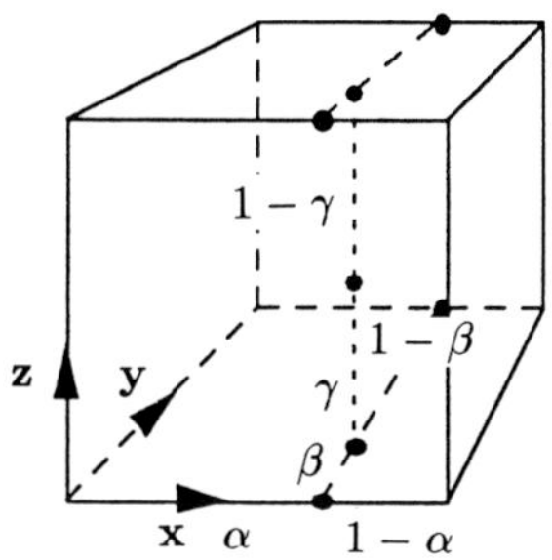

Fig. 11. Tri-linear interpolation; interpolation in the x-direction gives four points in the plane $x = \alpha$, interpolation in the y-direction gives two points on the line $x = \alpha$, $y = \beta$, interpolation in the z-direction gives the interpolated value at (α, β, γ).

accuracy as a result of interpolation. Use of higher order interpolation techniques can reduce the loss of accuracy.

3.3.4 Interpolation

Flow quantities are usually given only at grid points, so that for other points values must be obtained by interpolation. Interpolations may be of zero, first, or higher order, depending on the accuracy required. In the following, we will assume a regular orthogonal grid (as often used in computational space C), in which the cells are cubes with (i, j, k) as integer indices; α, β, and γ are the fractional $(0 \ldots 1)$ offsets of a point within a cell; $E(i, j, k) = E_{ijk}$ is a grid point value.

For zero order interpolation, the value within each grid cell is assumed to be constant. The value for a point inside the cell is either the average of the surrounding eight grid points, or equal to the value in the nearest grid point. This interpolation is discontinuous over the grid cell boundaries.

In first order interpolation, a linear variation of the value in the grid cells is assumed. With tri-linear interpolation, a linear interpolation is performed in x-, y- , and z-direction using the fractions α, β, and γ, respectively (see Fig. 11).

In a regular orthogonal grid, α, β, and γ can be easily determined from the coordinates. The interpolation can then be described for a single cell (with $i, j, k = 0, 1$), with two linear basis functions: $\Phi_0(s) = 1 - s$, and $\Phi_1(s) = s$; the product of three basis functions is defined on each of the eight corner grid points of the cell:

$$\Phi_{ijk}(\alpha, \beta, \gamma) = \Phi_i(\alpha) \cdot \Phi_j(\beta) \cdot \Phi_k(\gamma), \text{ with } i, j, k = 0 \text{ or } 1. \tag{3.4}$$

Now the interpolated value is obtained as:

$$E_{\alpha\beta\gamma} = \sum_{i,j,k=0,1} E_{ijk} \cdot \Phi_{ijk}(\alpha, \beta, \gamma). \tag{3.5}$$

Higher order interpolations work in a similar way, using higher order basis functions. The order of continuity of the interpolated values across grid cells is lower than the order of the interpolation. Thus, C^0 (positional) continuity can be obtained with first order interpolation, and for C^1 (tangent) continuity at least second order interpolation is needed.

Interpolation in a regular curvilinear grid, as is often the case in physical space P, is much more complex. The fractions (α, β, γ) are unknown in this case, and must be computed. We can use the same interpolation method described above; starting from a given point $\mathbf{x}_p(x, y, z)$ in cell (i, j, k), we have:

$$\mathbf{x}_p = \sum_{i,j,k=0,1} \mathbf{x}_{ijk} \cdot \Phi_{ijk}(\alpha, \beta, \gamma), \tag{3.6}$$

giving three non-linear equations, which can be solved numerically to find α, β, and γ. Alternatively, α, β, and γ can be determined using the method described below.

3.3.5 Point Location in Computational and Physical Space

It is often necessary to determine the location of a point with respect to a grid, that is, given the coordinates of a point, we must find the grid cell indices (i, j, k), and fractional offsets (α, β, γ) within the cell containing that point. If we have a point $\mathbf{x}_c(\xi, \eta, \zeta)$ in C, it is easy to find the cell, if the grid in C is regular and orthogonal.

If we have a point $\mathbf{x}_p(x, y, z)$ in a structured curvilinear grid in P, and we must find the corresponding point $\mathbf{x}_c$ in C, an incremental search through the cells must be performed, starting from an initial point, each time taking a step towards a neighbouring cell. One algorithm for this search is the *stencil walk* (Buning [7]).

We have a point $\mathbf{x}_p$ in P, and we want to find the corresponding cell (i, j, k) and offsets (α, β, γ) in C. An initial guess for the right cell in P can be made by finding the nearest grid point, and choosing one of the cells adjacent to that grid point. In the corresponding initial cell in C, we choose $\mathbf{x}_c^*$, with $\alpha = \beta = \gamma = 0.5$, the center of the cell. This point is transformed to $\mathbf{x}_p^*$ in P, determined by trilinear interpolation between the P-coordinates of the corner points of the cell; for the center point, this amounts to taking the average.

The difference vector $\Delta\mathbf{x}_p = \mathbf{x}_p^* - \mathbf{x}_p$ is again transformed from P to C, using the inverse Jacobian matrix $\mathbf{J}^{-1}$, which is found by interpolation between the values of the corner points of the cell, again with $\alpha = \beta = \gamma = 0.5$. In C, we thus find a difference vector $\Delta\mathbf{x}_c = (\Delta\alpha, \Delta\beta, \Delta\gamma)) = \mathbf{J}^{-1}\Delta\mathbf{x}_p$, which is used to determine new values for α, β, and γ by adding the old values (0.5) to $\Delta\alpha$, $\Delta\beta$, and $\Delta\gamma$.

If at least one of the new values (α, β, γ) is outside $[0, 1]$, the point is not in the current cell, and the search continues. If $\alpha > 1$, then i is incremented, and if $\alpha < 0$, i is decremented; similar rules apply for β and γ. The center of the cell is used as the new starting point, and this continues until (α, β, γ) are within $[0, 1]$, which means we have found the right cell. Then we can continue in the same way within this cell, until $|\Delta \mathbf{x}_c|$ is small enough.

In cases when a grid contains holes or cavities, this procedure may fail when the search path crosses a boundary of the grid. This must be detected, and the search can be continued by following boundary cells to get around the obstacle.

Point location is greatly simplified if a previous location of the point is known, as often happens with particle tracing. In that case, the cell of the previous location (or a neighbouring cell in the direction of $\mathbf{v}$) can be used as an initial guess. In an unstructured grid, it is necessary for this purpose to store adjacency information of the cells.

3.3.6 Computing Derived Scalar Quantities

Several scalar quantities can be computed from a velocity field for each grid cell, and may be used to indirectly visualize the flow field. We give only a few examples here.

- The magnitude of all velocity vectors $||\mathbf{v}||$ defines a scalar field.
- The kinetic energy density is $\frac{1}{2} \cdot \rho \cdot ||\mathbf{v}||^2$.
- The scalar product of two vectors is a measure of the angle ϕ between them:

  $$\mathbf{u} \cdot \mathbf{v} = ||\mathbf{u}|| \cdot ||\mathbf{v}|| \cos \phi \tag{3.7}$$

  This can be used to find the components of all velocity vectors in a given direction, or to find the changes in direction at two neighbouring points.
- The magnitude of the vorticity ω (see 3.3.7) may be used to find vortices. Using ω, *helicity density* is computed as: $H_d = \mathbf{v} \cdot \omega$. (Buning [7])
- The scalar $f_{\mathbf{shock}}$ is defined for compressible media as:

  $$f_{\mathbf{shock}} = \frac{\nabla p}{|\nabla p|} \cdot \frac{\mathbf{v}}{c}, \tag{3.8}$$

  in which c is the speed of sound. The iso-surface for $f_{\mathbf{shock}} = 1$ shows shock waves (Buning [7]).

3.3.7 Computing Particle Path Lines and Other Vector Quantities

Particle path lines, or integral curves, can be computed from the velocity field by computing a series of consecutive particle positions in the field, and fitting a curve through them. The motion of a particle is governed by the equation:

$$\frac{d\mathbf{x}}{dt} = \mathbf{v}(\mathbf{x}), \tag{3.9}$$

with $\mathbf{x}$ the position vector of a particle at a given instant t, and $\mathbf{v}(\mathbf{x})$ the velocity field. Integration of this equation yields position $\mathbf{x}$ at any instant. As the velocity is usually only given at the grid point positions, the velocity at other positions must be obtained by interpolation.

The problem can thus be stated as determination of particle positions in physical space P, starting from an initial position $\mathbf{x}(x, y, z)$. This is performed in computational space, in the following three steps (Buning [7]):

1. find the position of $\mathbf{x}$ in computational space C, as a grid cell (i, j, k) and offsets (α, β, γ) in the cell;
2. interpolate within the grid cell to find the velocity vector at that point;
3. integrate equation (3.9) to find the next particle position.

The first two steps have already been discussed above (3.3.5 and 3.3.4). For the third step, several integration schemes can be used; each scheme gives an approximation of:

$$\mathbf{x}(t + \Delta t) = \mathbf{x}(t) + \int \mathbf{v}(\mathbf{x}(t))dt, \tag{3.10}$$

for a time step Δt. The simplest is the first order Euler technique. The integral is approximated as $\mathbf{v}(\mathbf{x}(t))\Delta t$, so that $\mathbf{x}(t + \Delta t) = \mathbf{x}(t) + \mathbf{v}(\mathbf{x}(t))\Delta t$. This approximation is too inaccurate; a second order Runge Kutta technique (also known as the Heun scheme) gives better results. Besides the velocity at $\mathbf{x}$ at instant t, we also use an estimation of the position $\mathbf{x}^*$ at instant $t + \Delta t$; for this *predictor* step, we use the Euler method:

$$\mathbf{x}^*(t + \Delta t) = \mathbf{x}(t) + \mathbf{v}(\mathbf{x}(t))\Delta t. \tag{3.11}$$

Now $\mathbf{x}^*(t + \Delta t)$ is used to estimate velocity at $t + \Delta t$, and $x(t + \Delta t)$ is computed in a *corrector* step:

$$\mathbf{x}(t + \Delta t) = \mathbf{x}(t) + \frac{1}{2}[\mathbf{v}(\mathbf{x}(t)) + \mathbf{v}(\mathbf{x}^*(t + \Delta t))]. \tag{3.12}$$

The choice of the time step Δt also affects accuracy. The optimum time step size is a compromise between accuracy and computational cost. Use of a variable time step, depending on the gradients in the velocity field, is the best solution. This may be done eg. with $\Delta t = \alpha / \mathbf{v}_a$, where α is the number of steps per cell, and $\mathbf{v}_a$ is the average velocity of the eight surrounding grid points.

For visualization of moving particles using animation, particle positions must be known at equal time intervals for each frame, to show velocity information. If particle positions were computed with variable time steps, each particle path must be resampled to find positions at equal intervals.

If we are able to compute particle traces, other objects can be easily derived. Stream surfaces (ribbons) can be constructed by fitting surfaces through particle positions, using polygons, or any type of smooth surface interpolation. Streak lines can be simulated by releasing a continuous flow of particles from a single point. Time lines (-surfaces) can be obtained by releasing a 1D (2D) array of particles at one instant t, and interpolating a curve (surface) through the positions of these particles at any given instant.

In a 3D incompressible flow, stream lines are curves that satisfy the equation:

$$\frac{dx}{u} = \frac{dy}{v} = \frac{dz}{w}, \tag{3.13}$$

where u, v, and w are the velocity components in x-, y-, and z-direction. For a time-independent flow field, the calculation of stream lines amounts to the same as computing particle paths (Strid et al. [53]).

Vorticity is another important vector quantity that can be computed from the velocity field. For this computation, the curl of the velocity field must be determined, since vorticity is defined as $\omega = \nabla \times \mathbf{v}$.

3.3.8 Contour Lines and Surfaces

Contour lines can be considered in 2D as intersection curves of a surface defined by a scalar function $z = f(x, y)$ and a (horizontal) plane $z = c$. Usually, no scalar function is known, but a field is defined by scalar data values at grid points. To determine contour lines, each grid cell is examined; if grid values both higher and lower than the contour value are found in a cell, it contains a part of the contour line. Intersection points can be determined with the edges of the cell, by linear or non-linear interpolation, and these points are connected by line segments, or a smooth curve is fitted through them. There are several methods to generate contour lines, for different types of grids, interpolations, and orders of curve generation. For an overview, see Sabin [46].

In the case of a 3D scalar field, a stack of planes, each with contour lines, may be determined for a number of positions in the i-, j- or k-direction of a regular rectangular grid. This can give some suggestion of 3D contour surfaces, but for properly shaded display other techniques must be applied. In volume rendering, contours are rendered directly by selective display of voxel values above a certain threshold; surface normals for the lighting calculations are estimated from gradients (Levoy [38]). This technique gives visually good results, without explicit calculation of the contour surface.

Another technique often applied in volume visualization is the Marching Cubes algorithm (Lorensen and Cline [40]). This algorithm finds an explicit polygonal representation of a contour surface. The principle is as follows:

mark all cells as undone;
find a starting cell, and put it into a queue;
WHILE the queue is not empty **DO**
  get the next cell from the queue;
  process the cell as described below;
  mark the cell as done;
  FOR EACH of the intersected edges **DO**
    put all unmarked neighbour cells sharing the edge into the queue.

In a structured grid, the edge-sharing neighbours can be easily found with the cell indices; in an unstructured grid, adjacency information is needed.

For a hexahedral (cubical) cell, all points on the edges with the contour value are determined by linear interpolation. As there can only be one point per edge, at most 12 points can be found in a cubical cell. Now the intersection of the cell with the contour surface is determined, consisting of one to four triangles. This is done by encoding the eight vertices of the cell as higher than, or lower than the contour value. The resulting 256 possible different binary code values were reduced to 14 different cases, each corresponding with a distinct configuration of triangles. The binary code for each cell is used as an index into a table of configurations, which is used to generate the output triangles, with the edge intersection points as vertices.

3.4 Flow Field Topology

Since 1987 the analysis and visualization of flow topology has been investigated thoroughly. In a number of papers, Helman and Hesselink [29, 30, 25, 24, 26, 27, 28] describe the development of a system to visualize flow topology. A general classification of flow fields is described by Chong et al. [9]. More recently, similar techniques to those of Helman and Hesselink were described. Dickinson [10] addresses the interactive aspect of flow topology visualization, and Globus et al. [20] give detailed information on how to implement a flow topology visualization module.

Flow topology analysis is based on critical point theory, which has been used widely to examine solution trajectories of ordinary differential equations. The topology of a vector field consists of critical points (where the velocity vector is zero) and integral curves and surfaces connecting these critical points. Images of a vector field topology display the topological characteristics of a vector field, without displaying too much redundant information.

The following steps are necessary to analyze and visualize vector field topologies:

- the location of the critical points must be calculated
- the critical points must be classified
- integral curves and surfaces must be calculated.

We will discuss these steps in more detail below. It should be noted that, although the techniques will be described for velocity fields, they can also be applied on any other vector field, such as vorticity fields or pressure gradient fields.

3.4.1 Critical points

The positions of the critical points can be found by searching all cells in the flow field. Critical points can only occur in cells where all three (or two, in 2D) components of the vector pass through zero. The exact position of a critical point can be calculated by interpolation in case of a rectangular grid. In case of a curvilinear grid, the position of a critical point can be calculated by recursively subdividing the cell, or by a numerical method such as Newton iteration (Globus et al. [20]).

Once the critical points have been found, they can be classified. This is done by approximating the velocity field in the neighbourhood of the critical point $\mathbf{x}_{cp}$ with a first order Taylor expansion. This gives the following formula for the velocity u:

$$u_i \approx u_{cp,i} + (x_j - x_{cp,j})\frac{\partial u_i}{\partial x_j}. \tag{3.14}$$

Because the velocity in a critical point is zero, the velocity field in the neighbourhood of a critical point is fully determined by the partial derivatives $\nabla\mathbf{u} = \partial u_i/\partial x_j$. The critical points can be classified according to the eigenvalues and eigenvectors of $\nabla\mathbf{u}$. Figure 12 shows some of the configurations for a three-dimensional vector field. In these figures, positive eigenvalues correspond to velocities away from the critical point (repelling nodes), and negative eigenvalues correspond to velocities towards the critical point (attracting nodes). Complex eigenvalues result in a focus; if the real part is non-zero, a spiral occurs, and if the real part is zero, concentric ellipses occur. If both negative and positive real values exist at a critical point, the critical point is a saddle.

3.4.2 Integral Curves and Surfaces

The classified critical points can be used as starting points for integral curves ("particle paths"), and the eigenvectors can be used as starting directions. This means that the starting point is a point on an eigenvector, very close to the critical point. The end points of the integral curves are critical points again, or are points on the boundary of the flow domain. Because of numerical errors, some integral curves might "miss" a critical point, or might enter an object in the flow. By attaching an integral curve to a critical point when it comes very close to the critical point, the first problem can be solved. To solve the second problem, integral curves that enter an object can be restricted to follow the surface of the object.

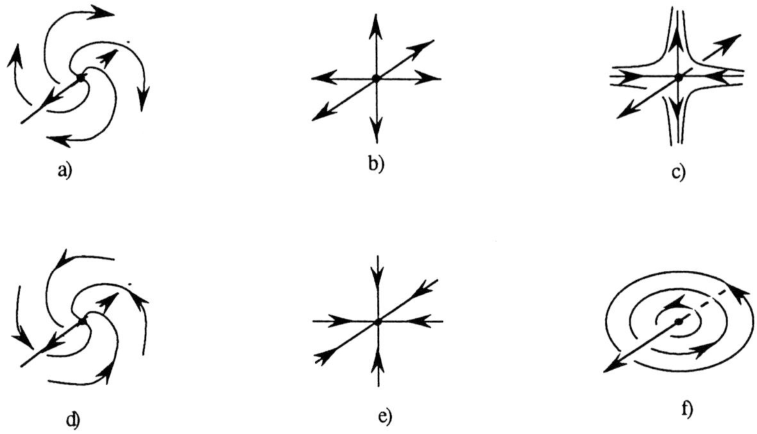

Fig. 12. Examples of three dimensional critical points a) repelling focus, also repelling in the third dimension, b) repelling node, c) saddle, repelling in the third dimension, d) attracting focus, repelling in third dimension, e) attracting node, f) center, repelling in the third dimension.

In two dimensions, critical points and integral curves completely describe the flow field in a qualitative way. In three dimensions, integral surfaces should also be calculated. Helman and Hesselink [28] describe a way to create some of these integral surfaces. When visualizing the flow around a hemisphere cylinder, they first calculate the topology of the vector field on the surface of the cylinder. This is, in fact, a two-dimensional topology. Points on the (two-dimensional) integral curves of this topology are used as starting points for a number of three-dimensional integral curves. By tessellating the space between two adjacent integral curves, integral surfaces can be visualized. In this way they are able to show surfaces of separation and reattachment.

4 Presentation Techniques

After data preparation and visualization mapping, the flow data have been cast into a form suitable for visual presentation. The visualization objects can be transformed into pictures, together with additional objects such as the environment of the flow, and auxiliary objects such as scales, pointers, colour bars, and numerical or textual annotations. First, we will turn to some aspects of human perception, and relevant visual cues to achieve optimum results for 3D flow visualization. Then we will discuss basic rendering techniques in 4.2, and some more advanced methods in 4.3.

4.1 Human Perception and Depth Cues

In 3D flow data visualization, often complex spatial structures must be shown, and for this a good orientation in 3-space is essential. Human visual perception can readily extract spatial information from a 2D picture, provided that enough supporting "depth cues" are available. Many depth cues are related to objects and surfaces, but field data (such as scalar and vector fields) do by nature not contain objects or surfaces and thus lack these cues. For visualization, extra depth cues must then be added, 3D objects and surfaces derived from the fields must be used, or environment geometry must be exploited to support localization in space.

We will give some examples of important cues for the perception of shapes, depth, and motion. (For a good introduction and further references, see Thorpe [55].)

3D *shapes* are mainly perceived by their contours, and the directional reflection of light on surfaces, showing as changes in reflected intensity. Gradual changes indicate a smoothly curved surface, while discontinuous changes indicate edges.

Depth or spatial information is perceived in many ways:

- perspective: parallel edges converge into the distance; the size of objects decreases linearly with their distance from the observer;
- stereoscopy: the disparity between images seen with two eyes gives a powerful depth effect, which is independent from other depth cues, and can be applied to virtually every type of image;
- occlusion: nearer objects cover other, more distant objects;
- surface texture density gradients: a continuous density change implies a surface expanding into depth, a discontinuity is perceived as a contour or an edge;
- shadows cast by one object onto the ground or onto other objects give information on spatial relationships;
- distance cues, such as atmospheric haze: the diminishing of colour contrast in the distance;
- artificial optical cues, such as depth of field;
- user control: if a user is given the possibility to navigate around the 3D display space, to probe and interrogate the data, this will help the user to build a 3D mental image of the data space. An example from flow visualization is the interactive positioning of particle sources (Wavefront Technologies [54]), and the virtual wind tunnel (Bryson and Levit [6]), where the user can move around in 3D data space, and can use data glove gestures to generate particle path lines in real time.

Depth ambiguity can easily occur when spatial cues are absent or too weak. This is quite common in line drawings (such as arrow plots), which contain no surface information. To support localization of an object floating in space, it can be related to the environment, or to coordinate planes by showing the object's coordinates or projection on two or three planes.

Motion can also support depth information. Changes in viewpoint resulting in gradual changes of occlusion relations and perspective, are known as *motion parallax*; this is readily perceived as extra depth information. The world as perceived by a moving observer has the characteristics of an *optic flow field* (Gibson [18, 19]). A static world "flows" around a moving observer in a standard pattern. If an object is perceived which moves in a way that is different from this pattern, it is detected to be in motion.

Moving objects can show direction and velocity information. A dense cloud of small objects can be perceived as a moving texture (Upson et al. [56]; Van Wijk [60, 62]), and can be given certain surface properties to improve localization in space. To preserve velocity information of moving objects (particles), the screen update time must be proportional to a simulation time step; in practice this means that both must be constant. Also, certain precautions must be taken to reduce the disturbing and possibly false effects caused by discretization (see 4.3).

In flow visualization, knowledge of human perception can be used to improve communicaton of spatial and motion information. Below we will show some examples where shape, depth, and motion cues are consciously applied for this purpose.

4.2 Basic Rendering Techniques

In the visualization mapping stage, visual primitives have been defined to represent the data. Generally speaking, the rendering process consists of four stages: viewing transformations, visible surface determination, lighting calculations, and scan conversion. We will not attempt to summarize the basic rendering techniques for these primitives. Many techniques for rendering lines, curves, polygons and curved surfaces can be found in any computer graphics textbook (eg. Foley et al. [13]). Rendering solids is described in Bronsvoort et al. [5], and volume rendering in Kaufman [34]. Often, a visualization will consist of a combination of several types of primitives, and therefore it is profitable if a visualization system permits the use of many different primitives in a single image.

We will concentrate here on the visualization mappings listed in 3.2, and the use of extra depth cues to improve spatial representation of 3D flow fields.

4.2.1 Arrows

Arrows can simply be drawn using only straight lines. In 2D, this can give reasonable results, provided that the arrows are scaled so they do not overlap (Fig. 13).

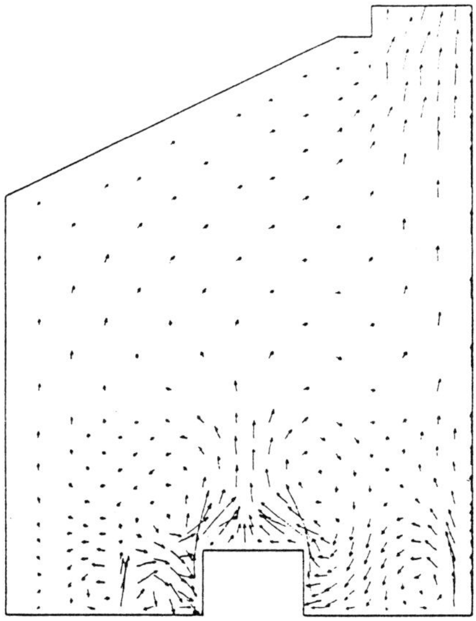

Fig. 13. 2D arrow plot (Bertrand and Tanguy [2], ©1988, John Wiley & Sons, Ltd. By permission).

In 3D, rendering arrows is much more complex. Perspective is the only depth cue available; there is no occlusion, or directional light reflection to assist depth perception. 3D line arrows are ambiguous in direction; for example, it is impossible to distinguish between arrows pointing towards and away from the observer (Fig. 14).

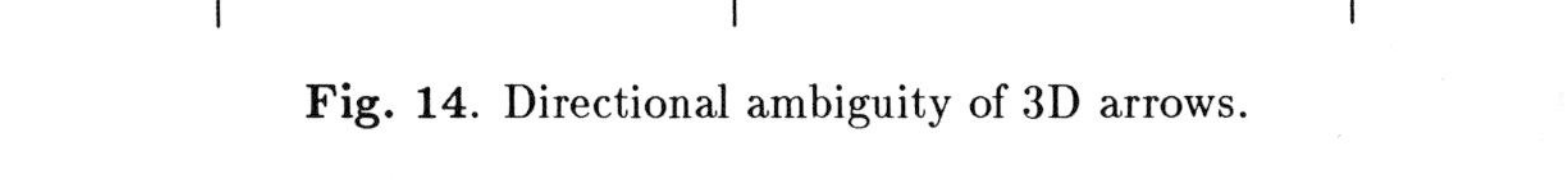

Fig. 14. Directional ambiguity of 3D arrows.

The size of the arrows is determined by three factors: velocity magnitude, direction with respect to the image plane, and perspective. Also, the display of a full 3D array of arrows very soon becomes cluttered. Various types of arrows have been devised to minimize cluttering, and to improve the directional effect (Kroos [37]), but none of these can fully solve these problems.

An improvement in this respect is to relate the arrows to a plane (Fig. 15), and to show projections ("shadows") of the arrows on the plane. The effect is similar to the use of *tufts* in experimental visualization.

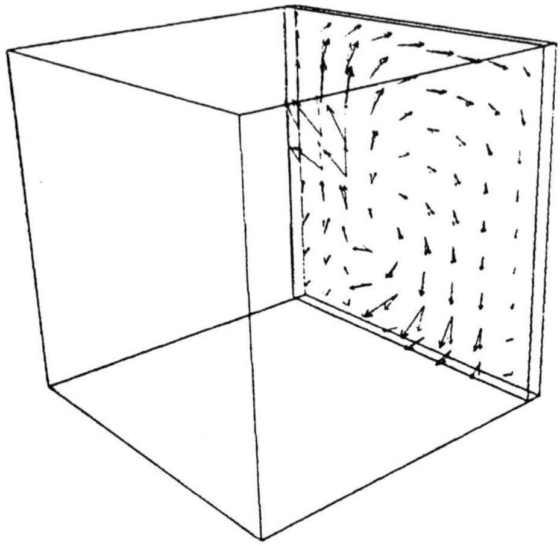

Fig. 15. Arrows connected to planes, with projections.

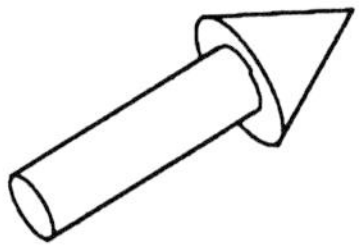

Fig. 16. Arrow as 3D object.

The flow field can be sliced, moving a section plane through the field, in the same style as applied in volume rendering. Observing the gradual changes of the arrows on the plane allows the user to mentally imagine the entire field. More spatial cues can be added by drawing arrows as 3D polygonal objects. The occlusion in a visible surface display gives a good depth cue, and reduces directional ambiguity. Also, directional light reflection and shading can be applied, giving some extra information on orientation in space (Fig. 16). But as polygonal arrows are larger, they can lead to an increase of cluttering.

4.2.2 Curves

Curves, such as stream lines, streak lines, path lines, and contour lines, can be visualized as a sequence of short line segments. Again, this gives good results in 2D, but in 3D spatial curves are hard to localize without further depth cues. Also, only a small number of curves can be displayed without confusion.

An improvement is showing projections of a curve in the main coordinate planes, allowing mental reconstruction of the 3D shape by the trained observer (Fig. B of the color illustrations of this chapter at the end of the book).

Another possibility is to display curves as 3D pipes, allowing occlusion and directional light reflection (Fig. C of the color illustrations).

It is often desirable to display curves (for example iso-curves or stream lines) on surfaces. If the curves are drawn as lines, their relation with the surface remains weak. A much better result is obtained when the lines are modelled as 3D strips of constant width pasted onto the surface (Van Wijk [61]). The strips can be rendered transparently, using texture mapping (Fig. D of the color illustrations of this chapter at the end of the book).

4.2.3 Surfaces

A good directional effect for stream lines is achieved by rendering *ribbons* (Hultquist [33]), which can be produced by connecting two adjacent curves with a maze of polygons. Rendered with proper lighting and shading, stream ribbons give a good impression of flow direction, also showing significant effects such as twist and divergence (Fig. 17).

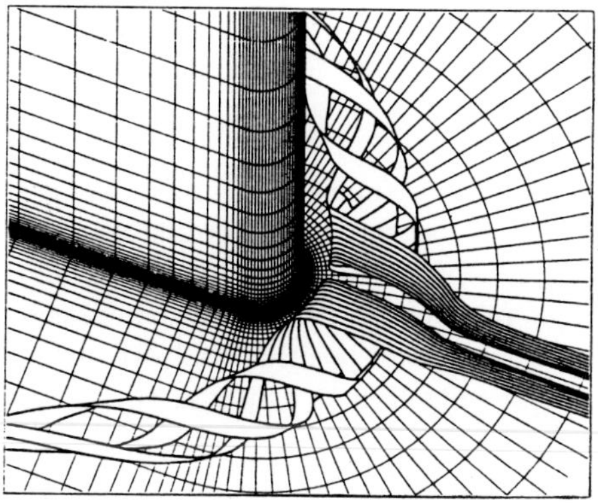

Fig. 17. Stream ribbons, showing twist and divergence (Hultquist [33], ©1990, Pergamon Press Ltd. By permission).

Surfaces, such as stream surfaces (Hultquist [33]), time surfaces, or iso-surfaces, can be represented as explicit geometric objects (see 3.3.8), or displayed directly from the data. In volume rendering, iso-surfaces can be displayed directly using segmentation techniques (Levoy [38]). Surface normals for shading are approximated by gradients of scalar values. In flow visualization, surfaces are often defined by points, eg. particle positions. An explicit surface can be obtained by interpolating a set of points, generating a polygonal mesh or a smooth surface. In 4.3.5 techniques will be described for direct display of surfaces in flow visualization.

Good shape and depth effects can be achieved by many techniques well known in computer graphics. Visible surface display must be applied, and directed lighting and shading with diffuse and specular reflections (see Foley et al. [13]).

A surface can also be used to display extra data pertaining to it, such as pressure and temperature. The object colour can be varied according to a scalar quantity, and contour lines can be shown on the surface (Van Wijk [61]; see Fig. D of the color illustrations of this chapter at the end of the book. Texture mapping allows display of directional information on the surface (Van Wijk [62]; see 4.3.3). Finally, surfaces can be rendered semi-transparently, to allow multiple layers of surfaces to be visualized at the same time (Fig. E of the color illustrations). A maximum of three layers can be viewed distinctly in this way.

A special use of a surface in flow visualization is the *stream polygon* (Schroeder et al. [48]). This is a regular, n-sided polygon, positioned in a flow field, oriented normal to the flow direction. The polygon can be scaled, sheared, and rotated in response to local strain and rotation in the field, or to other quantities. By sweeping the polygon along a stream line, an n-sided cylindrical *stream tube* is obtained. The effects of deformation and sweeping can be combined, yielding a complex cylinder that is twisted and tapered. The surfaces of the polygon and the stream tube can be shaded using directed lighting, and can also show associated data. Figure F of the color illustrations shows an example of this technique.

4.2.4 Particles

A particle may be rendered as a small-sized object, as a single point, or as a special type of primitive that has some characteristics of both. As an object, a circle (in 2D) or a sphere (in 3D) would be an obvious choice. But these objects do not show any changes in orientation. Even with directed light reflection, a sphere only shows its shape, not its position and orientation in space. To take advantage of changing light reflection as a spatial cue, a flat or elongated shape should be used.

To improve the directional effect, especially in still images, a tail can be attached to a particle. This can be the motion path in the preceding time interval, with the length indicating its velocity (Fig. G of the color illustrations of this chapter at the end of the book).

A related way to render a particle is the *stream arrow*. The head of the arrow points to the particle's position, and indicates its direction of motion. The tail again is a part of the motion path (Fig. 18).

Particles can also be rendered as points. This is of special interest to visualize the global structure of a flow field, using large numbers of particles. Although a point seems the simplest of all geometric primitives, rendering points is by no means trivial (Van Wijk [63]). If a point is simply rendered by setting the pixel onto which the point is projected, regardless of its position within the pixel, serious errors occur due to aliasing, caused by the limited resolution of the display screen. We will return to this problem in 4.3.2; we will discuss other particle rendering methods, including techniques to add extra spatial information to particles in 4.3.5.

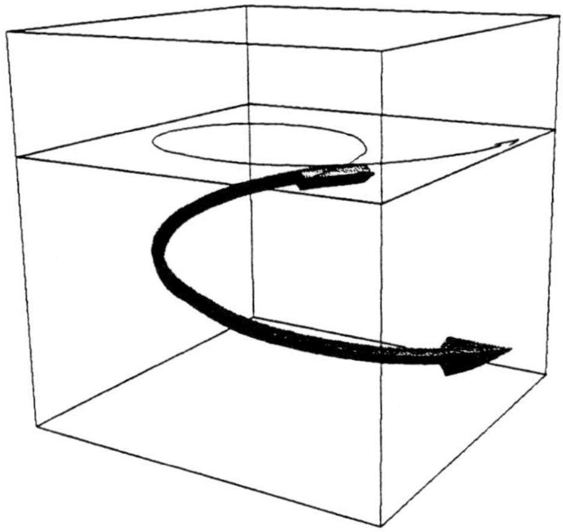

Fig. 18. Stream arrow.

4.2.5 Environment Geometry

Display of environment geometry is often desirable in flow visualization, to show the context of the particular flow problem studied, and also to support spatial orientation (Fig. H of the color illustrations of this chapter at the end of the book). The environment can usually be shown as a collection of surfaces or solid objects, allowing the use of many spatial cues mentioned before. Techniques for rendering are mainly standard computer graphics methods. But it is important to note that objects of the environment must be displayed together with the data, and therefore the rendering technique must be used in combination with other rendering techniques (see 4.3.4).

4.2.6 Volume Rendering

If scalar data fields are associated with flow fields, or when scalar quantities are derived from a flow field (see 3.3.6), volume rendering techniques can be used to visualize these data (for a survey of volume rendering techniques, see Kaufman [34]). In volume rendering, a scalar data field (eg. a density field) is usually considered as a regular 3D array of cubical volume elements (voxels), with one scalar data item associated with each volume element. A segmentation is made of the data, by selecting only the voxels that satisfy a selection criterion (such as a given minimum scalar value), or by applying other 3D image segmentation techniques. The selected voxels are then projected onto the screen and rendered using special techniques for visible surface display and shading.

Although in volume rendering surfaces are often displayed directly, iso-surfaces can also be generated explicitly from volume data as a polygon mesh, using the method described in 3.3.8.

If a surface is defined that intersects the data volume, the scalar values on the surface can be determined by interpolation, and these can be shown using colour coding and iso-lines. The same can be done with a surface at the model boundary, such as an airplane wing or a ship hull. The surface may be shaded using directed light reflection, to provide good spatial cues.

Plane sections are often used in volume visualization to inspect a 3D data volume. This can be achieved by moving a plane through the volume and watching the changing patterns projected onto the section plane, to build a mental image of the whole field.

4.3 Special Rendering Techniques

4.3.1 Animation

Animation can be used for many purposes in flow visualization. As mentioned before, particle motion can be very effectively visualized using animation sequences. The velocity of particles can be directly visualized, provided that particle positions are known at constant intervals, and the update time of the images on the screen is also constant.

Animations can either be produced in real time, or simply by playback of a series of pre-computed images (frames). In real time animation, rendering of each frame is done during display. This has the advantage that the animation may be interactively controlled by the user. A disadvantage is that the screen update time usually depends on image complexity (for example, the number of visible particles), which may vary per frame. Also, the choice of rendering techniques is severely restricted. If we want to achieve the update rate of at least 5-10 frames per second required to produce the illusion of motion, only the most simple and fast rendering techniques can be used. To accelerate the process, some parameters such as particle paths can be pre-computed, but this reduces the possibilities for interactive control.

Pre-computed frames may be displayed one at a time and recorded on a video tape, or played back directly on the screen. In the latter case, all frames must be displayed from main memory, as transfer from disk is too slow. The size of main memory thus restricts the length of an on-screen animation. As the screen update rate is independent of the contents of the frames, playback speed can be constant. Interactive control of most viewing parameters (such as viewing direction) or positioning of particle sources is not possible.

A problem in animated flow visualization is the number of frames needed to get a clear view of the flow patterns. For an unsteady flow, a new velocity field must be computed for each frame, and new particle positions in this field. For one minute of animation, this must be performed for about 1500 frames. With a steady flow, particle paths are derived from a single velocity field; as the paths do not change in time, *cyclic animation* can be applied (Van Wijk [60]; Stolk and Van Wijk [52]), showing a smooth motion by continuously repeated display of a limited number of frames (typically 10-30).

Animation can be used for many other purposes in flow visualization. A changing viewpoint (or a rotating object) provides a powerful depth cue (motion parallax), that is quite independent of other cues, and can be used to resolve depth ambiguity even in wire frame drawings, such as arrow plots. Other applications of animation are inspection by moving a section plane through an area, or changing iso-values.

4.3.2 Aliasing and Anti-Aliasing

An important visual effect caused by the discrete nature of the display screen is known as aliasing. It is obvious from the jagged edges and silhouettes, Moiré patterns, irregular and granular textures, and missing small details. In animation, the effects are even more disturbing.

Aliasing can be explained from signal theory (Blinn [3]; Foley et al. [13]). The colour of a pixel is often determined by taking only one point sample. Better results can be achieved by considering a pixel as a small area to which one colour is assigned, and determining this colour from all objects that are visible within the pixel area. The contributions of the objects can be estimated by taking several point samples for one pixel (supersampling), or by geometric calculations (area sampling). The contributions are then combined to determine the colour of the pixel.

When combining several samples to determine the colour for a given pixel, digital low-pass filtering can be applied. The samples are weighted according to a filter function, which specifies the weights dependent on the distance of a sample point to the pixel center. From the sample values, a weighted average is then computed.

Analog to the spatial aliasing described above, *temporal* aliasing is caused by taking only one point sample for each time step. This results in jerky motion, flicker, "strobing" effects, and even false movements that are unacceptable in animated flow visualization.

To reduce this, the time interval between two frames should be considered. For a moving particle, this would be its trajectory over the time interval. Smooth motion can then be achieved using digital filtering, using this trajectory or a number of samples on it. The result is a blurred image of the moving particle, extended in the direction of motion, with a fuzzy shape. This effect, known as *motion blur*, produces smooth motion without false effects (Sims [50]).

4.3.3 Texture Synthesis and Texture Mapping

A texture is a pattern that can be mapped onto a surface, and used to modify the surface's optical properties. The most common form is modification of surface colour, resulting in a coloured pattern pasted onto the surface; other properties, such as reflectivity, transparency, or normal vector direction, can also be modified by texture (Heckbert [23]). In texture mapping, obvious adverse effects are

Fig. 19. Aliasing in texture mapping - left: with point sampling; right: with digital frequency filtering (Heckbert [23], ©1986, IEEE Computer Society Press. By Permission).

caused by the aliasing problems mentioned above, and therefore digital filtering is usually necessary to obtain good results (Fig. 19).

Existing images, such as digitized photographs, can be used as texture, but texture can also be generated synthetically. Techniques for synthesis of texture has been mainly developed to achieve naturalistic effects: to suggest certain materials or rough surfaces. As a primitive for visualization, texture has been little explored, but some of its potential is shown by Van Wijk ([61, 62]).

A good example is the synthesis of *spot noise*, a type of texture especially designed for data visualization (Van Wijk [62]). Spot noise is a stochastic texture generated as an addition of many randomly distributed and weighted 2D patterns (spots). The texture can be globally and locally controlled by varying the attributes of the spots. The spot size controls the granularity of the texture: with small spots, it has the nature of white noise, while with large spots it has fractal characteristics. This effect can be used to reflect a 2D data field, by scaling spot size according to a scalar field. If the spots are scaled non-uniformly, the texture becomes non-isotropic, and thus gets a directional nature. If the spots are locally scaled according to a 2D vector field, the texture clearly shows the directional pattern. Other effects may be obtained by taking spots with different shapes or patterns. In Fig. I of the color illustrations of this chapter at the end of the book, two sets of lines and a velocity field are mapped onto the hull of a ship.

4.3.4 Hybrid Rendering

In scientific visualization it can be of particular importance to display several types of data in a single picture. Data sets often contain combinations of scalar and vector data, and several types of derived quantities may be computed from these data. Also, display of environment geometry is necessary. Exploring the data may involve trying several combinations of visualization mappings, which should not be restricted to rendering in separate modes. Quick and crude rendering should be available as well as high-quality images. All these considerations lead to the concept of hybrid rendering.

There are three approaches to achieve this. First, before rendering, all geometric primitives can be converted to a single type. Examples of this are conversion of polygons and solids to volumetric representations using 3D scan-conversion algorithms (Kaufman and Shimony [35]), or to convert volume data to polygons by surface reconstruction (Lorensen and Cline [40]) or to represent volume elements by transparent tetrahedra (Shirley and Tuchman [49]). A second approach is using a technique that can be adapted for rendering volumetric and surface data directly, such as ray casting (Levoy [39]). A third approach to hybrid rendering is rendering each type of primitive with its own rendering technique, and then either merging the results using one or more pixel buffers (Kaufman et al. [36], Ebert and Parent [12], Van Walsum et al. [59], Frühauf [14]). Most of these techniques combine only volume and surface data (see eg. Figure H of the color illustrations of this chapter at the end of the book), and do not consider particles as a separate type of primitive; an exception is Sims [50]. Other types of primitives may have to be added as well.

4.3.5 Advanced Particle Rendering

As observed in 4.2.4, particles may be considered as a special type of visual primitives, for which special rendering techniques should be used. A first recognition of this idea was the development of *particle systems* (Reeves [44], Reeves and Blau [45]), where particles were used for modelling and rendering "fuzzy objects" with irregular, complex geometry, such as fire, trees and grass. An early application of particle systems in 2D flow visualization was the animation of the atmosphere of Jupiter in the film *2010* (Yaeger et al. [64]). A more recent example was shown by Van Wijk [60]).

A particle system is a collection of particles representing a fuzzy object. The particles have a certain life cycle: they are born, have a limited lifetime, and die; during the lifetime, various attributes such as motion dynamics (position, speed, motion direction), and visual appearance (shape, size, colour, reflectivity, transparency) may vary as a function of time. They can also have certain collective properties, which allows modelling of structured objects. In flow visualization, the motion dynamics of the particles is determined by a flow field, but the other attributes can be used to convey spatial information or to visualize associated data.

The rendering of the particles depends on their attributes. If particles are considered as light emitting points, then rendering is simply done by adding intensities of particles projected onto a single pixel, without any visibility calculations. These particles do not carry much spatial information. If the particles are considered as light-reflecting, shading calculations must be performed; also, self-shadowing (shadows cast by one particle onto others) is desirable as a depth cue. Precise shading and shadowing calculations are prohibitive for very large numbers (more than 10^6), but then probabilistic shading models can be used. The position and orientation of a particle determine the probability that it is lighted directly, and ambient, diffuse, and specular reflection components are assigned, based on these probabilities (Reeves and Blau [45]). The particles are depth sorted for visibility, and rendered in back-to-front order. Transparency can thus be achieved by blending the colours of the particles appearing in the same pixel.

To display more spatial information, a particle can be modelled as a very small surface element that reflects directed light; it is then called a *surface-particle* (Stolk and Van Wijk [52]). This has the additional advantage that a (large) collection of surface-particles may show a stream surface or a time surface (or indeed, any surface moving with and deformed by the flow), without determination of explicit surface geometry. As an example, a stream surface consisting of surface-particles is shown in Fig. J of the color illustrations of this chapter at the end of the book.

Shading calculations are performed for each particle, and for this purpose a normal vector is attached to the particle. The direction of this vector is dependent on the desired type of surface. The type of surface shown is determined by the shape of the particle source and the release time of the particles. The particle source is a zero to three-dimensional geometric object, on which particle starting positions are defined. The starting positions are distributed regularly or randomly over the length of a line segment, the area of a polygon, or the surface of a solid. Particles are released either continuously, or at discrete instants, at regular or random intervals. As space between the particles is not filled, the surfaces are naturally transparent, depending on the density of the particles.

A stream surface results if a line segment is used as a particle source, and particles are released at random intervals. If a polygon, or a rectangle is used for the source, and particles are all released at the same instant, a time surface is shown.

This technique is also very suitable to be used in animation. Particle release time and lifetime are chosen such that no discontinuities show up in cyclic animation of steady flow. Animated surface particles can give a very compelling view of 3D flow patterns.

Particle systems are often used in animations, and usually a kind of both spatial and temporal anti-aliasing must be applied. The particles can be rendered as short line segments or small circles, exploiting hardware facilities (Reeves and Blau [45]). A more elaborate particle model has been used by Sims [50]. He used rounded, elongated shapes, with a distinct position and radius for head and tail. The particles are scan converted, and spatial anti-aliasing and motion blur are accomplished by decreasing opacity near the edges, and setting the head and tail positions in accordance with velocity and shutter speed. Rendering of the particles is done without light reflection.

5 Conclusions and Research Directions

In the preceding sections, we have reviewed different aspects of flow visualization, including experimental and computer graphics flow visualization techniques. The connection between experimental and computer-aided flow visualization is now beginning to develop. The current strong demand for new flow visualization techniques, especially for large scale 3D numerical flow simulations, can only be satisfied by combining the efforts of fluid dynamics specialists, numerical analysts, and computer graphics experts. Additional knowledge will be required from perceptual and cognitive psychology, and artists and designers can also contribute to this effort.

Flow visualization will not be restricted to techniques for giving an intuitively appealing, general impression of flow patterns, but will increasingly focus on more specific physical flow phenomena, such as turbulence, separations and reattachments, shock waves, or free liquid surfaces. Also, purely visual analysis of flow patterns will be increasingly complemented by algorithmic techniques to extract meaningful patterns and structures, that can be visualized separately.

To take its place as a common research tool in fluid dynamics, computer graphics flow visualization will also have to deal with more specific questions, such as validation of new flow analysis techniques, involving comparison of data from CFD simulation results with experimental data. This may be done visually by displaying both types of data with the same techniques, but may also be supported by algorithmic similarity checking. Image processing techniques are often applied to images acquired from experimental visualization (Yang [65]) to derive data that can also be computed in the CFD simulation or the visualization stage.

Before they can be used as reliable research tools, the visualization techniques themselves must also be carefully tested and validated. As we have seen in the previous sections, visualization involves a sequence of many processing steps, where approximations are frequently used and numerical errors can easily occur. Even at the visual level, interpolation artifacts, aliasing, or contradictory depth cues may cause incorrect interpretations.

An important issue following development of visualization techniques, is the design and implementation of flow visualization systems. Here, there are many other issues that should be explored further, such as distributed processing, data management, standardization of data formats, and user interaction. Many styles and modes of interaction are possible, depending on the extent of control of the visualization (and flow simulation) process (Hearn and Baker [22]). The interaction style provided by general purpose visualization systems such as AVS (Upson et al. [56]) and apE (Dyer [11]), enabling the user to construct data flow networks, allows manipulation of the visualization process rather than the displayable objects representing the data. The concept of "direct manipulation" of the data needs clarification; desirable facilities would certainly include interactive probing and interrogation (Speray and Kennon [51]).

Research in computer graphics flow visualization is still in its early stages, and especially 3D flow field visualization is still very much an open problem. At present, this is one of the great challenges of scientific visualization. This calls for a cooperative effort in the development of new techniques at all stages of the flow visualization process.

Acknowledgements

We would like to thank Delft Hydraulics, the Netherlands Energy Research Foundation (ECN) and the High-Speed Laboratory of the Department of Aerospace Engineering at Delft University of Technology for their cooperation and for permission to use several illustrations. Also, we want to express thanks to Andrea Hin, Arthur Mynett, Guus Segal, and Jack van Wijk for their valuable criticism of earlier versions of this paper. Finally, we thank the editor Heinrich Müller for his encouragement and patience.

References

[1] G.K. Batchelor (1967). *An Introduction to Fluid Dynamics.* Cambridge University Press.

[2] F.H. Bertrand, P.A. Tanguy (1988). *Graphical Representation of Two-dimensional Fluid Flow by Stream Vectors. Communications in Applied Numerical Methods,* 4:213–217.

[3] J.F. Blinn (1989). *What We Need Around Here Is More Aliasing / Return of the Jaggy. IEEE Computer Graphics & Appl.,* 9(1/2):75–79 and 82–89.

[4] M. Briscolini, P. Santangelo (1991). *Animation of Computer Simulations of Two-Dimensional Turbulence and Three-Dimensional Flows. IBM Journal of Research and Development,* 35(1/2):119–138.

[5] W.F. Bronsvoort, F.W. Jansen, F.H. Post (1991). *Design and Display of Solid Models.* In I. Hermann G. Garcia, editor, *Advances in Computer Graphics VI,* pp. 1–57. Springer-Verlag.

[6] S. Bryson, C. Levit (1991). *The Virtual Windtunnel: An Environment for the Exploration of Three-Dimensional Unsteady Flows.* In G.M. Nielson, L. Rosenblum, editors, *Proceedings Visualization '91,* pp. 17–24. IEEE Computer Society Press.

[7] P.G. Buning (1989). *Numerical Algorithms in CFD Post-Processing.* von Karman Institute for Fluid Dynamics, Lecture Series 1989-07.

[8] H.H. Chen, T.S. Huang (1988). *A Survey of Construction and Manipulation of Octrees. Computer Vision, Graphics and Image Processing,* 43:403–431.

[9] M.S. Chong, A.E. Perry, B.J. Cantwell (1990). *A General Classification of Three-dimensional Flow Fields. Physics of Fluids A,* 2(5):765–777.

[10] R.R. Dickinson (1991). *Interactive Analysis of the Topology of 4D Vector Fields. IBM Journal of Research and Development,* 35(1/2):59–66.

[11] D.S. Dyer (1990). *A Dataflow Toolkit for Visualization. IEEE Computer Graphics & Appl.,* 10(4):60–69.

[12] D.S. Ebert, R.E. Parent (1990). *Rendering and Animation of Gaseous Phenomena by Combining Fast Volume and Scanline A-buffer Techniques. Computer Graphics,* 24(4):357–366.

[13] J.D. Foley, A. van Dam, S. Feiner, J. Hughes (1990). *Computer Graphics: Principles and Practice, second edition.* Addison-Wesley Publishing Company.

[14] M. Frühauf (1991). *Combining Volume Rendering with Line and Surface Rendering.* In F.H. Post, W. Barth, editors, *Eurographics '91*, pp. 21–32. North Holland.

[15] R.S. Gallagher (1991). *Span Filtering: an Optimization Scheme for Volume Visualization of Large Finite Element Models.* In G.M. Nielson, L. Rosenblum, editors, *Proceedings Viualization '91*, pp. 68–75. IEEE Computer Society Press.

[16] R.S. Gallagher, J.C. Nagtegaal (1989). *An Efficient 3-D Visualization Technique for Finite Element Models and Other Coarse Volumes. Computer Graphics*, 23(3):185–194.

[17] M. Geiben, M. Rumpf (1992). *Visualization of Finite Elements and Tools for Numerical Analysis.* In F.H. Post, A.J.S. Hin, editors, *Advances in Scientific Visualization.* Springer-Verlag.

[18] J.J. Gibson (1950). *The Perception of the Visual World.* Houghton Mifflin Co.

[19] J.J. Gibson (1979). *The Ecological Approach to Visual Perception.* Houghton Mifflin Co.

[20] A. Globus, C. Levit, T. Lasinski (1991). *A Tool for Visualizing the Topology of Three-Dimensional Vector Fields.* In G.M. Nielson, L. Rosenblum, editors, *Proceedings of Viualization '91*, pp. 33–40. IEEE Computer Society Press.

[21] R.B. Haber, D.A. McNabb (1990). *Visualization Idioms: A Conceptual Model for Scientific Visualization Systems.* In L.J. Rosenblum G.M. Nielson, B. Shriver, editor, *Visualization in Scientific Computing*, pp. 74–92. IEEE Computer Society Press.

[22] D.D. Hearn, P. Baker (1991). *Scientific Visualization.* Eurographics Technical Report Series EG91 TN6.

[23] P.S. Heckbert (1986). *Survey of Texture Mapping. IEEE Computer Graphics & Appl.*, 6(11):56–67.

[24] J. Helman, L. Hesselink (1989). *Analysis and Visualization of Flow Topology in Numerical Data Sets.* In *IUTAM Symposium on Topological Fluid Mechanics, Cambridge, England.*

[25] J. Helman, L. Hesselink (1989). *Automated Analysis of Fluid, Flow Topology, 3D Visualization and Display Technologies.* In *Proc. SPIE 1083*, pp. 825–835. SPIE, Bellingham, Wash.

[26] J. Helman, L. Hesselink (1989). *Representation and Display of Vector Field Topology in Fluid Flow Data Sets. IEEE Computer*, 22(8):27–36.

[27] J. Helman, L. Hesselink (1990). *Surface Representations of Two- and Three-Dimensional Fluid Flow Topology.* In A. Kaufman, editor, *Proceedings Visualization '90*, pp. 6–13. IEEE Computer Society Press.

[28] J. Helman, L. Hesselink (1991). *Visualizing Vector Field Topology in Fluid Flows. IEEE Computer Graphics & Appl.*, 11(3):36–46.

[29] L. Hesselink, J. Helman (1987). *Evalution of Flow Topology from Numerical Data.* Invited AIAA-paper, 87-1181-CP.

[30] L. Hesselink, J. Helman, K. Wu (1988). *Visualization and Interpretation of 3-D Scientific Data Sets.* In *ICALEO '88-Conference, Santa Clara CA.*

[31] W. Hibbard, D. Santek (1989). *Visualizing Large Data Sets in the Earth Sciences. IEEE Computer*, 22(8):53–57.

[32] A.J.S. Hin, E. Boender, F.H. Post (1990). *Visualization of 3D Scalar Fields using Ray Casting.* In Y. Le Lous M. Grave, editor, *Proceedings of the Eurographics Workshop on Visualization in Scientific Computing.* to be published by Springer-Verlag.

[33] J.P.M. Hultquist (1990). *Interactive Numeric Flow Visualization Using Stream Surfaces. Computing Systems in Engineering*, 1(2–4):349–353.

[34] A. Kaufman, editor (1990). *Volume Visualization.* IEEE Computer Society Press.

[35] A. Kaufman, E. Shimony (1986). *3D Scan Conversion Algorithms for Voxel-Based Graphics.* In F. Crow, S.M. Pizer, editors, *Proceedings 1986 Workshop on Interactive 3D Graphics*, pp. 45–75. ACM.

[36] A. Kaufman, R. Yagel, D. Cohen (1990). *Intermixing Surface and Volume Rendering.* In K.H. Höhne, H. Fuchs, S.M. Pizer, editors, *3D Imaging in Medicine: Algorithms, Systems, Applications*, pp. 217–227. Springer-Verlag.

[37] K.A. Kroos (1984). *Computer Graphics Flow Visualization Techniques for Three-Dimensional Flow Visualization.* In T.L. Kunii, editor, *Frontiers in Computer Graphics*, pp. 129–145. Springer-Verlag.

[38] M. Levoy (1988). *Display of Surfaces from Volume Data. IEEE Computer Graphics & Appl.*, 8(3):29–37.

[39] M. Levoy (1990). *A Hybrid Ray Tracer for Rendering Polygons and Volume Data. IEEE Computer Graphics & Appl.*, 10(2):33–40.

[40] W.E. Lorensen, H.E. Cline (1987). *Marching Cubes: a High Resolution 3D Surface Construction Algorithm. Computer Graphics*, 21(4):163–169.

[41] W. Merzkirch (1987). *Flow Visualisation, second edition.* Academic Press Inc.

[42] G.M. Nielson, B.S. Shriver, L.J. Rosenblum, editors (1990). *Visualization in Scientific Computing.* IEEE Computer Socitety Press.

[43] T.V. Papathomas, J.A. Schiavone, B. Julesz (1988). *Applications of Computer Graphics to the Visualization of Meteorological Data. Computer Graphics*, 22(4):327–335.

[44] W.T. Reeves (1983). *Particle Systems - a Technique for Modelling a Class of Fuzzy Objects. ACM Transactions on Graphics*, 2(2):91–108.

[45] W.T. Reeves, R. Blau (1985). *Approximate and Probabilistic Algorithms for Shading and Rendering Structured Particle Systems. Computer Graphics*, 19(3):313–322.

[46] M.A. Sabin (1985). *Contouring – the State of the Art.* In R.A. Earnshaw, editor, *Fundamental Algorithms for Computer Graphics*, pp. 411–482. Springer-Verlag.

[47] H. Samet (1990). *The Design and Analysis of Spatial Data Structures and Applications of Spatial Data Structures.* Addison-Wesley Publishing Company.

[48] W.J. Schroeder, C.R. Volpe, W.E. Lorensen (1991). *The Stream Polygon: a Technique for 3D Vector Field Visualization.* In L. Rosenblum G.M. Nielson, editor, *Proceedings Visualization '91*, pp. 126–132. IEEE Computer Society Press.

[49] P. Shirley, A. Tuchman (1990). *A Polygonal Approximation to Direct Scalar Volume Rendering. Computer Graphics*, 24(5):63–69.

[50] K. Sims (1990). *Particle Animation and Rendering Using Data Parallel Computation. Computer Graphics*, 24(4):405–413.

[51] D. Speray, S. Kennon (1990). *Volume Probes: Interactive Data Exploration on Arbitrary Grids. Computer Graphics*, 24(5):5–12.

[52] J. Stolk, J.J. van Wijk (1992). *Surface-particles for 3D Flow Visualization.* In A.J.S. Hin F.H. Post, editor, *Advances in Scientific Visualization.* Springer-Verlag.

[53] T. Strid, A. Rizzi, J. Oppelstrup (1989). *Development and Use of some Flow Visualization Algorithms.* von Karman Institute for Fluid Dynamics, Lecture Series 1989-07.

[54] Wavefront Technologies, editor (1990). *The Data Visualizer Version 1.0, Users Guide, 1990.* Wavefront Technologies.

[55] S.J. Thorpe (1990). *Image Processing by the Human Visual System.* Eurographics Technical Report Series EG90 TN4.

[56] C. Upson, et al. (1989). *The Application Visualization System: a Computational Environment for Scientific Visualization. IEEE Computer Graphics & Appl.*, 9(4):30–42.

[57] C. Upson, M. Keeler (1988). *V-Buffer: Visible Volume Rendering. Computer Graphics*, 22(4):59–64.

[58] M. van Dyke (1982). *An Album of Fluid Motion.* The Parabolic Press.

[59] T. van Walsum, A.J.S. Hin, J. Versloot, F.H. Post (1992). *Efficient Hybrid Rendering of Volume Data and Polygons.* In A.J.S. Hin F.H.Post, editor, *Advances in Scientific Visualization.* Springer-Verlag.

[60] J.J. van Wijk (1990). *A Raster Graphics Approach to Flow Visualization.* In D.A. Duce C.E. Vandoni, editor, *Eurographics '90*, pp. 251–259.

[61] J.J. van Wijk (1990). *Rendering Lines on Curved Surfaces.* In Y. Le Lous M. Grave, editor, *Proceedings of the Eurographics Workshop on Visualization in Scientific Computing.* Springer-Verlag.

[62] J.J. van Wijk (1991). *Spot Noise – Texture Synthesis for Data Visualization. Computer Graphics*, 25(4):309–318.

[63] J.J. van Wijk (1992). *Rendering Surface Particles. submitted for publication.*

[64] L. Yaeger, C. Upson, R. Myers (1986). *Combining Physical and Visual Simulation – Creation of the Planet Jupiter for the Film "2010". Computer Graphics*, 20(4):85–93.

[65] W.J. Yang, editor (1989). *Handbook of Flow Visualization.* Hemisphere Publishing Corporation.

Volume Visualization in Medicine: Techniques and Applications

Andreas Pommert, Michael Bomans, Martin Riemer, Ulf Tiede, Karl Heinz Höhne

Three-dimensional visualization of medical objects from tomographic volume data is increasingly considered useful in various fields. This paper reviews methods for all steps of the 3D imaging pipeline from data preprocessing to object definition and display, with special emphasis on advanced segmentation methods and surface- and voxel-based rendering techniques. Furthermore, multimodality matching, data manipulation, and aspects of image fidelity and implementation are discussed. Methods are illustrated with applications in craniofacial surgery, traumatology, neurosurgery, radiotherapy, and medical education.

1 Introduction

1.1 Objectives

Medical imaging technology has experienced a dramatic change over the past two decades. Previously, only X-ray radiographs were available which showed the depicted organs as superimposed shadows on photographic film. These images suffered from poor contrast and — even more important — gave no information about the depth of an object. With the advent of modern computers, new *tomographic* imaging modalities like computed tomography (CT) and magnetic resonance imaging (MRI) could be developed which deliver cross-sectional images of a patient's anatomy (*tomography*, from greek: $\tau o \mu \eta$ cut, $\gamma \rho \alpha \varphi \varepsilon \iota \nu$ to record). These images show different organs free of overlays in an unprecedented precision (Fig. 1). Even the three-dimensional (3D) structure of organs can be recorded if a sequence of parallel cross-sections is taken.

In current practice, the individual cross-sectional images of a tomographic study are visually inspected in order to establish a diagnosis. This procedure is suitable for typical radiological investigations like the detection of a tumor. For many clinical tasks like surgical planning, however, it is necessary to understand complex and often disordered 3D structures. Experience has shown that a "mental reconstruction" of objects from cross-sectional images is extremely difficult and strongly depends on the observer's training and imagination. For these cases, it is certainly desirable to present the human body as a surgeon or an anatomist would see it.

The aim of *3D imaging in medicine* is to create precise and realistic views of objects from medical volume data. The resulting images, even though they are of course two-dimensional, are often called *3D images* or *3D reconstructions*,

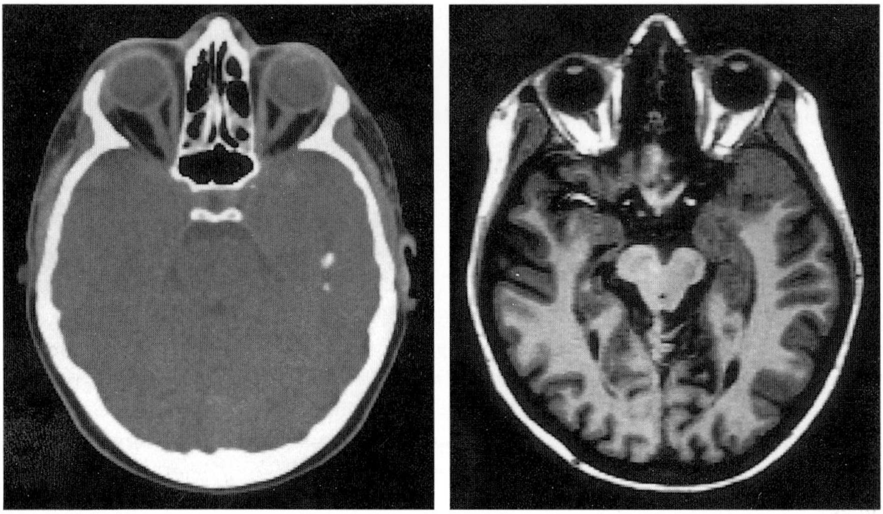

Fig. 1. Tomographic images of a head. Left: X-ray computed tomography (CT). Right: magnetic resonance imaging (MRI). These modalities show widely complementary aspects of the human anatomy.

to distinguish them from 2D cross-sections or conventional radiographs. The first attempts date back to the late 1970s [39, 40, 99], with the first clinical applications reported on visualization of bone from CT in craniofacial surgery and orthopedics [37, 42, 101, 111]. Methods and applications have since been extended to other subjects and imaging modalities. Recently, the same principles have also been applied to sampled and simulated data from other domains, such as fluid dynamics, geology, and meteorology. As a general expression, the term *volume visualization* is now widely accepted [54].

1.2 Related Fields

3D imaging has its roots in three other fields of computer science, which are image processing, computer vision, and computer graphics, respectively. *Image processing* deals with any image-to-image transformations, such as filters or geometric transformations [91]. Most steps in 3D imaging can therefore be considered as special image processing methods.

Computer vision, also known as *image understanding*, is the construction of symbolic descriptions from input images [5, 100]. In 3D imaging, the more low-level functions of image segmentation and interpretation are used in order to identify different parts of a volume which may be displayed or removed. They have to be strictly distinguished from higher level functions such as automatic detection of lesions or even computer aided diagnosis, which are investigated in *artificial intelligence* [96].

Computer graphics provides methods to synthesize images from numerical descriptions [29, 115]. These techniques were originally developed for realistic display of human-defined objects, such as models from *computer aided design* (CAD). Objects in 3D space are usually represented by infinitely thin surface patches such as triangles or higher order curves. Contributions of computer graphics to 3D imaging include data structures, projection techniques and shading models.

2 Imaging Modalities

Medical imaging technology is based on various physical phenomena, such as X-ray attenuation in *computed tomography* (CT) [53, 63], relaxation of magnetized hydrogen nuclei in *magnetic resonance imaging* (MRI) [98], sound reflections in *ultrasonography* (US) [51, 116], or radioactive decay of injected markers in *positron emission tomography* (PET) and *single photon emission computed tomography* (SPECT) [7]. The resulting images show widely complementary aspects of a patient's anatomy (structure) and physiology (function). CT is especially suitable to image high density objects such as bone; MRI, in contrast, is very sensitive to variations in soft tissue (Fig. 1). Compared to these, US offers a rather low image quality, the depicted structures are mainly borders between different organs. PET and SPECT visualize the metabolism inside a patient.

All mentioned imaging modalities are fully computerized and deliver the images already in digital form as a matrix of currently typical 128^2 to 1024^2 picture elements. The *intensity* of a picture element or *pixel* represents the physical property, measured in a small rectangular volume element or *voxel* in 3D space. In CT and MRI, intensity resolution is typically 12 bits, equivalent to 4096 different gray values. This exceeds by far the abilities of a human observer who can distinguish at most between 100 different intensity levels.

For 3D imaging, it is important whether different tissue types can be automatically identified. In CT, major classes such as background, soft tissue and bone show characteristic intensity ranges. At their borders, however, the so-called *partial volume effect* occurs where two or more tissues are present within one voxel. The resulting intermediate intensity value indicates the percentages of the different materials.

In contrast to CT, the intensity ranges of different tissue types in MRI images typically overlap or are even identical. However, MRI can generate multi-parameter images of the same anatomy, which somewhat compensates for this problem. Each parameter shows a different aspect, such as proton density or various relaxation times (so-called T_1- and T_2-weighted images). The recently introduced *MR angiography* (MRA) emphasizes flow effects and is thus especially suitable to visualize blood vessels.

3 Methods

3.1 Overview

This chapter presents a rather detailed description of 3D imaging methodology. An overview is shown in Fig. 2. After acquisition of a series of parallel cross-sections, the data usually undergoes some preprocessing for data conversion (Sect. 3.2.1) and image filtering (Sect. 3.2.2). From this point, one of several paths may be followed.

The dotted line in fig. 2 represents an early approach where an object is reconstructed from its contours on the cross-sectional images. This method is shortly reviewed in Sect. 3.4.1. All other methods, represented by the solid line, start from a contiguous *data volume*. If required, equal spacing in all three directions can be achieved by interpolation (Sect. 3.2.3). Like a 2D image, a 3D volume can be filtered to improve image quality.

The next step is to identify the different objects represented in the data volume so that they can be removed or selected for visualization. This task breaks down into segmentation and interpretation (Sect. 3.3). The simplest way is to binarize the data with an intensity threshold, e.g. to distinguish bone from other tissues in CT. Especially for MRI data, however, more sophisticated segmentation methods are required. This field is still subject to basic research, the major directions are presented in Sect. 3.6.

At this point, there is a choice which *rendering* technique is to be used. The more traditional surface-based methods first create an intermediate surface representation of the object to be shown (Sect. 3.4). It may then be rendered with any standard computer graphics method. More recently, voxel-based methods have been developed which create a 3D view directly from the volume data (Sect. 3.5). These methods use the full gray level information to render surfaces, cuts, or transparent and semi-transparent volumes. Both surface- and voxel-based methods have their merits; the decision which one should be taken for a particular application depends both on the available memory and computing power and on the visualization goals.

The next two sections discuss some extensions to the 3D imaging pipeline presented so far. The objective of multimodality matching is to register data volumes from different sources (Sect. 3.7). Manipulation of volume data is used e.g. for surgical simulation systems (Sect. 3.8). Finally, sections 3.9 and 3.10 take a look at the accuracy of the resulting 3D images, and some hard- and software considerations for implementing a 3D imaging system.

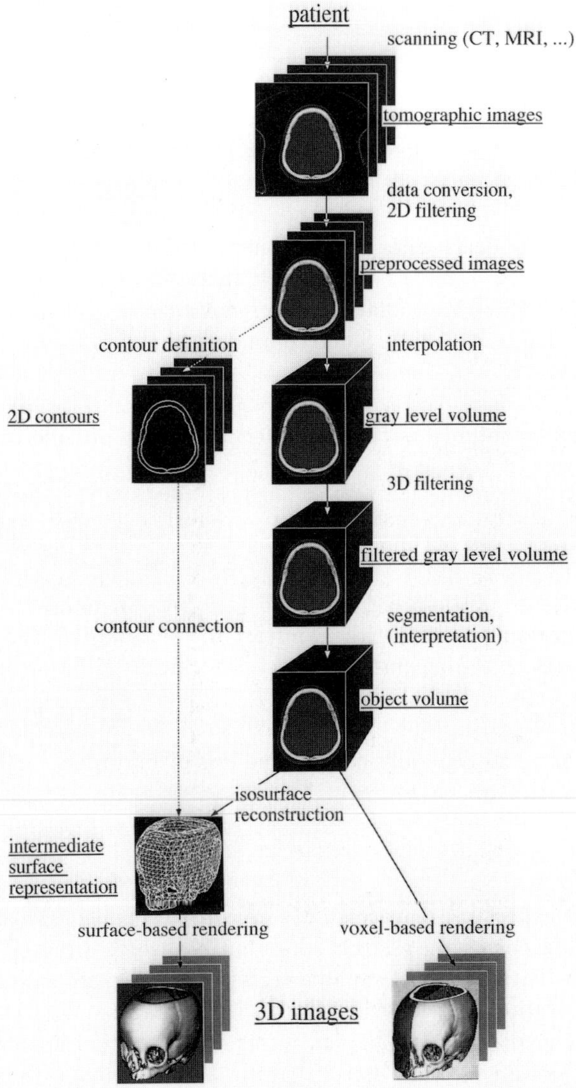

Fig. 2. General sketch of the 3D imaging pipeline. Individual processing steps may be left out, combined, or reversed in order by a particular method.

3.2 Preprocessing

3.2.1 Data Conversion

The first step in the 3D processing pipeline after image acquisition is usually data conversion. Besides a change of the data format which may be required, this involves a number of measures for data reduction to save storage space and processing time.

3D imaging usually deals with huge amounts of data. A typical CT study of 80 cross-sections with 512×512 pixels each takes 40 megabytes of memory. Furthermore, if an explicit interpolation step is performed, the size of the data may be multiplied (see Sect. 3.2.3). Some common techniques for data reduction are

- Cutting: a region of interest is chosen; other parts (e.g. outside the body) of the images are cut.
- Reduced spatial resolution: matrix size is reduced, e.g. by averaging from 512^2 to 256^2 pixels.
- Reduced intensity resolution, e.g. from 16 to 8 bit: an intensity window is chosen which represents most of the contrast in the images. This is usually done on the basis of a *histogram* which shows the distribution of the gray values.

The latter two points will generally cause some loss of information, they should therefore be used with care.

3.2.2 Filtering

Another important aspect of preprocessing is image *filtering* [91]. This is a rather general term for all kinds of image processing routines which are used to smooth or enhance the information contents of an image. A typical example is to improve the signal-to-noise ratio, especially in MRI and US images. Well-known *noise filters* are average, median, and Gaussian filters. These filters however tend to smooth out small details as well. Better results are obtained with *anisotropic diffusion* methods [82]. Other filter types are applied to emphasize special aspects of an image, e.g. to enhance edges.

Filters can be designed to work on 1D lines, 2D images, 3D volumes, or higher dimensional data. In principle, a 1D filter can also be applied to the individual rows or columns of a 2D image. In general, however, results are better if an image is filtered with a 2D filter and a volume with a 3D filter.

3.2.3 Interpolation

At this point, the data is still a stack of 2D images. If they are put on top of each other, a contiguous *gray level volume* is obtained. The resulting data structure is an orthogonal 3D array of voxels, each representing an intensity value. It is called the *voxel-model.*

Many algorithms for 3D imaging are working on *isotropic* volumes where the sampling density is equal in all three dimensions. In practice, however, only very few data sets have this property. CT images are often taken with considerable and varying spacing between the cross-sections so that the resolution in the image plane is much better than perpendicular to it. In these cases, the missing information has to be reconstructed in an *interpolation* step. A very simple method is image replication [111]. Better results are obtained with linear interpolation of the intensities between adjacent images. Higher order functions such as splines may be used as well [80]. An alternative approach are shape-based methods [89]; these are however wasting most of the original gray level information.

3.2.4 Data Structures

There are a number of different data structures for volume data. The most important are

- *binary voxel-model:* voxel values are either 1 (object) or 0 (no object). This very simple model is not much used any more. In order to reduce storage requirements, binary volumes may be recursively subdivided into homogeneous subvolumes; the resulting data structure is called an *octree* [74].
- *gray level voxel-model:* each voxel holds an intensity information.
- *generalized voxel-model:* besides a gray value, each voxel contains further information such as an object membership label ("this voxel belongs to brain"), material percentages ("this voxel contains 60 % bone"), or data from other sources (e.g. MRI and PET). This data structure is the basis for many advanced applications [46].

3.3 Object Definition

A gray level volume usually represents a large number of different organs. To display a particular one, we thus have to decide which parts of the data we want to use or ignore. In an ideal case, selection would be done with a command like "show only the brain". This however requires that the computer knows which parts of the volume (or, more precisely, which voxels) constitute the brain and which do not. This information is also needed for morphometric measurements of distances, angles and volumes.

The aim of *object definition* is to establish relationships between voxels and meaningful (anatomical) terms. This task breaks down into segmentation and interpretation. The result is an *object volume* which may be organized e.g. using the generalized voxel-model.

3.3.1 Segmentation

The first step towards object definition is to partition the gray level volume into different regions which are homogeneous with respect to some formal criteria and corresponding to real anatomical objects. This process is called *segmentation* [5].

A very simple but nevertheless important example which is used throughout the next sections is to specify a certain intensity range with lower and upper *threshold values*. A voxel belongs to the selected class if and only if its intensity value is within the specified range. Thresholding is a common method to select bone or soft tissues in CT. It is often performed during the rendering process itself so that no explicit segmentation step is required.

A drawback of any binary decisions (such as thresholding) is that small objects which take only a small fraction of a voxel and cases of uncertainty cannot be handled properly. To model these cases, *fuzzy* segmentation techniques have been developed. Here, a set of probabilities is assigned to every voxel, indicating the evidence for different materials. For CT data, Drebin et al. use a maximum likelihood classifier which calculates material percentages from a-priori known distributions, according to Bayes' rule [24]. Fuzzy segmentation is closely related to the so-called volume rendering methods discussed in Sect. 3.5.5.

The segmentation task gets much more complicated if different organs with similar gray level characteristics are to be distinguished. Advanced segmentation methods for these cases are still subject to basic research, the major directions are presented in Sect. 3.6.

3.3.2 Interpretation

Segmentation delivers a set of regions in a data volume, but in many cases it is not clear which anatomical objects they represent. If required, an *interpretation* step can be performed in which the various regions are *identified* and *labeled* with meaningful terms such as "white matter" or "ventricle". This can be done interactively or with an automatic system. For simple applications such as visualization of bone from CT, interpretation is usually left out.

3.4 Surface-Based Rendering

The first techniques for volume visualization which evolved in the late 1970s were strongly based on conventional computer graphics methods. Key idea of these so-called *surface-based* methods is to extract an *intermediate surface description* of the relevant objects from the volume data. Only this information is then used for rendering.

A clear advantage of surface-based methods is the possibly very high data reduction from volume to surface representations. This affects both memory requirements and computing times. Views from different angles can thus quickly be generated. Computing times can be further reduced if the surface representations are based on common data structures such as triangle meshes which are supported by computer graphics workstations. Another advantage in this case is that standard computer graphics software can be used.

On the other hand, the *surface reconstruction* step throws away most of the valuable information on the cross-sectional images [102]. Once the surface representation is created, there is no way to get back to the original intensity values. Even simple cuts cannot be done because there is no information about the interior of an object. Furthermore, every change of surface definition criteria requires recalculation of the whole data structure.

The following subsections focus on methods of how to reconstruct surface descriptions from tomographic images. The subsequent rendering is largely based on standard computer graphics methods and will thus only shortly be reviewed.

3.4.1 Surface Reconstruction from Contours

In 1975, Keppel presented an algorithm to reconstruct a surface representation from a stack of planar contours [59]. This method has since been modified by several other authors [9, 18, 32, 99]. In a first step, a set of object contours is defined on every tomographic image (Fig. 2, dotted line). This may be done interactively or with an edge detecting operator. In a second step, the contours from adjacent cross-sections are connected to form a 3D structure. If triangles are used as surface elements, this process is called *triangulation.*

The crucial step with this method is to connect properly the different contours. Especially for medical data, shapes are often extremely complex and vary greatly from one cross-section to the next. With general solutions lacking, contour connection is based on heuristic rules; it is however questionable whether they will apply in every case. Therefore, methods for surface reconstruction from 2D contours are not much used in practice today.

3.4.2 Surface Reconstruction from Volumes

In contrast to contour connection, surface reconstruction from volumes is a true 3D operation. Given a certain intensity value, its goal is to create an *isosurface*, representing all points where this intensity is found in the original gray level volume. Alternatively, the surface can be defined using object membership labels.

The first method to be widely used in clinical practice was developed by Herman and coworkers [3, 17, 23, 37, 40]. It is known as the *cuberille model* (Fig. 3). The gray level volume is first binarized with an intensity threshold. Then, a list of square voxel faces is created which denote the border between voxels in- and outside the object, using a surface tracking algorithm. It can be shown that the resulting surfaces are always well-defined and closed.

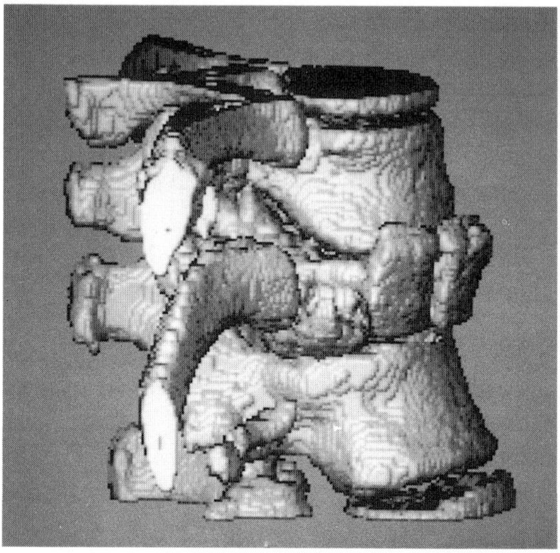

Fig. 3. 3D image of a fractured vertebra produced from CT scans using the cuberille model.

A surface description created with this algorithm is quite simple in the sense that all faces are of the same size and shape, with only six different orientations. Of course, this is only a rough approximation of the actual object form. The resulting 3D images thus miss a lot of fine details.

More recently, methods have been described which utilize the full gray level information. The *marching cubes* algorithm developed by Lorensen and Cline basically considers a cube of 2×2×2 contiguous voxels in the data volume [21, 69]. Depending on whether one or more of these voxels are inside the object (i.e. above a threshold value), a surface representation of up to four triangles is placed within the cube. The exact location of the triangles is found by linear interpolation of the intensities at the voxel vertices. The result is a highly detailed surface representation with subvoxel resolution. Surface orientations are calculated from gray level gradients (see sections 3.4.3 and 3.5.2). Some erroneous details of the algorithm are discussed in [4, 118].

Applied to clinical data, the marching cubes algorithm typically creates hundreds of thousands of triangles. Most of these triangles however are so small that they hardly contribute to the final 3D image. A somewhat simplified approach developed by the same group uses points instead of triangles [21]. This method, called *dividing cubes*, subdivides a group of 2×2×2 contiguous voxels into smaller cubes, whereby the intensities are interpolated. The surface description is made from the cubes that approximate the threshold value. As with the marching cubes algorithm, surface orientations are calculated from gray level gradients.

3.4.3 Shading

After the surface representation has been created with one of the above described methods, it is mapped to a raster image display to make the final 3D image. This so-called *rasterization* step breaks down into the subtasks of scan conversion, hidden surface removal and shading [29]. Scan conversion and hidden surface removal are standard problems of computer graphics and will thus not be covered here. For surface shading, however, a number of non-standard methods are used. A more detailed survey of shading methods for 3D imaging is found in [57].

In general, *shading* is the realistic display of an object, based on position, orientation and characteristics of its surface and the light sources illuminating it [29, 36]. The reflective properties of a surface are described with an *illumination model* such as the Phong model, which uses a combination of ambient light, diffuse (such as chalk) and specular (such as polished metal) reflections. A key input to these models is the local surface orientation, described by a *normal vector* perpendicular to the surface.

In principle, the surfaces created with the cuberille method can be rendered with any of the methods developed in computer graphics. Due to the low dynamic range of only six different surface orientations, however, the images appear more or less jagged. An alternative approach is to use the information in the so-called *z-buffer*. This 2D array describes the local depth of a scene, i.e. the distance between image plane and object surface. In *distance shading*, the intensity of a pixel is a function only of the corresponding value in the z-buffer [41, 111]. A more realistic impression is obtained if the z-buffer is used to estimate the local surface normal vectors. This *distance gradient shading* method was first used by Gordon and Reynolds [35], a number of variations have since been published [17, 57, 103]. Still, image quality is low, as compared to other methods.

The original marching cubes algorithm calculates the surface normal vectors from the gray level gradients in the data volume. This method was first used for voxel-based surface rendering (see Sect. 3.5.2). Alternatively, the surface normal vectors of the created triangles can be taken. Both versions deliver highly detailed images, where the latter variation shows some staircase artifacts. Images produced with both methods are compared in [87, 103].

3.5 Voxel-Based Rendering

In *voxel-based rendering*, images are created directly from the volume data. No intermediate surface representations are needed. After some early experiments by Oswald et al. [79], Tuy and Tuy [106], and Lenz et al. [64], voxel-based rendering of gray level volumes evolved in the second half of the 1980s and has since gained an enormous popularity. Compared to surface-based methods, the major advantage is that all gray level information which has originally been acquired is kept during the rendering process. As shown by Höhne et al. [45, 47, 49], this makes it an ideal technique for interactive data exploration. Threshold values and other parameters which are not clear from the beginning can interactively be changed. Furthermore, voxel-based rendering allows a combined display of different aspects such as opaque and semi-transparent surfaces, cuts, and maximum intensity projections.

A current drawback of voxel-based techniques is that the large amount of data which have to be handled does not allow real-time applications on present day computers. With dedicated hardware, however, rendering times are already down to less than a second. With computing power further increasing, this problem will be overcome in a few years.

3.5.1 Projection Techniques

In voxel-based rendering, we basically have the choice between two scanning strategies: pixel by pixel (image order) or voxel by voxel (volume order). These strategies correspond to the image and object order rasterization algorithms used in computer graphics [29].

Image order strategies scan the data volume on rays along the view direction [64, 106]. These methods are commonly known as *ray casting*:

FOR each pixel on image plane **DO**
  FOR each sampling point on associated ray **DO**
    compute contribution to pixel

The principle is illustrated in Fig. 4. Ray casting can also be considered as a non-recursive variant of the ray tracing methods used in computer graphics. Along the ray, visibility of surfaces and objects is easily determined. After an opaque surface has been found, the ray can stop.

Ray casting is a very flexible and intuitive scanning strategy. Integration of opaque, semi-transparent and transparent rendering methods is comparatively easy. Furthermore, image order scanning can be used to render both voxel and polygon data at the same time [58, 67]. All images presented in this section were rendered with a ray casting technique.

On the other hand, performance of ray casting algorithms is limited both by high memory (random access to the input volume) and computing (interpolation for oblique rays) requirements. A strategy to reduce computation times is to prerotate the whole volume so that the rays for a given view angle scan along the lines of the array [24, 49]. Even if several images are rendered from this view, interpolation is thus required only once.

A different technique for speed-up is to start with a coarse sampling density to generate a view quickly. If the user does not specify any changes, sampling density is adaptively refined to full resolution [68].

Volume order strategies scan the input volume along lines or columns of the 3D array, projecting a chosen aspect onto the image plane in the direction of view:

FOR each sampling point in volume **DO**
  FOR each pixel projected onto **DO**
    compute contribution to pixel

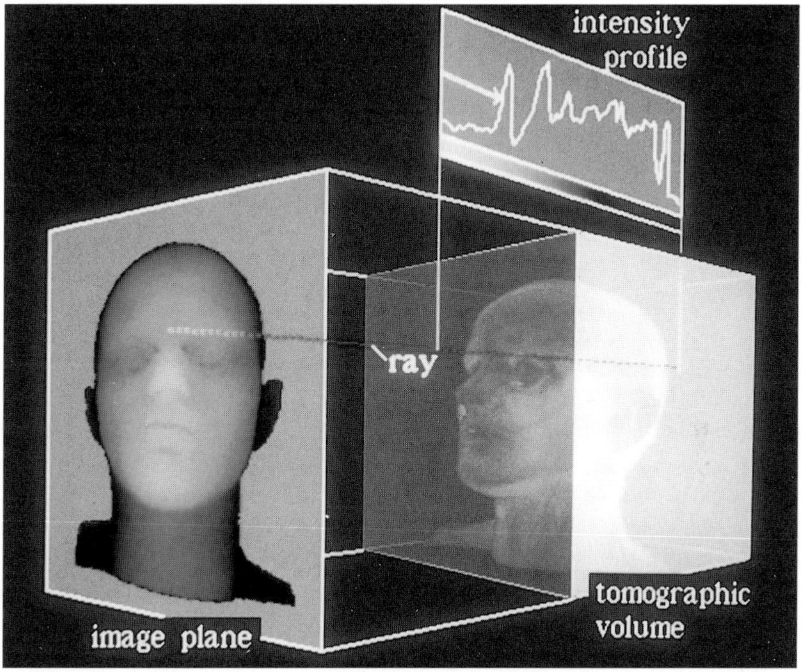

Fig. 4. Principle of ray casting for volume visualization.

The volume can either be traversed in back-to-front (BTF) order [30] from the voxel with maximal to the voxel with minimal distance to the image plane, or vice versa in front-to-back (FTB) order [35, 79]. In both cases, several voxels may be projected to the same pixel. If an opaque surface is to be shown, the visible parts thus have to be determined. In BTF, pixel values are simply overwritten so that only the visible surface appears. In FTB, pixels which have already been written are protected using a z-buffer.

Scanning the input data as they are stored, these techniques are reasonably fast even on computers with small main memory, and especially suitable to be parallelized. So far, ray casting algorithms still offer a higher flexibility in combining different display techniques. However, volume rendering techniques (see Sect. 3.5.5) working in volume order have already been developed [117].

3.5.2 Surfaces

Using one of the described projection methods, the visible surface of an object can be determined with a threshold or an object membership label. For shading, any of the methods developed for the cuberille model, such as distance or distance gradient shading, can be applied (see Sect. 3.4.3).

As shown by Höhne and Bernstein [43] and independently by Barillot et al. [6], a much more realistic and detailed presentation is obtained if the gray level information present in the data is taken into account. As a consequence of the tomographic acquisition process, the gray levels in the 3D neighborhood of a surface voxel are representing the relative proportions of different materials inside these voxels. The resulting *gray level gradients* can thus be used to calculate the surface inclinations. The simplest variant is to calculate the components of a gradient G for a surface voxel at (i, j, k) from the gray levels g of its six neighbors along the main axes as

$$
\begin{aligned}
G_x &= g(i+1, j, k) - g(i-1, j, k) \\
G_y &= g(i, j+1, k) - g(i, j-1, k) \\
G_z &= g(i, j, k+1) - g(i, j, k-1)
\end{aligned}
$$

Normalization of G yields the surface normal [104]. The gray level gradient may also be calculated from all 26 neighbors in a $3 \times 3 \times 3$ neighborhood, weighted according to their distance from the surface voxel [104, 121]. Aliasing patterns are thus almost eliminated.

In case of very small objects like thin bones, the gray level gradient does not correspond to the actual surface inclination any more. Pommert et al. proposed an *adaptive gray level gradient method* which chooses only 3–6 meaningful neighbors [86, 87, 103]. Basic idea is to maximize the gradient magnitude. This algorithm gives smooth images even for thin objects.

3.5.3 Cut Planes

Once a surface view is available, a very simple and effective method to visualize interior structures is cutting. If the original intensity values are mapped onto the cut plane, they can be better understood in their anatomical context [45, 47, 49]. A special case is selective cutting, where certain objects are excluded (Fig. 5).

3.5.4 Integral and Maximum Intensity Projection

A different way to look into an object is to integrate over the intensity values along the viewing ray. If applied to the whole data volume, this is a step back to the old X-ray projection technique [47, 64]. If applied in a selective way, it is nevertheless helpful in special cases [49, 103].

For small bright objects such as vessels from MR angiography, *maximum intensity projection* is a suitable display technique. Along each ray through the data volume, the maximum gray value is determined and projected onto the image plane[27]. The advantage of this method is that neither segmentation nor shading are needed, which may fail for very small vessels. But there are also some drawbacks: as light reflection is totally ignored, maximum intensity projection does not give a realistic 3D impression. Spatial perception can be improved by rotating the object or by a combined presentation with other surfaces or cut planes [45].

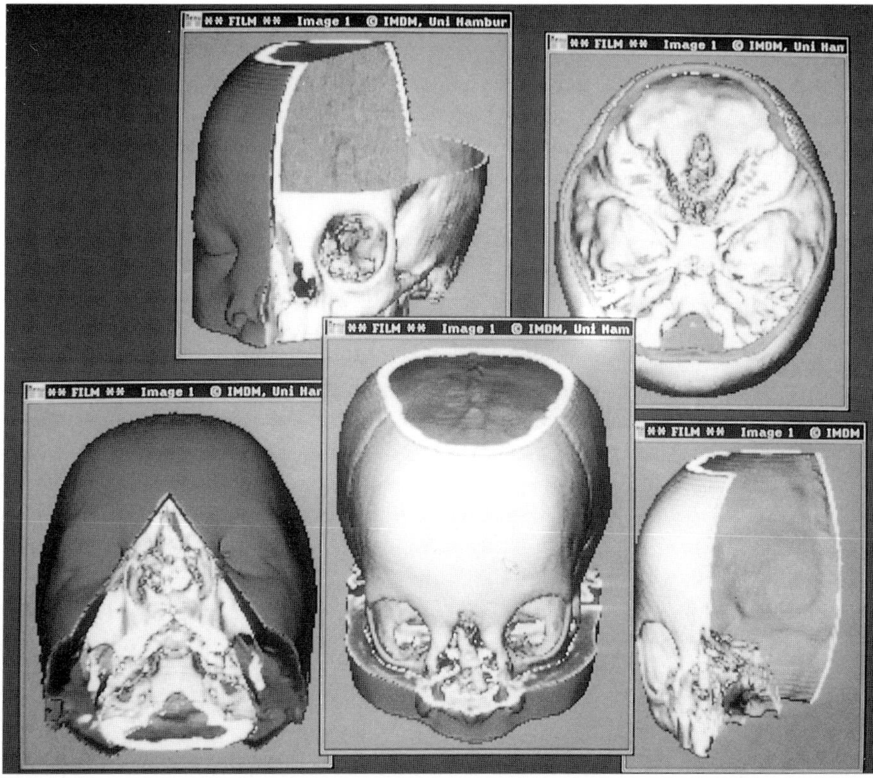

Fig. 5. 3D image of a child with a congenital facial cleft from CT (ray casting, gray level gradient shading). Soft tissues and bone were detected with threshold values and rendered as opaque surfaces. Cuts are visualizing the original intensity values.

3.5.5 Volume Rendering

Volume rendering is the visualization equivalent to fuzzy segmentation (Sect. 3.3.1). These methods were first described in 1988 by Drebin et al. [24], Levoy [65], Sabella [92], and Upson and Keeler [109], and since modified by various groups [45, 62, 75, 90, 117]. A commonly assumed underlying model is that of a colored, semi-transparent gel with suspended reflective particles [8]. Illumination rays are partly reflected and change color while travelling through the volume.

Each voxel is assigned a color and an opacity. This opacity is a product of an "object weighting function" and a "gradient weighting function". The object weighting function is usually dependent of the gray value, but it can also be the result of a more sophisticated fuzzy segmentation algorithm. The gradient weighting function emphasizes surfaces for 3D display. All voxels are shaded, using the gray level gradient method. The shaded values along a viewing ray are weighted and summed up. A somewhat simplified basic equation is given as follows (frontal illumination):

I	intensity of reflected light	
p	index of sampling point on ray	$(0 \ldots$ maximum depth of scene)
l	fraction of incoming light	$(0.0 \ldots 1.0)$
α	local opacity	$(0.0 \ldots 1.0)$
s	local shading component	

$$I(p,l) = \alpha(p) \cdot l \cdot s(p) + (1 - \alpha(p)) \cdot I(p+1, l - \alpha(p))$$

The total reflected intensity as displayed on the 3D image is given as $I(0, 1.0)$.

Since binary decisions are avoided in volume rendering, the resulting images are very smooth and show a lot of fine details (Fig. A. of the color illustrations of this chapter at the end of the book). Another important advantage is that even coarsely defined objects can be rendered with acceptable quality. The 12 week old fetus shown in Fig. B of the color illustrations was rendered from rather noisy ultrasonography data for which a binary segmentation proved worthless. Other examples can be found in [103].

On the other hand, the more or less transparent images produced with volume rendering methods are often hard to understand such that their clinical use is limited [87, 103]. Spatial perception can however be improved by rotating the object. Another serious problem is the large number of parameters which have to be specified to define the weighting functions. A good mapping is difficult to find, and even small variations can completely change the image. Finally, volume rendering is comparably slow because weighting and shading operations are performed for many voxels on each ray. If certain values such as gradients are pre-calculated, a substantial speed-up can be achieved at the cost of higher memory requirements [24].

3.6 Advanced Segmentation Methods

For all 3D images shown so far, the interesting regions were selected with a threshold or a fuzzy classification. Unfortunately, these simple segmentation methods are not suitable if different structures have overlapping or even identical gray level ranges. This situation frequently occurs e.g. for soft tissues from CT and MRI.

Currently, a large number of advanced segmentation methods for 3D medical data are being developed. The major directions of research are presented in the following subsections. Division into three classes – point- edge- and region-based methods – roughly follows that in [5]. The described methods often have been tested successfully on a number of cases; experience has shown, however, that they should be used with extreme care.

3.6.1 Point-Based Segmentation

Point-based segmentation methods classify a voxel only depending on its intensity, no matter where it is located. A simple example is thresholding, which can be applied to single- or multi-parameter data, such as T_1- and T_2-weighted images in MRI. In the latter case, individual threshold values are used for every parameter.

To somewhat generalize this concept, voxels in an n-parameter data set can be considered as n-dimensional vectors in an n-dimensional *feature space*. In *pattern recognition* [26], the feature space is partitioned into arbitrarily shaped subspaces, representing different tissue classes or organs. This is called the *training phase*: in supervised training, the partition is derived from feature vectors which are known to represent particular tissues [20, 33, 77, 113]. In unsupervised training, the partition is automatically generated [33]. In the subsequent *test phase*, a voxel is classified according to the position of its feature vector in the partitioned feature space.

Pattern recognition methods have successfully been applied to considerable numbers of two- or three-parametric MRI data volumes of head [20, 33, 113] and chest [77]. Quite frequently, however, isolated voxels or small regions are incorrectly classified (e.g. subcutaneous fat in the same class as white matter). To get rid of these errors, a connected component analysis is often applied (see region-based segmentation).

A closely related method based on recent *neural network* methodology has been developed by Kohonen [60]. Instead of an n-dimensional feature space, a so-called *topological map* of $m \times m$ n-dimensional vectors is used. During the training phase, the map iteratively adapts itself to a set of training vectors which may either represent selected tissues (supervised learning, [114]) or the whole data volume (unsupervised learning, [75]). Finally, the map develops several relatively homogeneous regions, which correspond to different tissues or organs in the original data. Performance of the topological map for 3D MRI data seems to be generally equivalent to pattern recognition methods [75, 114].

3.6.2 Edge-Based Segmentation

The aim of edge-based segmentation methods is to detect intensity discontinuities in a gray level volume. These edges (in 3D, they are actually surfaces; it is however common to speak about edges) are assumed to represent the borders between different organs or tissues. Regions are subsequently defined as the enclosed areas.

A common strategy for edge detection is to locate the maxima of the first derivative of the 3D intensity function. A method which very accurately locates the edges was developed by Canny [16]. All algorithms using the first derivative however have the drawback that the detected contours are usually not closed, i.e. they do not properly separate different regions. To repair broken edges, combinations with region-based methods are currently being investigated [13].

An alternative approach is to detect zero-crossings of the second derivative. The Marr-Hildreth operator convolves the input data with the Laplacian of a Gaussian; the resulting contour volume describes the locations of the edges [73]. With a 3D extension of this operator, the complete human brain could be segmented and visualized from MRI for the first time [11, 12]. Similar approaches were developed by other groups [61]. Occasionally, however, this operator creates erroneous "bridges" between different materials which have to be interactively removed. Furthermore, curved edges are dislocated outwards. A method to correct this error, based on morphological filters[97], is presented in [11].

3.6.3 Region-Based Segmentation

Region-based segmentation methods are considering whole regions instead of individual voxels or contours. Since we are actually interested in regions, this approach appears to be the most natural. Properties of a region are e.g. its size, shape, location, variance of gray levels, and its spatial relation to other regions.

A typical application of region-based methods is to post-process the results of a previous point-based segmentation step. A *connected component analysis* is applied to determine whether the voxels which have been classified as belonging to the same class are part of the same (connected) region. If not, the voxels in the smaller regions are often wrongly classified. On the other hand, *region growing* algorithms can be used to split and merge greater regions, according to certain criteria [5].

Region-based methods often combine segmentation and interpretation steps into a single algorithm. The knowledge required may be provided interactively by a human user, or automatically by a model. Cline et al. have developed an algorithm which grows a region from a user-selected seed voxel [19]. Höhne and Hanson [48] propose an interactive segmentation system based on *mathematical morphology* [97]. Regions are initially defined with thresholds; the user can subsequently apply operations such as "erosion" (to remove small "bridges" between erroneously connected parts), "dilation" (to close small gaps), region fill, or Boolean set operations. After each step, segmentation results are instantaneously inspected on a 3D image (Fig. C of the color illustrations of this chapter at the end of the book).

A different interactive method developed by Pizer et al. first creates a multiscale hierarchy of volume representations, based on higher-order features such as symmetry axes [83, 84]. Finally, the user can interactively select, add or subtract regions, or move to larger "parent" or smaller "child" regions in the hierarchy.

There is also a number of automatic systems for region-based segmentation and interpretation. Raya and Udupa use a rule-based system to successively generate a set of thresholds [88]. Brummer et al. encode the knowledge required to detect brain contours in a fixed sequence of morphological operations [15]. Bomans generates a set of hypotheses for every voxel, depending on its gray value [10]. Location, surface-volume ratio etc. of the resulting regions are compared to some predefined values, and the regions are modified accordingly (Fig. D of the color illustrations of this chapter at the end of the book). Menhardt uses a rule-based system which models the anatomy with relations such as "brain is inside skull" [76]. Regions are defined as fuzzy subsets of the volume, and the segmentation process is based on fuzzy logic and fuzzy topology.

The problem with automatic segmentation and interpretation systems is of course that the results may be wrong if the underlying model does not properly represent the data. The models used so far are not yet adequate to handle e.g. various pathologies.

3.7 Multimodality Matching

It is often desirable to combine information from different imaging modalities to improve the information available to the clinician. For example, PET images show only physiological aspects; for their interpretation, it is necessary also to know the anatomy, as shown in MRI.

In general, different data sets do not match geometrically. It is therefore required to *register* the volumes in relation to each other. This is quite a difficult task; to make it somewhat easier, external markers can be attached to the patient which will be visible on the different modalities. Their positions in the respective volumes are defining the geometric transformation. Without this additional information, it is necessary to define corresponding features in both data sets. A very robust method which registers the 3D skin surface of the patient has been developed by the group of Levin [52, 81]. Other approaches use corresponding landmarks, which are e.g. interactively defined on 3D images (Fig. F of the color illustrations of this chapter at the end of the book) [93].

3.8 Manipulation

So far, focus has been on mere visualization of the data. A step further is to manipulate the data at the computer for surgery simulation [2, 14, 119]. These techniques are especially useful for craniofacial surgery where a skull is dissected into small pieces, and then rearranged to achieve a desirable shape. In the system developed by Yasuda et al. [119], a cut is defined by interactively drawing a closed contour onto a 3D image. The resulting segments can be inspected from other view directions, and individually moved and rearranged in 3D space. The system can even roughly predict the resulting face.

3.9 Image Fidelity

For clinical applications, it is of course important to assure that the resulting 3D images really show the true anatomical situation, or at least to know about their limitations. A common approach is to compare 3D images rendered by means of different algorithms [108]. This method however is of limited value since the "truth" usually is not known.

A different approach to investigate image fidelity is to apply 3D imaging techniques to simulated data [70, 86, 87, 103], and to data acquired from cadavers [25, 38, 85]. In both cases, the actual situation is available for comparison. Using the first technique, the accuracy e.g. of different shading algorithms could be shown. Results of the latter studies are e.g. visibility of sutures or fracture gaps as a function of acquisition parameters and object size. Diagnostic performance of 3D imaging and conventional radiographs has been compared using the *receiver operator characteristic* (ROC) analysis [112].

3.10 Implementation Aspects

Acceptance of volume visualization systems in a clinical environment strongly depends both on computing times used and the availability of user-friendly interfaces. In order to speed up image generation, a number of dedicated hardware systems have been developed which allow near-real-time applications even for volume data (e.g. voxel processor by Goldwasser et al. [34], Cube by Kaufman and Bakalash [55], or Pixel-Planes 5 by Fuchs et al. [31, 66]). A key idea of these systems is to support massively parallel operation. A survey of major systems, both research and commercial, is presented in [56].

While dedicated hardware systems are not yet used in large numbers, a number of software packages have been developed which run directly on scanner or general purpose workstations (e.g. Voxel-Man [45], ANALYZE [90], 3D98[107], AVS[109]). While the use of advanced window and menu techniques is already widespread, handling of the whole process from data conversion to 3D imaging is still too complicated for the non-technical user.

4 Applications

At first glance, one might expect diagnostic radiology to be the major field of application for 3D imaging. This is however not the case. One of the reasons is clearly that radiologists are especially skilled in reading cross-sectional images. Another reason is that many diagnostic tasks such as tumor detection and classification can well be done from tomographic images. Furthermore, 3D visualization of these objects from MRI requires robust segmentation algorithms which are not yet available.

The situation is generally different in all fields where therapeutical decisions have to be made by non-radiologists on the basis of radiological images [44]. A major field of application for 3D imaging methods is *craniofacial surgery* [23, 72, 110, 120]. A typical case is shown in Fig. 5.

Another important field of application is *traumatology*. Due to the emergency situation, planning times are usually very short. With new faster imaging modalities available and computing power ever increasing, 3D imaging techniques are being introduced for difficult cases. Especially in pelvic surgery where the morphology is difficult to assess, 3D imaging is considered a great help (Fig. E of the color illustrations of this chapter at the end of the book) [28, 78].

An application that becomes more and more attractive with the increasing resolution and specificity of MRI is *neurosurgery planning.* Here the problem is to choose a proper access path to a lesion. 3D visualization of brain tissue from MRI and blood vessels from MR angiography before surgical intervention allows the surgeon to find a path with minimal risk in advance (Fig. D of the color illustrations of this chapter at the end of the book) [22, 45]. In combination with a 3D coordinate digitizer, the acquired information can even be used to guide the surgeon during the intervention [1]. In conjunction with functional information from PET images, localization of a lesion is facilitated [52].

Another potential application that reduces the risk of a therapeutical intervention is *radiotherapy planning.* Here, the objective is to focus the radiation to the target volume while avoiding radiation on healthy organs. 3D visualization of target volume, organs at risk and simulated radiation dose allows the realistic rehearsal of the treatment procedure (Fig. G of the color illustrations of this chapter at the end of the book). Work in this field is done at several places [94, 95].

Applications apart from clinical work are e.g. *medical research and education* [71, 105]. The 3D brain atlas shown in Fig. H of the color illustrations is based on a high quality MRI data set where every voxel has interactively been marked with a label, describing its membership to an anatomical or functional constituent of the brain [50]. On the basis of the thus acquired description, various teaching programs can be written that allow dissection and surgical training at the computer screen.

5 Conclusions

3D imaging in medicine has come a long way from the first experiments to the current, highly detailed renderings. As the rendering algorithms are improved and the fidelity of the resulting images is investigated, 3D images are not only pretty pictures, but a powerful source of information to the clinician. In certain areas such as craniofacial surgery or traumatology, 3D imaging is increasingly becoming part of the standard preoperative procedures.

A number of problems still hinder an even broader use of 3D imaging methods. First, current workstations are not yet able to deliver 3D images fast enough. For the future, it is certainly desirable to be able to interact with the workstation in real time, instead of just looking at static images or pre-calculated movies. With further increase of computing power, this problem will be overcome in a few years.

The second major problem is the design of a user interface which is suitable for the clinician. Currently, there are still a large number of rather technical parameters e.g. for control of segmentation and shading. Acceptance in the medical community will certainly depend heavily on progress in this field.

A third problem is the segmentation of a volume into meaningful parts, representing different objects. To date, there are no robust methods which perform well in every case; especially for MRI. As has been shown, there is research in different directions going on which might eventually succeed.

In the near future, 3D imaging is likely to see the surface-based rendering methods widely replaced by the much more powerful and flexible voxel-based rendering methods. Furthermore, a number of applications beyond mere visualization will become operational, e.g. surgical simulation systems and educational systems. In all these cases, experimental setups are already available.

The more distant future may see incorporation of functional information, e.g. of motion, force, etc., even dynamically changing over time (*4D imaging*). Another intriguing idea is to combine 3D imaging with current efforts in *virtual reality* systems, which will enable the clinician to walk around or even fly through a virtual patient [31]. The future will show which of these new techniques are really useful for clinical work.

Acknowledgements

The authors would like to thank B. Pflesser, T. Schiemann (both IMDM) and Dr. D. Friboulet (Institut National des Sciences Appliquées, Lyon) for many discussions and practical assistance. We also thank I. Gulens for her help in preparing the manuscript.

Tomographic raw data were kindly provided by Dr. F. Hottier, Philips Paris (Fig. B), Dr. J. P. Mugler, University of Virginia (Fig. C), and Siemens, Erlangen (Figs. D, H). Applications are in cooperation with Prof. W.-J. Höltje, Dept. of Craniofacial Surgery (Fig. 5), Dr. A. Wening, Dept. of Traumatology (Fig. F), Dr. R. Schmidt, Dept. of Radiotherapy (Fig. G), and Prof. W. Lierse, Dept. of Anatomy (Fig. H).

Work at the IMDM is supported by the Werner Otto Foundation, Hamburg.

References

[1] L. Adams, W. Krybus, D. Meyer-Ebrecht, R. Rueger, J.M. Gilsbach, R. Moesges, G. Schloendorff (1990). *Computer-Assisted Surgery. IEEE Computer Graphics & Appl.*, 10(3):43–51.

[2] S.R. Arridge (1990). *Manipulation of Volume Data for Surgical Simulation.* In K.H. Höhne, et al., editors, *3D-Imaging in Medicine: Algorithms, Systems, Applications*, pp. 289–300. Springer-Verlag, Berlin.

[3] E. Artzy, G. Frieder, G.T. Herman (1981). *The Theory, Design, Implementation and Evaluation of a Three-Dimensional Surface Detection Algorithm. Computer Graphics and Image Processing*, 15(1):1–24.

[4] H.H. Baker (1989). *Building Surfaces of Evolution: the Weaving Wall. Comput. Vis.*, 3:51–71.

[5] D.H. Ballard, C.M. Brown (1982). *Computer Vision.* Prentice-Hall Inc., Englewood Cliffs, NJ.

[6] C. Barillot, B. Gilbaud, L.M. Luo, J.M. Scarabin (1985). *3-D Representation of Anatomic Structures from CT Examinations.* In *Proc. SPIE 602: Biostereometrics '85*, pp. 307–314.

[7] D.R. Bernier, P.E. Christian, J.K. Langman, L.D. Wells (1989). *Nuclear Medicine: Technology and Techniques.* C. V. Mosby Co., St. Louis, MO.

[8] J.F. Blinn (1982). *Light Reflection Functions for Simulation of Clouds and Dusty Surfaces. Computer Graphics*, 16(3):21–29.

[9] J.D. Boissonnat (1988). *Shape Reconstruction from Planar Cross Sections. Computer Vision, Graphics and Image Processing*, 44(1):1–29.

[10] M. Bomans (in preparation). *Vergleich verschiedener Verfahren und Entwicklung eines kombinierten Verfahrens zur Segmentation von Kernspintomogrammen des Kopfes.* Ph. D. thesis, Dept. of Computer Science, University of Hamburg.

[11] M. Bomans, K.H. Höhne, U. Tiede, M. Riemer (1990). *3D-Segmentation of MR-Images of the Head for 3D-Display. IEEE Trans. Med. Imaging*, MI-9(2):177–183.

[12] M. Bomans, M. Riemer, U. Tiede, K.H. Höhne (1987). *3D-Segmentation von Kernspin-Tomogrammen.* In E. Paulus, editor, *Mustererkennung 1987, Proc. 9. DAGM-Symposium*, pp. 231–235. Springer-Verlag, Berlin.

[13] G.J. Brelstaff, M.C. Ibison, P.J. Elliot (1990). *Edge-Region Integration for Segmentation of MR Images.* In *Proc. British Machine Vision Conference, BMVC '90*, pp. 139–144.

[14] L.J. Brewster, S.S. Trivedi, H.K. Tuy, J.K. Udupa (1984). *Interactive Surgical Planning. IEEE Computer Graphics & Appl.*, 4(3):31–40.

[15] M.E. Brummer, R.M. Mersereau, R.L. Eisner, R.R.J. Lewine (1991). *Automatic Detection of Brain Contours in MRI Data Sets.* In A.C.F. Colchester, D.J. Hawkes, editors, *Information Processing in Medical Imaging, Proc. IPMI '91*, pp. 188–204. Springer-Verlag, Berlin.

[16] J. Canny (1985). *A Computational Approach to Edge Detection. IEEE Trans. Pattern Anal. Machine Intell.*, PAMI-8(6):679–698.

[17] L.S. Chen, G.T. Herman, R.A. Reynolds, J.K. Udupa (1985). *Surface Shading in the Cuberille Environment. IEEE Computer Graphics & Appl.*, 5(12):33–43.

[18] S.-Y. Chen, W.-C. Lin, C.-C. Liang, C.-T. Chen (1990). *Improvement on Dynamic Elastic Interpolation Technique for Reconstructing 3-D Objects from Serial Cross Sections. IEEE Trans. Med. Imaging*, MI-9(1):71–83.

[19] H.E. Cline, C.L. Dumoulin, H.R. Hart, W.E. Lorensen, S. Ludke (1987). *3D Reconstruction of the Brain from Magnetic Resonance Images Using A Connectivity Algorithm. Magn. Reson. Imaging*, 5:345–352.

[20] H.E. Cline, W.E. Lorensen, R. Kikinis, F. Jolesz (1990). *Three-Dimensional Segmentation of MR Images of the Head Using Probability and Connectivity. J. Comput. Assist. Tomogr.*, 14(6):1037–1045.

[21] H.E. Cline, W.E. Lorensen, S. Ludke, C.R. Crawford, B.C. Teeter (1988). *Two Algorithms for Three-Dimensional Reconstruction of Tomograms. Med. Phys.*, 15(3):320–327.

[22] H.E. Cline, W.E. Lorensen, S.P. Souza, F.A. Jolesz, R. Kikinis, G. Gerig, T.E. Kennedy (1991). *3D Surface Rendered MR Images of the Brain and its Vasculature. J. Comput. Assist. Tomogr.*, 15(2):344–351.

[23] D.J. David, D.C. Hemmy, R.D. Cooter (1990). *Craniofacial Deformities: Atlas of Three-Dimensional Reconstruction from Computed Tomography.* Springer-Verlag, New York.

[24] R.A. Drebin, L. Carpenter, P. Hanrahan (1988). *Volume Rendering. Computer Graphics*, 22(4):65–74.

[25] R.A. Drebin, D. Magid, D.D. Robertson, E.K. Fishman (1989). *Fidelity of Three-dimensional CT Imaging for Detecting Fracture Gaps. J. Comput. Assist. Tomogr.*, 13(3):487–489.

[26] R.O. Duda, P.E. Hart (1973). *Pattern Classification and Scene Analysis.* John Wiley and Sons, New York.

[27] H.-H. Ehricke, G. Laub (1990). *Combined 3D-Display of Cerebral Vasculature and Neuroanatomic Structures in MRI.* In K.H. Höhne, et al., editors, *3D-Imaging in Medicine: Algorithms, Systems, Applications*, pp. 229–239. Springer-Verlag, Berlin.

[28] E.K. Fishman, D.R. Ney, D. Magid (1990). *Three-Dimensional Imaging: Clinical Applications in Orthopedics.* In K.H. Höhne, et al., editors, *3D-Imaging in Medicine: Algorithms, Systems, Applications*, pp. 425–440. Springer-Verlag, Berlin.

[29] J.D. Foley, A. van Dam, S.K. Feiner, J.F. Hughes (1990). *Computer Graphics: Principles and Practice, Second edition.* Addison-Wesley Publ. Comp., Reading, MA.

[30] G. Frieder, D. Gordon, R.A. Reynolds (1985). *Back-to-Front Display of Voxel-Based Objects. IEEE Computer Graphics & Appl.*, 5(1):52–59.

[31] H. Fuchs (1990). *Systems for Display of Three-Dimensional Medical Image Data.* In K.H. Höhne, et al., editors, *3D-Imaging in Medicine: Algorithms, Systems, Applications*, pp. 315–331. Springer-Verlag, Berlin.

[32] H. Fuchs, Z.M. Kedem, S.P. Uselton (1977). *Optimal Surface Reconstruction from Planar Contours. Commun. ACM*, 20(10):693–702.

[33] G. Gerig, J. Martin, R. Kikinis, O. Kübler, M. Shenton, F.A. Jolesz (1991). *Automating Segmentation of Dual-Echo MR Head Data.* In A.C.F. Colchester, D. Hawkes, editors, *Information Processing in Medical Imaging, Proc. IPMI '91*, pp. 175–187. Springer-Verlag, Berlin.

[34] S.M. Goldwasser, R.A. Reynolds, T. Bapty, D. Baraff, J. Summers, D.A. Talton, E. Walsh (1985). *Physicians Workstation With Real Time Performance. IEEE Computer Graphics & Appl.*, 5:44–57.

[35] D. Gordon, R.A. Reynolds (1985). *Image Space Shading of 3-Dimensional Objects. Computer Vision, Graphics and Image Processing*, 29:361–376.

[36] R. Hall (1986). *A Characterization of Illumination Models and Shading Techniques. Visual Comput.*, 2:268–277.

[37] D.C. Hemmy, D.J. David, G.T. Herman (1983). *Three-Dimensional Reconstruction of Craniofacial Deformity Using Computed Tomography. Neurosurgery*, 13:534–541.

[38] D.C. Hemmy, P.L. Tessier (1985). *CT of Dry Skulls with Craniofacial Deformities: Accuracy of Three-Dimensional Reconstruction. Radiology*, 157(1):113–116.

[39] G.T. Herman, H.K. Liu (1977). *Display of Three-Dimensional Information in Computed Tomography. J. Comput. Assist. Tomogr.*, 1(1):155–160.

[40] G.T. Herman, H.K. Liu (1979). *Three-dimensional display of human organs from computed tomograms. Computer Graphics and Image Processing*, 9:1–21.

[41] G.T. Herman, J.K. Udupa (1981). *Display of Three-Dimensional Discrete Surfaces.* In *Proc. SPIE 283*, pp. 90–97.

[42] G.T. Herman, J.K. Udupa (1983). *Display of 3-D Digital Images: Computational Foundations and Medical Applications. IEEE Computer Graphics & Appl.*, 3(5):39–46.

[43] K.H. Höhne, R. Bernstein (1986). *Shading 3D-Images from CT Using Gray Level Gradients. IEEE Trans. Med. Imaging*, MI-5(1):45–47.

[44] K.H. Höhne, M. Bomans, B. Pflesser, A. Pommert, M. Riemer, T. Schiemann, U. Tiede (1992, in press). *Perspectives of 3D Imaging. Diagn. Imaging*, 1.

[45] K.H. Höhne, M. Bomans, A. Pommert, M. Riemer, C. Schiers, U. Tiede, G. Wiebecke (1990). *3D-Visualization of Tomographic Volume Data Using the Generalized Voxel-Model. Visual Comput.*, 6(1):28–36.

[46] K.H. Höhne, M. Bomans, U. Tiede, M. Riemer (1988). *Display of Multiple 3D-Objects Using the Generalized Voxel-Model.* In R.H. Schneider, S.J. Dwyer, editors, *Proc. SPIE 914, Medical Imaging II, Part B, Newport Beach*, pp. 850–854.

[47] K.H. Höhne, R.L. DeLaPaz, R. Bernstein, R.C. Taylor (1987). *Combined Surface Display and Reformatting for the 3D-Analysis of Tomographic Data. Invest. Radiol.*, 22:658–664.

[48] K.H. Höhne, W.A. Hanson (1992, in press). *Interactive 3D-Segmentation of MRI and CT Volumes Using Morphological Operations. J. Comput. Assist. Tomogr.*

[49] K.H. Höhne, M. Riemer, U. Tiede. (1987). *Viewing Operations for 3D-Tomographic Gray Level Data.* In H.U. Lemke, et al., editors, *Computer Assisted Radiology, Proc. CAR '87*, pp. 599–609. Springer-Verlag, Berlin.

[50] K.H. Höhne, M. Riemer, U. Tiede, W. Lierse (submitted). *A 3D Anatomical Atlas Based on a Volume Model. IEEE Computer Graphics & Appl.*

[51] F. Hottier, A. Collet Billon (1990). *3D Echography: Status and Perspective.* In K.H. Höhne, et al., editors, *3D-Imaging in Medicine: Algorithms, Systems, Applications*, pp. 21–41. Springer-Verlag, Berlin.

[52] X. Hu, K.K. Tan, D.N. Levin, C.A. Pelizzari, G.T.Y. Chen (1990). *A Volume-Rendering Technique for Integrated Three-Dimensional Display of MR and PET Data.* In K.H. Höhne, et al., editors, *3D-Imaging in Medicine: Algorithms, Systems, Applications*, pp. 379–397. Springer-Verlag, Berlin.

[53] A.C. Kak, M. Slaney (1988). *Principles of Computerized Tomographic Imaging.* IEEE Press Inc., New York.

[54] A. Kaufman, editor (1991). *Volume Visualization.* IEEE Computer Society Press, Los Alamitos, CA.

[55] A. Kaufman, R. Bakalash (1988). *Memory and Processing Architecture for 3D Voxel-Based Imagery. IEEE Computer Graphics & Appl.*, 8(11):10–23.

[56] A. Kaufman, R. Bakalash, D. Cohen, R. Yagel (1991). *Architectures for Volume Rendering.* In A. Kaufman, editor, *Volume Visualization*, pp. 311–320. IEEE Computer Society Press, Los Alamitos, CA.

[57] A. Kaufman, D. Cohen, R. Yagel (1991). *Volumetric Shading Techniques.* In A. Kaufman, editor, *Volume Visualization*, pp. 169–173. IEEE Computer Society Press, Los Alamitos, CA.

[58] A. Kaufman, R. Yagel, D. Cohen (1990). *Intermixing Surface and Volume Rendering.* In K.H. Höhne, et al., editors, *3D-Imaging in Medicine: Algorithms, Systems, Applications*, pp. 217–227. Springer-Verlag, Berlin.

[59] E. Keppel (1975). *Approximating Complex Surfaces by Triangulation of Contour Lines. IBM J. Res. Develop.*, 19(1):2–11.

[60] T. Kohonen (1988). *Self-Organisation and Associative Memory, second edition.* Springer-Verlag, Berlin.

[61] O. Kübler, J. Ylä-Jääski, E. Hiltebrand (1987). *3-D Segmentation and Real Time Display of Medical Volume Images.* In H.U. Lemke, et al., editors, *Computer Assisted Radiology, Proc. CAR '87*, pp. 637–641. Springer-Verlag, Berlin.

[62] D. Laur, P. Hanrahan (1991). *Hierarchical Splatting: A Progressive Refinement Algorithm for Volume Rendering. Comput. Graphics*, 25(4):285–288.

[63] J.K.T. Lee, S.S. Sagel, R.J. Stanley, editors (1989). *Computed Body Tomography with MRI Correlation, second edition.* Raven Press, New York.

[64] R. Lenz, P.E. Danielsson, S. Cronström, B. Gudmundsson (1986). *Presentation and Perception of 3-D Images.* In K.H. Höhne, editor, *Pictorial Information Systems in Medicine*, pp. 459–468. Springer-Verlag, Berlin.

[65] M. Levoy (1988). *Display of Surfaces from Volume Data. IEEE Computer Graphics & Appl.*, 8(3):29–37.

[66] M. Levoy (1989). *Design for a Real-Time High-Quality Volume Rendering Workstation.* In C. Upson, editor, *Proc. Chapel Hill Workshop on Volume Visualization*, pp. 85–92. Dept. of Computer Science, University of North Carolina, Chapel Hill, NC.

[67] M. Levoy (1990). *A Hybrid Ray Tracer for Rendering Polygon and Volume Data. IEEE Computer Graphics & Appl.*, 10(?):33–40.

[68] M. Levoy (1990). *Volume Rendering by Adaptive Refinement. Visual Comput.*, 6(1):2–7.

[69] W.E. Lorensen, H.E. Cline (1987). *Marching Cubes: A High Resolution 3D Surface Construction Algorithm. Comput. Graphics*, 21(4):163–169.

[70] M. Magnusson, R. Lenz, P.E. Danielsson (1988). *Evaluation of Methods for Shaded Surface Display of CT-Volumes.* In *Proc. 9th International Conference on Pattern Recognition, ICPR '88*, pp. 1287–1294. IEEE Computer Society Press, Washington, DC.

[71] I. Mano, Y. Suto, M. Suzuki, M. Iio (1990). *Computerized Three-Dimensional Normal Atlas. Radiat. Med.*, 8(2):50–54.

[72] D. Marchac, editor (1987). *Craniofacial Surgery: Proc. of the First International Congress of the International Society of Cranio-Maxillo-Facial Surgery.* Springer-Verlag, Berlin.

[73] D. Marr, E. Hildreth (1980). *Theory of Edge Detection. Proc. R. Soc. Lond.*, B 207:187–217.

[74] D.J. Meagher (1982). *Geometric Modeling Using Octree Encoding. Computer Graphics and Image Processing*, 19(2):129–147.

[75] H.P. Meinzer, U. Engelmann, D. Scheppelmann, R. Schäfer (1990). *Volume Visualization of 3D Tomographies.* In K.H. Höhne, et al., editors, *3D-Imaging in Medicine: Algorithms, Systems, Applications*, pp. 253–259. Springer-Verlag, Berlin.

[76] W. Menhardt (1988). *Image Analysis Using Iconic Fuzzy Sets.* In *European Conference on Artificial Intelligence, Proc. ECAI'88*, pp. 672–674. Pitman Publ., London.

[77] M.B. Merickel, T. Jackson, C. Carman, J.R. Brookeman, C.R. Ayers (1990). *A Multispectral Pattern Recognition System for the Noninvasive Evaluation of Atherosclerosis Utilizing MRI.* In K.H. Höhne, et al., editors, *3D-Imaging in Medicine: Algorithms, Systems, Applications*, pp. 133–146. Springer-Verlag, Berlin.

[78] D. Ney, E.K. Fishman, D. Magid, R.A. Drebin (1990). *Volumetric Rendering of Computed Tomography Data: Principles and Techniques. IEEE Computer Graphics & Appl.*, 10:24–32.

[79] H. Oswald, W. Kropatsch, F. Leberl (1982). *A Perspective Projection Algorithm with Fast Evaluation of Visibility for Discrete Three-Dimensional Scenes.* In *Proc. ISMIII '82, International Symposium on Medical Imaging and Image Interpretation*, pp. 464–468. IEEE Computer Society Press, Silver Spring, MD.

[80] J.A. Parker, R.V. Kenyon, D.E. Troxel (1983). *A Comparison of Interpolating Methods for Image Resampling. IEEE Trans. Med. Imaging*, MI-2(1):31–39.

[81] C.A. Pelizzari, G.T.Y. Chen, D.R. Spelbring, R.R. Weichselbaum, C. Chen (1989). *Accurate Three-Dimensional Registration of CT, PET, and/or MR Images of the Brain. J. Comput. Assist. Tomogr.*, 13(1):20–26.

[82] P. Perona, J. Malik (1987). *Scale Space and Edge Detection using Anisotropic Diffusion.* In *Proc. IEEE Workshop on Computer Vision, Miami*, pp. 16–22.

[83] S.M. Pizer, T.J. Cullip, R.E. Fredericksen (1990). *Toward Interactive Object Definition in 3D Scalar Images.* In K.H. Höhne, et al., editors, *3D-Imaging in Medicine: Algorithms, Systems, Applications*, pp. 83–105. Springer-Verlag, Berlin.

[84] S.M. Pizer, H. Fuchs, M. Levoy, J. Roseman, R.E. Davis, J.B. Renner (1989). *3D Display with Minimal Predefinition.* In H.U. Lemke, et al., editors, *Computer Assisted Radiology, Proc. CAR '89*, pp. 723–736. Springer-Verlag, Berlin.

[85] A. Pommert, W.-J. Höltje, N. Holzknecht, U. Tiede, K.H. Höhne (1991). *Accuracy of Images and Measurements in 3D Bone Imaging.* In H.U. Lemke, et al., editors, *Computer Assisted Radiology, Proc. CAR '91*, pp. 209–215. Springer-Verlag, Berlin.

[86] A. Pommert, U. Tiede, G. Wiebecke, K.H. Höhne (1989). *Image Quality in Voxel-Based Surface Shading.* In H.U. Lemke, et al., editors, *Computer Assisted Radiology, Proc. CAR '89*, pp. 737–741. Springer-Verlag, Berlin.

[87] A. Pommert, U. Tiede, G. Wiebecke, K.H. Höhne (1990). *Surface Shading in Tomographic Volume Visualization: A Comparative Study.* In *Proc. First Conference on Visualization in Biomedical Computing, VBC '90*, pp. 19–26. IEEE Computer Society Press, Los Alamitos, CA.

[88] S.P. Raya, J.K. Udupa (1990). *Low-Level Segmentation of 3-D Magnetic Resonance Brain Images — A Rule-Based System. IEEE Trans. Med. Imaging*, MI-9(3):327–337.

[89] S.P. Raya, J.K. Udupa (1990). *Shape-Based Interpolation of Multidimensional Objects. IEEE Trans. Med. Imaging*, MI-9(1):32–42.

[90] R.A. Robb, C. Barillot (1989). *Interactive Display and Analysis of 3-D Medical Images. IEEE Trans. Med. Imaging*, MI-8(3):217–226.

[91] A. Rosenfeld, A.C. Kak (1982). *Digital Picture Processing, second edition.* Academic Press, New York.

[92] P. Sabella (1988). *A rendering algorithm for 3D scalar fields. Computer Graphics*, 22(4):51–58.

[93] C. Schiers, U. Tiede, K.-H. Höhne (1989). *Interactive 3D-Registration of Image Volumes from Different Sources.* In H.U. Lemke, et al., editors, *Computer Assisted Radiology, Proc. CAR '89*, pp. 666–670. Springer-Verlag, Berlin.

[94] W. Schlegel (1990). *Computer Assisted Radiation Therapy Planning.* In K.H. Höhne, et al., editors, *3D-Imaging in Medicine: Algorithms, Systems, Applications*, pp. 399–410. Springer-Verlag, Berlin.

[95] R. Schmidt, T. Schiemann, K.-H. Hübener K.H. Höhne (1991, in press). *3-D Treatment Planning for Fast Neutrons.* In *Proc. Advances in Radiation Treatment (ART '91), München.*

[96] W.B. Schwartz, R.S. Patil, P. Szolovits (1987). *Artificial Intelligence in Medicine: where do we stand? New Engl. J. Med.*, 316(11):685–688.

[97] J. Serra (1982). *Image Analysis and Mathematical Morphology.* Academic Press, New York.

[98] D.D. Stark, W.G. Bradley (1988). *Magnetic Resonance Imaging.* C. V. Mosby Co., St. Louis, MO.

[99] A. Sunguroff, D. Greenberg (1978). *Computer Generated Images for Medical Applications. Computer Graphics*, 12(3):196–202.

[100] S.L. Tanimoto (1987). *The Elements of Artificial Intelligence.* Computer Science Press, Rockville, MD.

[101] A.W. Templeton, J.A. Johnson, W.H. Anderson (1985). *Computer Graphics for Digitally Formatted Images. Radiology*, 152:527–528.

[102] P. Tessier, D. Hemmy (1986). *Three Dimensional Imaging in Medicine: A Critique by Surgeons. Scand. J. Plast. Reconstr. Surg.*, 20:3–11.

[103] U. Tiede, K.H. Höhne, M. Bomans, A. Pommert, M. Riemer, G. Wiebecke (1990). *Investigation of Medical 3D-Rendering Algorithms. IEEE Computer Graphics & Appl.*, 10(2):41–53.

[104] U. Tiede, M. Riemer, M. Bomans, K.H. Höhne (1988). *Display Techniques for 3D-Tomographic Volume Data.* In *Proc. NCGA '88, Vol. III, Anaheim,*, pp. 188–197.

[105] A.W. Toga (1990). *Three-Dimensional Neuroimaging.* Raven Press, New York.

[106] H.K. Tuy, L.T. Tuy (1984). *Direct 2-D Display of 3-D Objects. IEEE Computer Graphics & Appl.*, 4(10):29–33.

[107] J.K. Udupa, G.T. Herman, L.S. Chen, P.S. Margasahayam, C.R. Meyer (1986). *3D98: A Turnkey System for 3D Display and Analysis of Medical Objects in CT Data.* In *Proc. SPIE 671*, pp. 154–168.

[108] J.K. Udupa, H.-M. Hung (1990). *Surface versus Volume Rendering: A Comparative Assessment.* In *Proc. First Conference on Visualization in Biomedical Computing, VBC '90*, pp. 83–91. IEEE Computer Society Press, Los Alamitos, CA.

[109] C. Upson, M. Keeler (1988). *V-BUFFER: Visible Volume Rendering. Computer Graphics*, 22(4):59–64.

[110] M.W. Vannier (1987). *Despite its Limitations, 3-D Imaging Here To Stay. Diagn. Imaging*, 9(11):206–210.

[111] M.W. Vannier, J.L. Marsh, J.O. Warren (1983). *Three Dimensional Computer Graphics for Craniofacial Surgical Planning and Evaluation. Computer Graphics*, 17(3):263–273.

[112] M.W. Vannier, T.K. Pilgram, C.F. Hildebolt, J.L. Marsh (1989). *Diagnostic Evaluation of Three-Dimensional CT Reconstruction Methods.* In H.U. Lemke et al., editor, *Computer Assisted Radiology, Proc. CAR '89*, pp. 87–91. Springer-Verlag, Berlin.

[113] M.W. Vannier, C.M. Speidel, D.L. Rickman, L.D. Schertz, L.R. Baker, C.F. Hildeboldt, C.J. Offutt, J.A. Balko, R.L. Butterfield, M.H. Gado (1988). *Multispectral Analysis of Magnetic Resonance Images.* In *Proc. 9th International Conference on Pattern Recognition, ICPR '88*, pp. 1182–1186. IEEE Computer Society Press, Washington, DC.

[114] E. Vaske (1991). *Segmentation of MRI Volume Data with the Topological Map for 3D Visualization.* In *IMDM Technical Report 91/1, Institute of Mathematics and Computer Science in Medicine, University of Hamburg.*

[115] A. Watt (1989). *Three-Dimensional Computer Graphics.* Addison Wesley Publ. Comp., Wokingham.

[116] P.N.T. Wells (1977). *Biomedical Ultrasonics.* Academic Press, New York.

[117] L. Westover (1990). *Footprint Evaluation for Volume Rendering. Computer Graphics*, 24(4):367–376.

[118] J. Wilhelms, A. van Gelder (1990). *Topological Considerations in Isosurface Generation. Computer Graphics*, 24(5):79–86.

[119] T. Yasuda, Y. Hashimoto, S. Yokoi, J.-I. Toriwai (1990). *Computer System for Craniofacial Surgical Planning Based on CT Images. IEEE Trans. Med. Imaging*, MI-9(3):270–280.

[120] F.W. Zonneveld, S. Lobregt, J.C.H. van der Meulen, J.M. Vaandrager (1989). *Three-Dimensional Imaging in Craniofacial Surgery. World J. Surg.*, 13:328–342.

[121] S.W. Zucker, R.A. Hummel (1981). *A Three-Dimensional Edge Detector. IEEE Trans. Pattern Anal. Machine Intell.*, PAMI-3(3):324–331.

Application of Visualization in Environmental Protection

Ralf Denzer

Like in other disciplines, visualization is a fundamental method for environmental protection tasks. Researchers, practitioners, politicians and the public use graphics to get deeper insights into environmental problems and processes. As users, data, processes and models differ a lot, many kinds of visualization and user interface techniques are used.

Environmental protection, planning and research raise questions of how decisions and proceedings in one field may involve other fields. These kinds of questions and the underlying complexity of the data and the models make visualization difficult and make it necessary to develop new visualization techniques.

This chapter gives a survey about the specific requirements, problems and applications in environmental data visualization today.

1 Introduction

Environmental protection, management and research today is only possible on the grounds of valid informations. These informations are gathered, stored and used in many computer systems. Environmental protection is also an interdisciplinary and international task. This involves that many people at many different sites need and use informations about the current state of the environment. It is obvious that computers play an important role in the every day work of environmental protection.

Mainly from the beginning of the 1980's, an increasing number of public authorities and research groups in Germany began to use software systems more extensively. Within the last years, it turned out that environmental software systems have to deal with a number of specific information problems, e.g. the size and the distribution of the data sets or the complexity of the models used.

In 1987, this lead to the foundation of the technical committee "Computer Science in Environmental Protection" and a special interest group of the same name in the German Computer Society. Practitioners and researchers (mainly from Germany and Austria, but also from other Central European countries) are organized in this group. Subgroups work on specific fields, e.g. on databases, expert systems and visualization. Since 1987, the group organizes a yearly conference [16, 17, 19].

The visualization SIG exists since the beginning of 1990. The goal of the SIG is to investigate the practical needs and the suitability of existing methods and to develop new visualization and graphical user interface techniques. For this purpose, we create cooperations and organize a yearly workshop which brings the SIG together [5, 7].

After 18 month of work, this contribution covers preliminary results and experiences and some personal insights into the domain. Because of the variety of disciplines and people working in environmental protection, it is hard to say what visualization of environmental data is. Therefore the text should be understood as an approach to show some of the specific requirements and problems.

2 Applications of Software Systems in Environmental Protection

Following [18], the applications of computer systems in environmental protection can be classified by the four categories

- monitoring and control systems
- information systems
- evaluation and interpretation systems
- decision support systems.

The following sections will give an introduction to each of them.

2.1 Monitoring and Control Systems

Monitoring systems provide the automatic measurement of environmental media. In general, these systems measure a set of environmental parameters at a number of distinct points spread all over the area to be observed. Most of the monitoring systems which have been built are air pollution monitoring systems. In Germany and Austria many of the federal states are currently operating or building such systems. A smaller number of water monitoring systems and radiosity control systems for nuclear power plants exist also.

Monitoring systems have in common, that they are operated as a distributed computer network. The measurement stations and the central database together build a complex distributed technical system.

An example of an air pollution monitoring system is the system built by the Austrian Research Centre Seibersdorf [22, 23]. Currently, there are installations for the Austrian provinces of Carinthia and Upper Austria. A third installation is used for the control of a steam power plant in Styria.

Another important area is the control of technical systems. Optimal strategies of guidance and control help to minimize pollutions and accidents. Many methods have been developed in the real time community for this purpose including simulation, observers, diagnose systems and knowledge based graphical user interfaces. An overview can be found in [6].

2.2 Information Systems

Information systems are the tools to store relevant environmental informations and to provide access to them. Applications cover basic informations like literature, research projects, toxical elements, geographical informations, hazardous sites, measurement data from monitoring systems, etc. Most of the information systems are local databases for a specific purpose.

Key problems managing the data are the size of the databases, their distribution and the fact, that most of the data relate to geographical locations or areas and therefore spatial queries are necessary [13].

2.3 Evaluation and Interpretation Systems

These are systems for the interpretation and scientific analysis of environmental informations. They use mathematical ans statistical analysis and graphical presentation to show specific entities of data and models. Interpretation systems include data analysis, simulation, image analysis and environmental scenarios.

2.4 Decision Support Systems

Decision support is always required when decisions about new industry sites, traffic, buildings, etc. have to be taken. It is also important if one has to decide how to care with a hazardous site. Decision support systems include knowledge about specific domains and must recombine many different aspects to answer questions about risks or an optimal strategy.

2.5 Environmental Information Systems

The fact that processes from one field of the environment influence other fields nowadays determines a lot the structure of information systems. To understand the relationships and correlations better, it is necessary that different computer systems work together and make environmental data available at many sites. This has lead to a new understanding of the term "environmental information system". The term is still being discussed and not well defined, but the discussions tend towards the following notion: the term "environmental information system" denotes a highly distributed, heterogeneous computer systems which is composed of many autonomous subsystems generating, storing and accessing environmental information.

There is no such system today. But the actual discussions show, that the development of a large nation-wide information concept will be one of the great questions of the near future [3].

3 Requirements of Visualization Systems

The requirements of visualization systems are determined by the data and models used and the tasks users want to perform. Many tasks in environmental practice in public authorities and government agencies may be supported by rather simple graphical output like line graphs or charts (e.g. the control of air pollution measurement values).

Other tasks involve the comparison and combination of many different parameters which may be related to different geometrical (geographical) objects and/or to different time scales. These tasks occur mostly in data analysis, interpretation and decision support, where correlations and intersections play an important role for the interpretation of the data. Many tasks also need a combination of many different styles of representations and methods of visualization.

An important fact is that most of the tasks are performed by people who are not specialists in the usage of computers. Therefore the systems must be extremely user-friendly.

The data which has to be visualized is of any possible kinds like

- scalar values (SO_2 at place P, time T)
- sets of scalar values (SO_2, NO, ... at place P, time T)
- vectors (wind direction and velocity at place P, time T)
- domain information ("company X produces hazardous element Y at place P")
- knowledge ("thirty years ago, ...").

These different types of data may be related to any kinds of possible geometries like

- points on a regular grid (air pollution flow simulation)
- points on a nonregular grid (finite element methods, groundwater flow simulation)
- scattered points (values from measurement networks or ground probes)
- rectangular areas (animal population statistics)
- non-rectangular areas ("catasters", land-registers)
- polygons (pollution statistics along streets)
- sets of polygons (e.g. watersheds)
- closed polygons (isolines)

- volumes (3D flow simulation)
- pixel locations (satellite data, maps).

Often, like in monitoring systems, a single value already consists of a number of items which describe not only the value but also the circumstances under which the value has been measured (e.g. a measurement method, validity flags, etc.). It is also most likely, that values do not exist for a certain time, because a measurement instrument was calibrating or not working. There may also exist an uncertainty about some of the values.

All these facts determine, that in environmental data visualization we do not deal with the nice data sets we often know as engineers. Visualization systems which intend to have a general approach, must be able to deal with all these problems.

Most of the visualization being used today addresses one or two of the possible data types related to one of the geometries mentioned above. Only in geographical information systems, several two-dimensional layers are used for multiple parameters. However, visualization is not yet used to investigate correlations, e.g. to support the process of building scientific hypothesis. This is mainly due to the fact that environmental data is still unavailable on a broad basis, which will be the case until we will have large distributed environmental networks [4].

With the increased availability of environmental data, visualization will more and more face complex questions which need to combine and correlate many different data sets and models. These are multi dimensional problems related to a 2D or 3D space and a time coordinate. We look forward to the results the visualization community will produce in this field within the next couple of years.

There is a necessity to look into the different methods whether they are suited for our domain and we need hints how to choose the right representations [20].

4 Methods and Applications

In this chapter we want to give an overview of methods and applications currently being developed and used. Visualization and environmental computer science are both areas which are at their beginnings. Therefore this chapter is more likely to be a list of examples than a methodology.

4.1 Data Analysis and Control in Monitoring Systems

The system built by the Austrian Research Centre Seibersdorf is a typical example of an air quality measurement network [22]. There are three main interactive visualization tasks for the measurement network:

- interactive control of the input data sets
- interactive control of the network itself and
- presentation of the information.

The latter is done by simple line graphs and therefore will be omitted in the sequel.

The network monitors more than 25 parameters at 20 places, generating half our mean values and 3 hour mean values. Because of possible data errors, every single curve of each parameter must be controlled by an expert before it is permanently stored in the database. There is only a partial possibility to process the input control automatically. It is obvious, that the manual control of these large datasets takes too much expert time.

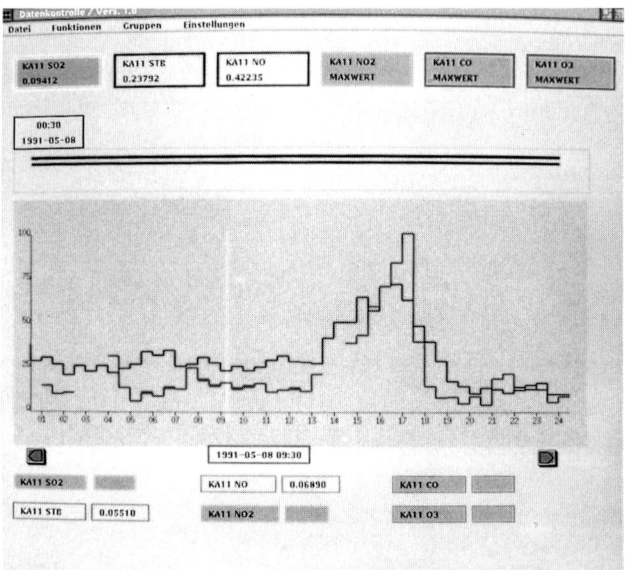

Fig. 1. Input data control (courtesy Forschungszentrum Seibersdorf, Austria).

Figure 1 shows the interactive tool for the control of the input data (for detailed information see [21, 22]). The interesting features in this figure are:

- at some times, measurement values do not exist;
- other values are highlighted; this happens, if less than 90% of the values for a mean hour value are valid;
- the expert can look into, control and correct single values or whole curves interactively using the slider.

The second interactive task is the control of the network itself. The network is a highly distributed technical system and therefore fault situations may occur, which have to be corrected immediately.

In a cooperation with the Forschungszentrum Seibersdorf, we implemented a direct manipulation interface for the control of the network [9]. We could show that, using direct manipulation techniques, it is possible to integrate the informations from several programs (network monitor, network visualization, data visualization and some more) into a single holistic user interface framework (see Fig. A and B of the color illustrations of this chapter at the end of the book). The user interface corresponds directly to the mental model of the system. We currently extend the work towards more complex sets of informations [8], e.g. if you want to visualize more than one network (which is analogue to thematic maps).

One of the main experiences of the research we have done in the last three years is, that very large and complex sets of informations may be visualized by interaction: display as little information as possible for the overall view and give access to the hidden information in the first or second interaction step using comfortable user interface techniques.

4.2 Information Systems

Information systems can help evaluating and predicting the impact of human activities into the environment. Interactive access of databases, in combination with model based methods and visualization, built a good work environment for people who have to take decisions.

The most important feature of such work places is that they must be integrated software systems with a common user interface. Most of the systems being used in public authorities today do not fulfil this requirement. Examples of really integrated environments can be found in [10, 11].

4.3 Scattered Data Methods

A common feature of environmental measurement data is that it is not measured on regular geometries. One reason for this fact is that many laboratory analysis are extremely expensive and one aims to work with as little probes as possible (for an example of such a case see [1]). Therefore many datasets are sets of scattered points.

Having sets of scattered points, one is interested of how the contamination in between those points may be (e.g. in the worst case). Scattered data interpolation methods in combination with 2 1/2 D or 3D visualization beyond the area observed are used for this purpose [15]. In order to determine where new probes should be located, it is especially interesting, how good the quality of the approximation is.

4.4 Particle Flow Visualization and Animation

Particle flow visualization is useful to show the spatial behavior of pollutants. This technique is mainly used for the visualization of air pollution, although it is not limited to this environmental medium. An example of such a system, built on top of a terrain modeling system, is described in [12].

In general the data being used is the result of a simulation. This gives the opportunity of an animation in order to show the temporal behavior of the pollutants. However animations like in [2] are still too expensive to use them in every day work.

4.5 Groundwater Protection and Finite Element Methods

The pollution of the groundwater is often calculated using finite element methods. This is an interesting case, as a combination of real measurement data and calculated values go into the simulation.

In this case visualization is not only necessary to display the results of the simulation (Fig. D of the color illustrations of this chapter at the end of the book) but also to edit the finite element mesh, to describe the ground parameters of the elements and the boundary conditions. The application described in [14] shows how important interactivity is for visualization.

4.6 Intelligent User Interfaces in Process Control

The management of a technical plant is a complex decision process under time pressure. In order to avoid constant pollution or pollution in case of accidents, it is necessary that the management is optimized.

Many of the problems come from the huge mass of data and messages and the complex relationships between the different states of the plant. Often the relationships are not exactly or only partially known. In critical situations, decisions about several hundred of alarms must be taken within a few minutes.

One possibility of operator decision support is the design of intelligent, knowledge-based visualization systems, which adapt to a given process state and which are highly interactive. In a three year research project, we designed and built a distributed visualization system for this purpose [3].

Building several user interfaces for very complex technical systems like a garbage burning test plant, we could show the advantages of using knowledge for the control of the visualization. Figure 2 shows such a user interface. The process of information filtering and adaptation is described in detail in [7].

5 Conclusions

Visualization is an important method for environmental research and practice. The goal of this chapter was to give an overview of visualization and interactive techniques as they are used in environmental practice today.

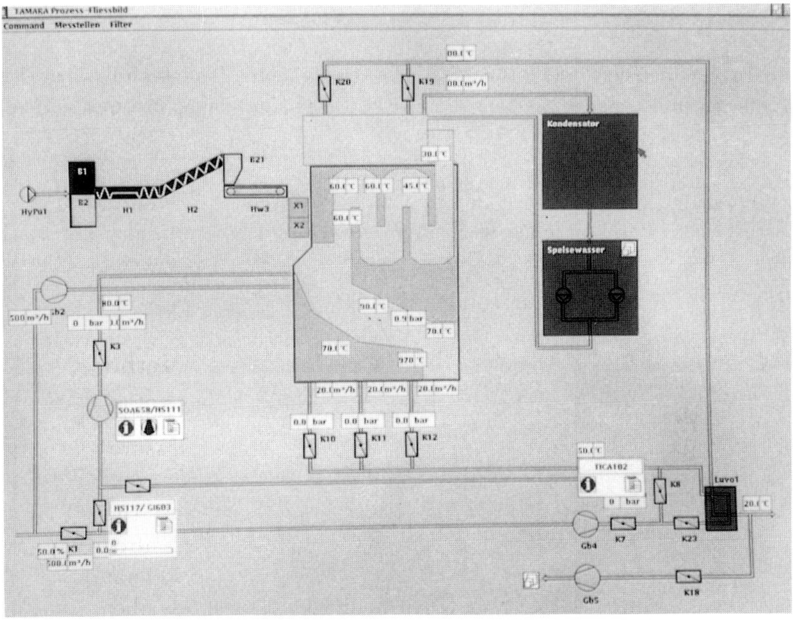

Fig. 2. Knowledge based visualization of a technical plant.

Environmental computer science is at a very early stage. At the moment, many of us are looking for more general principles of our domain. They are difficult to define because of the variety of our domain.

This is also true for the application of visualization and interaction. We hope that during the coming years, we will be able to provide decision takers, researchers and the public with even better visualization techniques to enhance the awareness and understanding of environmental processes. For many of the future questions, we will need new methods which are capable of analyzing many parameters on many geometries simultaneously. We believe that advances in visualization, together with other methods, will enhance the possibilities to aid safeguarding our natural environment.

Acknowledgements

Aknowledgements are due to my colleagues and friends who kindly permitted the publication of their material and allowed me to use their photographs: Gerald Schimak and Heinrich Humer from the Austrian Research Centre Seibersdorf and Werner Haas and Barbara Gruber-Geymayer from Joanneum Research Graz, Austria.

References

[1] J. Burnetti, et al. (1991). *Visualizing Environmental Data for Program Decision Support.* In *Visualization '91, San Diego, California, October 1991*, pp. 398–404. IEEE Computer Society Press.

[2] T. Culviner (1991). *Winning Animation. IEEE CG & Appl.*, 11(3):9–11.

[3] R. Denzer (1991). *Towards Open Environmental Information Systems, position paper.* In *6. Symposium "Computer Science for Environmental Protection".* Springer-Verlag, Berlin.

[4] R. Denzer (1991). *User Interface Management in Distributed Systems.* In R. Denzer, H. Hagen, K.-H. Kutschke, editors, *Visualisierung von Umweltdaten, (Visualization of Environmental Data), Workshop, Rostock, Germany*, pp. 83–97. Springer-Verlag, Berlin.

[5] R. Denzer, R. Güttler, R. Grützner, editors (1992). *Visualisierung von Umweltdaten (Visualization of Environmental Data) 2. Workshop, Schloß Dagstuhl, Germany.* Springer-Verlag, Berlin.

[6] R. Denzer, H. Hagen, G. Kira, F. Koob (1991). *Using Process Knowledge for Adaptive User Interfaces.* In *6. Conference on the Applications of Artificial Intelligence in Engineering (AIENG VI), Oxford, UK*, pp. 583–596. Computational Mechanics Publications, Elsevier.

[7] R. Denzer, H. Hagen, K.-H. Kutschke, editors (1991). *Visualisierung von Umweltdaten, (Visualization of Environmental Data), Workshop, Rostock, Germany.* Springer-Verlag, Berlin.

[8] R. Denzer, F. Koob, G. Kira (1992). *Interactive Visualization of Environmental Measurement Networks.* In R. Denzer, R. Güttler, R. Grützner, editors, *Visualisierung von Umweltdaten (Visualization of Environmental Data) 2. Workshop, Schloß Dagstuhl, Germany*, pp. 77–85. Springer-Verlag, Berlin.

[9] R. Denzer, G. Schimak (1991). *Visualization of an Air Quality Measurement Network.* In M. Hälker, A. Jaeschke, editors, *Computer science for environmental protection Informatik für den Umweltschutz, 6. Symposium, Munich*, pp. 78–86. Springer-Verlag, Berlin.

[10] K. Fedra (1990). *Interactive Environmental Software: Integration, Simulation and Visualization.* In W. Pillmann, A. Jaeschke, editors, *Informatik für den Umweltschutz, 5. Symposium, Vienna, Austria*, pp. 733–744. Springer-Verlag, Berlin.

[11] K. Fedra, E. Weigkricht (1990). *Environmental Software Featuring Interactive Interfaces.* In A. Jaeschke, B. Page, editors, *Informatik im Umweltschutz, 2. Symposium, Karlsruhe, Germany*, pp. 38–51. Springer-Verlag, Berlin.

[12] M. Groß (1991). *Effiziente Visualisierungstechniken für den Umweltschutz.* In R. Denzer, H. Hagen, K.-H. Kutschke, editors, *Visualisierung von Umweltdaten, (Visualization of Environmental Data), Workshop, Rostock, Germany*, pp. 63–75.

[13] O. Günther (1990). *Data Management in Environmental Information Systems.* In W. Pillmann, A. Jaeschke, editors, *Informatik für den Umweltschutz, 5. Symposium, Vienna, Austria*, pp. 57–66. Springer-Verlag, Berlin.

[14] W. Haas, R. Brantner (1991). *FE Analysis and Visualization Utilizing X-Windows in a Distributed Supercomputing Environment.* In G. Beer, J.R. Booker, J.P. Carter, editors, *Computer Methods and Advances in Geomechanics*, pp. 65–68. Balkerna, Rotterdam.

[15] H. Hagen (1991). *Scattered Data Algorithmen zur Umweltdatenvisualisierung.* In R. Denzer, H. Hagen, K.-H. Kutschke, editors, *Visualisierung von Umweltdaten, (Visualization of Environmental Data), Workshop, Rostock, Germany*, pp. 22–28. Springer-Verlag, Berlin.

[16] A. Jaeschke, W. Geiger, B. Page, editors (1989). *Informatik im Umweltschutz, 4. Symposium, Karlsruhe, Germany.* Springer-Verlag, Berlin.

[17] A. Jaeschke, B. Page, editors (1987). *Informatik im Umweltschutz, 2. Symposium, Karlsruhe, Germany.* Springer-Verlag, Berlin.

[18] B. Page, A. Jaeschke, W. Pillmann (1990). *Angewandte Informatik im Umweltschutz. Informatik-Spektrum*, 13(1):6–16.

[19] W. Pillmann, A. Jaeschke, editors (1990). *Informatik für den Umweltschutz, 5. Symposium, Vienna, Austria.* Springer-Verlag, Berlin.

[20] P. Robertson (1991). *A Methodology for Choosing Data Representations. IEEE CG & Appl.*, 11(3):56–67.

[21] G. Schimak (1992). *Benutzeroberfläche zur Datenkontrolle und Datenkorrektur in einem Luftmeßnetz.* In R. Denzer, R. Güttler, R. Grützner, editors, *Visualisierung von Umweltdaten (Visualization of Environmental Data) 2. Workshop, Schloß Dagstuhl, Germany*, pp. 86–94. Springer-Verlag, Berlin.

[22] G. Schimak, G. Ernst, H. Humer, H. Vrabl (1992). *Automation and Visualization in Monitoring Systems.* In *WMO Tecchnical Conference on Instruments and Methods of Observation (TECO-92), Vienna, May 1992, WMO Report No. 49*, pp. 31–35. World Meteorological Organization.

[23] Austrian Research Centre Seibersdorf . *UWEDAT - A System for Aquisition of Meteorological, Hydrological and Environmental Data.*

Data Structures in Scientific Visualization

Ulrich Lang, Michel Grave

This chapter presents a collection and compilation of informations about how scientific data is structured under the requirements of storing, processing and visualizing these data. Many different software packages from different application areas as well as scientific visualization packages are available. While they differ in many points, the commonalities in describing and structuring data are dominant.

1 Introduction

The amount of scientific data produced by different sources is dramatically increasing. The large amounts of data are not an aim in itself, the aim is to get insight into the behaviour of systems, to predict future developments or to control and influence them [9]. Together with the increase in data volume, the need for better tools to get insight into the data rises too. A major means to get insight is to visualize the data. Various mappings of visualization methods on scientific datasets are possible. To design the possible visualization tools for all datasets in the different application areas the analysis and systematic structuring of data is required.

Scientific data has its origin either in calculations, done on computers, or data is acquired using measuring instruments. Independent of its origin large quantities of data are often stored in archives or databases. For further processing it makes no difference where the data came from [1] (see Fig. 1). This chapter tries to derive and describe the necessary data structures, their hierarchies and relationships.

Different people have written papers and given presentations about structuring of data. Some of these papers, e.g. [6, 17, 1, 3], have influenced this chapter.

2 Computer Based Problem Handling

Scientific data is used to pass information between the different processing steps. Whether the origin of data is a measuring device or a computer simulation, the data has to be stored, processed, transported and visualized. Appropriate data structures are needed to support all these steps in an optimal manner. The optimal data structure might differ for the same dataset during the storage,

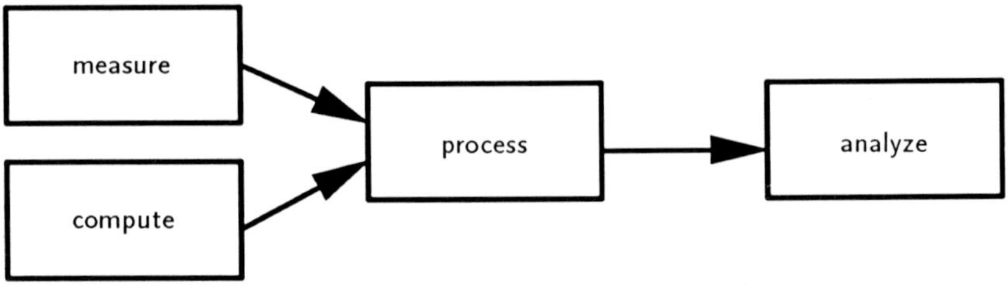

Fig. 1. Data handling steps.

processing, transport and visualization. Some requirements of the different steps for the definition of data structures are:

- Storage
  - To be appropriately accessible, data has to be structured into basic entities, required by the operations.
  - Data structures have to resemble and support the abstract model world and thinking of the application scientist.
  - The storage space requirements of the data structures have to be minimal. They should be suited for storage in memory and on disc.
  - In the case of large datasets a compact or compressed data representation should be available, at least on disc, possibly also in memory.
- Processing
  - The operations, which have to be applied to the data, have to be supported with a minimal interface to the data structures.
  - Data must be accessed and stored as fast as possible during the required operations on a dataset.
  - Operations should be applicable as directly as possible on the data without additional preparation steps.
  - It might be desirable to store a history about processing and conversion operations to track peculiarities seen during the visualization step.
- Transport
  - Basic data types have to be known at every level of abstraction to allow conversion of data during transfer between different machine types.
  - Data should be organized in large entities of homogeneous basic data types to allow optimized conversion of basic data representations.
  - To transfer large quantities of data through low bandwidth connections it might be necessary to compress data.

- Visualization
  - Some visualization algorithms only operate on certain data structures. Therefore fast conversion algorithms are needed to resample, interpolate or copy data into the appropriate data structures.

As a general rule data structures should be as general as possible, while still meeting the requirements of a specific application case. Software engineering methods should be applied to define data structures. Data should be described by defining abstract data type modules for all the operations being applied on it. This will give data a consistent selfcontained representation. In some applications this is called strongly typed datasets.

To further analyze what an optimal support means, we will have a closer look into the stages for treating scientific problems using computers. From an application scientist's point of view problem handling can be described in three stages [5]: modeling, solution and interpretation.

2.1 The Modeling Phase

In the modeling phase a problem described in a very informal manner is converted into a mathematical formulation. Mathematical laws such as partial differential equations define the behaviour of the system. The mathematical formulation is discretized to make it suitable for computer processing, using methods such as finite differences or finite elements. Assumptions are made to idealize the problem and reduce its complexity. Wrong decisions made here cannot be changed during the simulation. One has to go back and remodel the problem.

The data structures to store and access data are defined in this phase. They primarily reflect the requirements of the numerical model. Special computer architectures such as vectorizing or parallel machines might impose specific requirements on the data structures. This can be in contradiction to the requirement of simplicity and readability of the resulting code.

2.2 The Simulation Phase

During the simulation phase the data structures implemented in the modeling phase are filled with data. From a programming point of view the data structures should be defined, so that they occupy a minimal storage space in memory and on disc, while giving a maximum in access speed during manipulations. To give an example of required storage space we examine a scalar volumetric data field with floating point values (see Fig. 2).

During manipulation and visualization additional auxiliary arrays are needed. Therefore either machines with large enough main memory are used, or special storage techniques are needed to reduce or avoid paging on virtual memory machines or organize swapping of data. Three techniques are possible [19]:

grid size	*	bytes/data value	=	memory requirement
$64 * 64 * 64$	*	4	=	1 MB
$128 * 128 * 128$	*	4	=	8 MB
$256 * 256 * 256$	*	4	=	64 MB

Fig. 2. Storage Space Requirements for Volumetric Datasets

1. Store and access data in smaller entities. E.g., store volumetric data plane wise and read it in planes from disc into memory. Write algorithms that step plane wise through the volumetric dataset.
2. Compact the data, e.g., using standard compression techniques.
3. Use hierarchical data structures to make non sequential access of data stored on disc possible.

2.3 The Evaluation Phase

In the evaluation phase visualization tools either specifically designed for the data structure, or general visualization tools are used to give insight into the meaning of the data. Specific tools have reduced overhead in accessing the data. They show an optimized performance in visualizing the specific data. General tools need additional conversion codes to interface specific data representations to the more general data representation of the tool. For most of the simpler cases such a tool shows reduced performance. But on the other hand general tools allow their usage for a major percentage of all application problems.

3 Data Types of Dependent Variables

Dependent variables, are used to describe the behaviour of scientific, technical, social or other systems. In the case of scientific calculations the behaviour of the system to be simulated is described by a set of equations. These may be partial or ordinary differential equations, integral equations, linear or nonlinear equations. The base space is the region, where the equations are defined. An analytical solution of the equations, if possible, would yield functional values at all positions of the computational space. As analytic solutions are only possible for very simple cases, an approximated numerical solution is necessary. For that purpose the base space is discretized defining volumetric elements, surfaces, edges or grid points. The equations are converted and replaced by equation systems better suited for the solution with computers.

3.1 Basic Data Types

Data at a discretized position is usually described using floating point numbers. These discretized positions need not always be grid points. Depending on the needed precision of the numerical calculation single or double precision values are used. In the case of measured data additional basic data types occur. Data from cat-scanners produce after preprocessing 16 bit values at every grid point. Other measuring devices such as gray scale paper scanners may produce 8 bit information per point. A boolean data type can be used for flags. Further basic data types might also be possible.

Therefore available application software packages support some or all of the following basic datatypes: byte, short, long, float, double, complex and double complex.

3.2 Dimensionality of Dependent Variables

The dimensionality of the dependent variable defines how many values are available at every data point. The dependent variables at a data point can have a relationship such as the x, y, z component of a vector or can be independent of each other like temperature, pressure, etc..

3.2.1 Scalar Fields

Scalar variables have a tensor rank of zero [17]. They do not have a direction. These variables do not vary under rotation or translation of a local reference frame. They are also called single-valued functions. A scalar variable in three dimensional space and time can e.g. be described by

$$f(x, y, z, t).$$

Examples for single-valued functions are temperature, pressure, mass, density, energy, etc.

3.2.2 Vector Fields

Vector fields are tensors of rank one. They have both magnitude and direction. They are usually represented as an n-tuple of numbers, where n is the dimension of the space. A three dimensional vector function on a three dimensional base space can e.g. be described by:

$$\vec{f} = \begin{pmatrix} f_1(x, y, z) \\ f_2(x, y, z) \\ f_3(x, y, z) \end{pmatrix}.$$

Often a vector fields is derived from a scalar field by differentiation with respect to each coordinate dimension of the base space. Such a field is called a gradient. Further examples for vector valued functions are: force, velocity, acceleration, surface normal, etc..

3.2.3 Tensor Fields

Tensor variables have arbitrary tensor ranks. A tensor field is a field, which has certain simple transformation properties under coordinate transformations. The tensor rank is a measure of the amount of information at each point in space. Tensor variables are represented by multi-component functions. Mathematically they can be described as

$$f_{ij...}^{kl...}(x, y, z, ...t).$$

The previous lower dimensional cases can be regarded as special cases of the tensor field:

- scalar (zero-order tensor),
- vector (first-order tensor),
- matrix (second-order tensor).

For example a rank two tensor can be represented by an $n * n$ matrix. Examples for second order tensors are stress, which associates a force vector to each direction. A fourth order tensor is e.g. the curvature in differential geometry. Stress and strain are examples of symmetric tensors. The vorticity of a flow is an asymmetric tensor. The symmetry property is preserved under transformations.

4 Coordinate Systems

Within the description of this chapter we concentrate on dependent variables and functions defined on a multi dimensional base space. This base space is described by independent variables, which are defined within a certain right hand coordinate system. The reason to choose between different coordinate systems is to take advantage of inherent problem symmetries, to ease the computation or to exploit visualization advantages [6]. Within each coordinate system a metric tensor g contains coefficients for measuring distance. For orthogonal coordinate spaces the off diagonal elements of g are zero. Graphical representations of coordinate systems are shown in Fig. 3.

In addition to the depicted coordinate systems curvilinear coordinate systems also occur. In most cases they are generated by distorting a Cartesian coordinate system along the surface of a given geometric object such as the wing of an airplane.

5 Connectivity

Data is located at certain points in space. If these points are arbitrarily distributed and no connectivity between them is defined, the data is called scattered. Examples are the measurement points of temperature over a landscape,

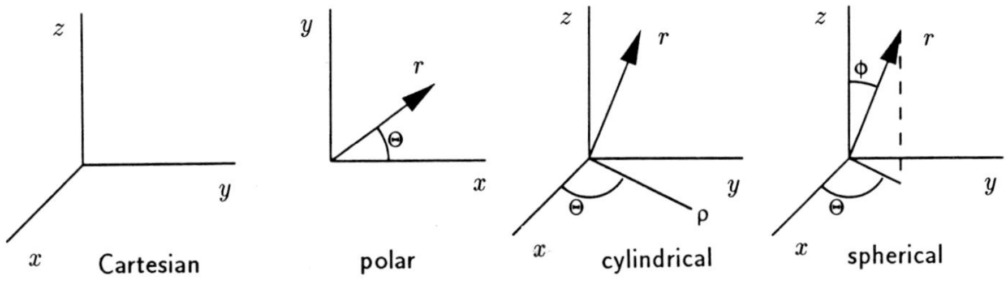

Fig. 3. Coordinate systems.

or the positions of oil drilling wholes. To include these cases into the further considerations of gridded data mathematical models are applied to map the data onto a grid. These models are defined over the entire domain and interpolate or approximate the given scattered data at grid points [10].

Within computer simulations numerical algorithms operate on a computational space. This computational space is mapped on a coordinate space. In most visualization software packages grids are expected to have a homeomorphic mapping from a regular uniform grid of the same dimension. The dependent variables are defined at grid points of the computational space. Through introduction of a connectivity, higher order objects are defined. Different authors use different even overlapping naming conventions for these objects. Some examples, sorted from zero to three dimensions, are:

- grid point, node,
- edge,
- cell, face, surface,
- cell, volume, element, mesh.

Grid points have certain relationships between each other such as closest neighbourhood. In a one dimensional regular grid every grid point has two neigbours. In the two dimensional regular case every point has four neighbours. The general rule for n dimensions is, that every point has 2 times n neighbours.

The different grid types are further explained in the following chapter. They are shown for the two dimensional case in Fig. 4. This simply extends to three and more dimensions.

5.1 Cartesian Grids

This type of grid is also called an array grid [15]. Coordinates are not explicitly stored or referenced. The addressing is done by integer subscripts (i, j, k). Examples for datasets defined on such grids are the electron density within a molecule or medical data from MRI or CT scanners after resampling. Some software packages use Cartesian grids to represent pixel images.

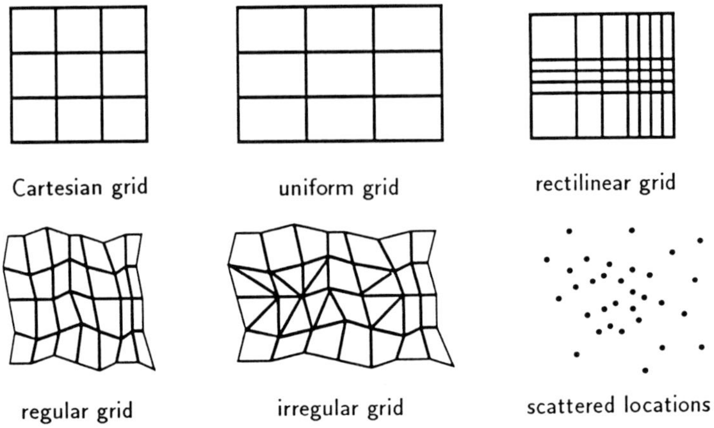

Fig. 4. Grid types.

5.2 Uniform Grids

A uniform grid is usually rectangular. It can be described through constant spacing between grid lines. Different types of coordinate systems are possible. Each axis has a separate distinct spacing. Coordinates are explicitly specified either by definition of a bounding box or by starting point and increment for every coordinate direction [7, 14]. Some authors call this a regular grid (see e.g. [15, 12]).

5.3 Rectilinear Grids

A rectilinear grid is usually rectangular with a nonuniform spacing along each coordinate axis. Different types of coordinate systems are allowed. It is described by a grid vector for every coordinate axis. Naming conventions vary across different authors. In apE [12] this is called a uniform grid with irregular spacing of axes. IRIS Explorer^{TM} [14] calls this a perimeter lattice.

5.4 Regular Grids

In a regular grid the coordinates for every grid point are stored explicitly. It is called regular, because the neighbouring relationship is regular, i.e., every interior point has the same number of neighbours. The connectivity is implicitly described by incrementing or decrementing the index in the coordinate or data fields. With this grid type one coordinate for a whole grid occupies the same amount of space as a scalar dependent variable does. To fully describe a three dimensional scalar field, four times the space of the dependent variable is needed. In [15] this is called a structured grid or curvilinear regular grid. Often, in 2D, cells are parallelograms and in 3D, they are hexahedra. A typical example for the usage of such a grid is in the aeronautics field for grids around the outer surface of an airplane wing.

5.5 Block-Structured Grids

A subcase of the regular grid is the block-structured grid, which is a collection or patchwork of multiple regular grids. In some application areas it is also called a multi domain grid. Blocking is useful in order to reduce the complexity of one big problem to multiple small ones and it is also useful during the meshing process. E.g., it is not always easy to describe the surrounding of a whole airplane with one structured block. Special care has to be taken at the connecting border between two grids. Preferably the number of grid points and their coordinates should be the same at the border between both grids. Otherwise interpolation is necessary to do processing across the border line or surface. In this case the same interpolation schemes should be used during simulation and visualization.

5.6 Irregular Grids

What is called an irregular grid here is called an unstructured grid by other authors. In irregular grids the coordinates for all grid points are also stored explicitly. In addition the connectivity information has to be stored. This is either just the direct relationship of grid points to edges, cells and volumes, or additionally it might be necessary to store neighbouring relationships between edges and cells, or cells and volumes. Finite element methods are usually based on these grid types. Techniques such as local refinement of grids to stay within error bounds of numerical algorithms might also lead to irregular grids, even if the algorithm started out with a regular grid. In this case the regular grid at the outset will already be described as an irregular grid.

In irregular grids specific questions arise, such as to devise fast algorithms to find the cells or volumes where arbitrary points fall into. These are trivial questions in regular or other simpler grids.

5.7 Hybrid Grids

Hybrid grids represent a mixture of different grid types such as uniform, rectilinear and irregular. Many of the currently available software packages do not support hybrid grids explicitly. Usually every single grid of the hybrid grids is mapped on an irregular grid. Thus a collection of irregular grids is used to describe hybrid grids. Special care has to be taken to merge these hybrid grids into one irregular grid. Interpolation schemes are used at the connecting borders of the irregular grids to map dependent values from the original grid on grid points of the new common grid.

5.8 Scattered Locations

As already mentioned scattered data can be included into the previous discussion of gridded data, if we find an ordering relationship between the spatial positions of the points where the data is defined. Some possibilities are to find pairs of closest neighbours, cells with minimal area or tetrahedra with minimal volume [10]. This would give a mapping on the case of an irregular grid in our previous description.

Another possibility is to use a regular or even uniform grid and map the irregular data using interpolation methods onto this grid. In this case it is very important to determine upper error bounds introduced by the mapping.

6 Zone Data

Complex simulations with small mesh sizes often lead to a large amount of grid points. To reduce the amount of stored information, such as connectivity, and speed up the storage and access of data regular grids are preferred in many simulation cases. As multiple materials might be situated in different regions of the computational domain, material zones are introduced [3]. They describe for every mesh of the grid, which type of pure material or which mixture of materials is stored in the mesh. The material properties are used in the calculation. An example is the conductivity, heat capacity and density of materials used in solving the heat conduction equation.

The zone information is also used to depict void areas. This might be to model interior regions or complicated outer boundaries of the computational domain. The material allocation to zones is a relevant information (see Fig. A of the color illustrations of this chapter at the end of the book), which should also be visualized, especially together with the values of dependent variables to reveal relationships. Even more important is the display of void regions. With measured data they can be caused by temporal unavailability of a measuring device. With computed data they might indicate obstacles, where the governing equations do not apply or represent boundary zones, where results do not have to be displayed (see Fig. B of the color illustrations of this chapter). Another possibility to show different zones is by usage of texturing such as in Fig. 5.

7 Auxiliary Information

During the modelling process certain assumptions have been made, which should remain known in the visualization process. Therefore this information should be passed on to the visualization. One of these assumptions is how values of dependent variables should be determined between grid points. Does a linear interpolation conform to the numerical algorithm? Has a cubic interpolation to be applied, or should a stepwise function be displayed?

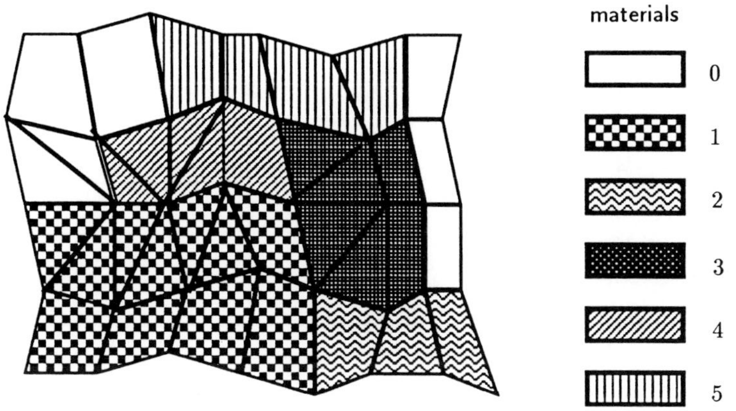

Fig. 5. Material zones.

Other auxiliary informations are e.g. the state of cell boundaries. For particle trace calculations they can be blocking, non blocking, reflective or even expose a cyclic behaviour. Auxiliary information will be explained further in the example of the following section.

8 Relationship between Data Sets

Within the same spatial domain multiple dependent variables are determined during a computer simulation. While they share the same domain the numerical methods applied to the original equations might lead to different grid points for the different variables within the same computational cells. The example shown in Fig. 6 will make this clear. For easier drawing purposes the example is in two dimensional space. T stands for temperature, u and v for the components of the velocity. As a result of a numerical algorithm u and v are not defined at the same grid points, they define the velocities at the input and output boundaries of the computational cell in their respective directions. The temperature is assumed to be cell centered. With finite difference methods often the assumption is made, that values are constant within a computational cell. In our example temperatures are assumed to be constant within every mesh (see Fig. 7). Velocity components are assumed to be constant along every connecting edge between meshes. Between opposite edges of the same cell a linear interpolation of the velocity values can be assumed (see Fig. 8). To show the relationship between the variables they should all be displayed in the same picture. To our knowledge, none of the available visualization packages are able to visualize exactly the spatial distribution of these values using the assumptions of the numerical algorithm. In most packages it is necessary to interpolate onto common grid points before displaying the different variables. Also interpolation methods other than linear interpolation between stored values are seldom available.

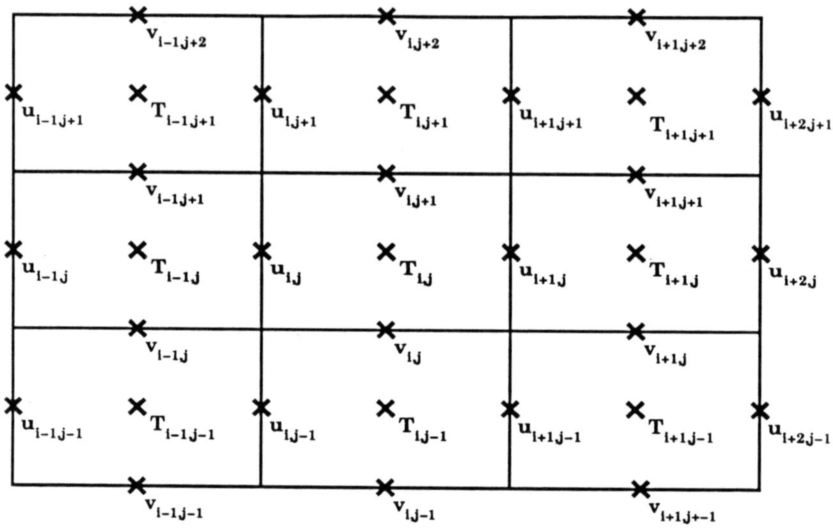

Fig. 6. Spatial relationship of multiple variables.

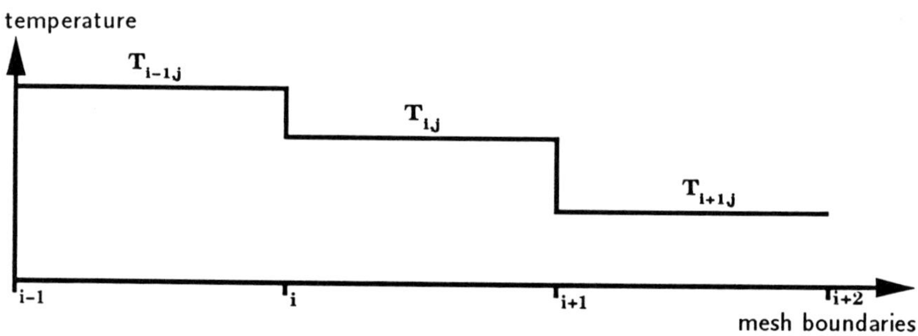

Fig. 7. Temperature distribution along a cutting plane at y_i.

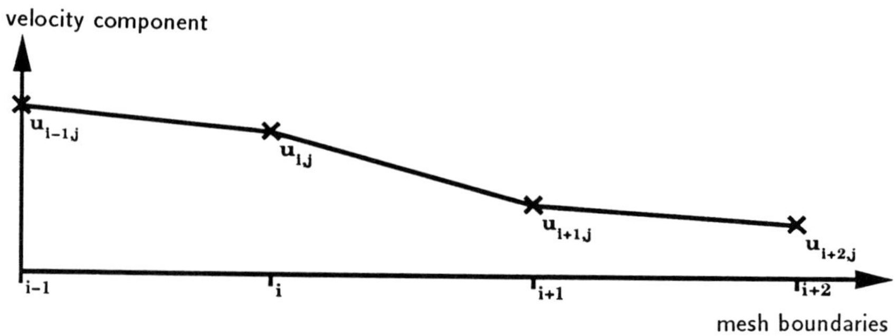

Fig. 8. Velocity Component u along Cutting Plane at y_i.

9 Consideration of Time

Most scientific simulations and measurements are done over a certain period of time. Time is a very important factor which has to be taken into account, when appropriate data structures have to be defined. As time can be regarded as a coordinate being orthogonal to space, it should be possible to avoid influence on the spatial structuring.

The introduction of time makes it necessary to store timing information. While in most cases timing information is global and the same for all parts of a simulation problem, there are cases, where it is necessary to deal with separate time granularity for different parts of the same simulation. This case has to be discussed in more detail.

Time dependent cases can be structured similar to the structuring of the previous spatial cases.

9.1 Handling of Time Information

9.1.1 Fixed Time Step Size

The easiest way to describe the discretized time is based on a constant time step size during the whole problem time period. This can be described by storing the initial and ending time together with the time step size. Alternatively the ending time can be replaced by the number of time steps.

9.1.2 Variable Time Step Size

If the time step size varies over the whole problem time, it is stored in a time vector. For every time step an entry exists in the vector. If different granularity in time for different parts of the overall problem exists, separate timing information is necessary for each part. This is stored in multiple time vectors.

The influence of varying time step size during time on the visualization is not recognized in most interactive visualization software packages. The usual approach tries to visualize data as fast as possible. The update speed of the graphical representation is determined only by the hardware capabilities and performance of the graphics library. Thus the display speed does not reflect the time step size of the stored or produced data. Non interactive animation software packages on the other hand try to represent time as well as possible. Instead of producing interactive output, they usually work with single framing methods, where the display of the produced animation is done at a later time. For proper analysis of results over time appropriate and consistent scaling of time due to time step size is also very important for interactive work and should be recognized in available packages.

9.2 Time Dependence of Data

Time dependence is introduced as soon as data at certain grid points changes over time. Therefore data has to be stored separately for every time step. This leads to an enormous increase in data volume to be dealt with.

9.3 Time Dependence of Grids

As soon as deformation of materials is introduced or non Eulerian grids are used in fluid flow calculations, the coordinates of grid points change over time. Therefore, in addition to data also grid coordinates have to be stored over time. This leads to an additional increase in data volume.

9.4 Time Dependence of Connectivity

Multiple reasons for the change of connectivity over time are possible. If meshes are deformed over time, it might be necessary to split them up and introduce additional meshes. Depending on the algorithm local refinement of meshes leads to new meshes and changes in the connectivity. In a regular grid this does not change the neighbouring scheme, but the actual neighbouring information is changed through introduction of new neighbours within a given mesh. In an irregular mesh the introduction of new neighbours requires even more changes in the coordinate list and connectivity table.

9.5 Different Time Steps in Separate Parts

If the time step sizes in different parts are not the same, appropriate methods are needed to visualize the data. One approach is to generate a common vector with identical intermediate time steps for all parts of the simulation and for every time step establish a pointer to the appropriate data, grid and connectivity values. The data values and grid positions are best interpolated over time for these intermediate time steps. To represent a change in connectivity between two time steps can not be done using interpolation. Therefore for these cases the most simple thing is to keep the values from the previous timestep and change values only at the new timestep.

10 Data Structures in Different Software Packages

The intention of this chapter is not to describe explicitly the used data structures within different visualization software packages, but to concentrate on common ideas or peculiarities of some of the available packages. As it would be very difficult to get access to all of the packages in such a dynamic segment of the software market, some packages, especially newer ones may not have been taken into account.

10.1 The Irregular Grid Approach

Packages such as Multi Purpose Graphics System (MPGS) [8] follow the idea, that every possible grid type can be mapped on an irregular grid. Therefore they just support irregular grids. Different widespread scientific codes are supported through available data conversion tools. These tools convert the code specific data representations, these are in most cases file formats, into the file formats of the visualization package. One file format to represent irregular grids which is supported by many packages is the MOVIE.BYU file format [16]. It is structured into a geometry file, a scalar file and a vector file. Just for the display of geometry the data files are not needed. The geometry file is organized into

- an element parts list, containing beginning and ending element numbers that describe which elements belong to which part,
- the nodal coordinate array consisting of an x, y, z triplet per node,
- and the element connectivity array, describing which nodes are linked to create elements.

The optional scalar function file contains just one scalar value for each node in the geometry file. The optional vector function file again contains x, y, z triplets for each node in the geometry function file.

10.2 Differentiated Data Structure Approach

Other packages such as Wavefront Data Visualizer[18], Animation Production Environment apE [12], Iris Explorer [14], or Application Visualization System AVS [7] follow a more differentiated approach.

The basic idea of these packages is to define locations through nodes, and to define grids as a set of nodes. Grids can be further separated into regular grids with evenly spaced nodes, irregular grids with separate x, y, z for every coordinate position.

Topology can be described through connectivity of the nodes into elements. Elements can either be 2D surface elements or 3D volume elements. A regular topology uses hexahedron volume elements, while in an irregular topology, node connectivity for data volume is either described by a hexahedron or tetrahedron. Geometric objects can be described by points, lines and polygons.

After introduction of basic components, additional hierarchical objects can be introduced. A mesh, e.g., is a combination of a grid and topology case. They can be even further separated into a regular mesh with a simple, rectangular array of volume elements, or into an irregular mesh with regular topology and x, y, z coordinates for every node. In the case of an unstructured mesh, an element node connectivity list is explicitly defined. In this case a grid may be regular or irregular.

11 Application Package Independent Data Access Software

In most cases, each visualization system has its own format for storing and manipulating visualization data. Even if many commonalities exist between them, there is always some work to be done when two different packages have to communicate. Some formats have become more popular than others, due to the wide distribution of the system they have been designed for (e.g. PLOT-3D, MOVIE-BYU, etc.), but there is a need of more standardized formats and access environments (see [2]).

In the recent years, several groups have worked on the definition of formats and tools able to be used in a wide range of applications, and portable on many hardware platforms. Three of them can be considered today as being well recognized in the academic community, even if their usage is far from being very large.

NASA's Common Data Format (CDF) can be seen as one of the first implementations of a scientific data model [4]. It provides tools for accessing multi-dimensional block-structures (all kinds of grids, except irregular ones, according to the vocabulary defined in this paper). A first version had a very simple interface (13 routines). The interest in the approach led another group (at Unidata Program Center) to use these concepts in its work.

Unidata has defined and implemented the netCDF software [13]. Like CDF, it also applies to block-structured data, but uses a machine independent binary representation (through the use of XDR), in order to facilitate transportation among heterogeneous machines. Another major enhancement to the initial CDF work has been the concept of hyperslab access for handling subparts of data structures, and the possibility to gather a complete dataset in a single file.

Both CDF and netCDF describe access mechanisms to Scientific Data Sets, and provide associated tools, but do not actually define data formats. In other words, they define the syntax of data sets, but semantics has still to be added, inside an application, or when two applications have to communicate.

The third important work has been done by NCSA (National Center for Supercomputer Applications at Urbana-Champaign). It is also a portable format, called HDF (Hierarchical Data Format) [11], for block-structured data, but in addition to netCDF, provides also image handling features. In the basic HDF format, user's are closer to the implementation than with netCDF, since they can see the location of bytes, offsets and other features, and data access is less general. However, with the introduction of the Vset concept, which includes an extension mechanism, all kinds of regular and irregular data sets, non Cartesian coordinates, as well as the hierarchical grouping of datasets can be described.

Those three data access systems are at present incompatible, but through the Vset extension mechanism a group at NCSA is working on the integration of netCDF objects in the HDF environment, providing a way to combine the rich and convenient data access mechanisms, with tools for handling images or irregular grids.

12 Conclusions

The intention of this chapter was not to give a survey of all available packages. Instead the commonalities between different packages are analyzed. Further analysis of available codes should give the possibility of a more detailed structuring. It would be a great advantage for the scientific community if an agreement on one interface for accessing and managing scientific data could be reached. This would improve the portability of modules between codes and thus support the exchange and reuse of these modules.

References

[1] D. Buttler, C. Hanson, editors (1991). *Visualization '91 Workshop on Scientific Visualization Environments, San Diego.*

[2] L. Carpenter (1992). *Mumford File Formats For Computer Graphics.* In *State of the Art Report.* Eurographics '92 conference, Cambridge.

[3] L.M. Cook (1990). *Mesh Types for Scientific Calculations.* Lawrence Livermore National Laboratory, July 30.

[4] M.L. Gough (1988). *NSSDC CDF Implementer's Guide (DEC VAX/VMS) version 1.1.* National Space Data Center 88-17, NASA Goddard Space Flight Center.

[5] R.B. Haber, D.A. McNabb (1990). *Visualization Idioms: A Conceptual Model for Scientific Visualization Systems.* In G.M. Nielson, B. Shriver, L.J. Rosenblum, editors, *Visualization in Scientific Computing*. IEEE Computer Society Press.

[6] D.D. Hearn, P. Baker (1991). *Scientific Visualization.* In *Eurographics '91 Tutorial Notes, Vienna, Sept 2.-6.*

[7] Stardent Computer Inc. (1991). *AVS User's Guide, Release 3.0.* Stardent Computer Inc.

[8] D. Laur, P. Hanrahan (1991). *Hierarchical Splatting: A Progressive Refinement Algorithm for Volume Rendering. Computer Graphics*, 25(4):285–288.

[9] B.H. McCormick, T.A. DeFanti, M.D. Brown (1987). *Visualization in Scientific Computing. Computer Graphics*, 21(6).

[10] G.M. Nielson, et al. (1991). *Visualizing and Modeling Scattered Multivariate Data. IEEE Computer Graphics & Appl.*, 11(3):20–27.

[11] National Center of Supercomputer Applications (1989). *NCSA: HDF Calling Interface and Utilities, NCSA HDF Version 3.1.* University of Illinois at Urbana-Champaign.

[12] The Ohio Supercomputer Graphics Project (November 1990). *apE Version 2.0 Programmer's Manual.* The Ohio Supercomputer Graphics Project.

[13] R. Russ, G. Davis (1990). *NetCDF: An interface for Scientific Data Access. IEEE Computer Graphics & Appl.*, 10(7):76–82.

[14] Inc. Silicon Graphics (1992). *IRIS-Explorer User's Guide.* Silicon Graphics, Inc., Mountain View, California.

[15] D. Sperey, S. Kennon (1990). *Volume Probes: Interactive Data Exploration on Arbitrary Grids.* ACM 0-89791-417-1.

[16] Brigham Young University (1987). *MOVIE.BYU, A General Purpose Computer Graphics System, January 1987 Edition, Version 6.* Brigham Young University.

[17] C. Upson (1989). *2D and 3D Visual Workshop.* In *Siggraph '89 Course Notes, Boston, MA, 31 July - 4 August.*

[18] Inc. Wavefront Technologies (Sept. 1991). *The Data Visualizer, Version 2.0 User's Guide.* Wavefront Technologies, Inc.

[19] J. Wilhelms (1990). *Visualizing Sampled Volume Data.* In D. Thalmann, editor, *Scientific Visualization and Graphics Simulation.* John Wiley & Sons.

A Visualization-Based Model for a Scientific Database System

Ted M. Sparr, R. Daniel Bergeron, Loren D. Meeker

The goal of this multi-disciplinary project is to design and prototype a new approach in database environments aimed at supporting collaborative scientific research. The design integrates scientific data visualization and mathematical and statistical analysis tools with database support in a highly interactive environment. A new schema model for scientific data is proposed. The new scientific database model is founded on the notion that a query of a scientific database conceptually creates new *derived* data whose relationship to the parent database is defined by the query. The system uses a process flow graph as the mechanism for representing queries. Each query, in principle, leads to the discovery of some form of structure in the data which is explicitly represented by the results of the query, or which is hypothesized by the scientist as a result of the current and previous queries.

1 Introduction

One of the most serious problems facing modern scientific investigations arises because we need to determine increasingly precise information with instruments of great cost and complexity. This demands ever more specialization of our scientists at a time when the most important problems facing society transcend the boundaries of our traditional disciplines. Modern scientific research is becoming increasingly dominated by the need to understand and interpret large amounts of complex data arising from the joint efforts of large interdisciplinary scientific teams. Traditionally such data is handled in an *ad hoc* manner. The absence of uniform interfaces to complex scientific data is particularly difficult in multidisciplinary research projects. In such environments scientists from different disciplines need to share data, tools, and resources in order to work effectively together.

We are particularly interested in cooperative research involving both scientists and analysts. We have initiated a multidisciplinary project to design and prototype a new approach in database environments to support collaborative scientific research. The environment should integrate analysis and data visualization tools within the context of an interactive interface to a scientific database system. The fundamental motivation is to encourage and facilitate cooperation between scientists with multi-dimensional domain-specific data and analysts with knowledge of generic analysis tools and techniques.

We focus especially on data interpretation problems that are not well-understood, and which require what Tukey has called Exploratory Data Analysis

(Tukey [42]). In this context scientists need to experiment with subsets of the data, with different analysis tools, and with different data visualization tools in an attempt to understand the relationships among the data. This often includes a "trial-and-error" component driven both by intuition and by the feedback the scientist receives from the analysis and visualization tools, which, in turn, stimulate additional intuition. From this interaction, the scientist may be able to postulate some hypothesis about the relationships within the data. Unfortunately, however, the selection and use of the analysis and visualization tools can have a significant impact on the interpretations that can be inferred about the data, and both can lead to interpretations based on artifacts created by either the analysis or visualization. Consequently, it is critical that the scientist's hypotheses be verified independently. Verification is most appropriately in the domain of the analyst. With the scientist's hypotheses and derivation processes, the analyst must try to validate the conclusions. In most cases, this may require additional analysis or experimentation.

We aim to achieve better cooperation between the analyst and scientist through the development of a data exploration environment that integrates a scientific database model with analysis and visualization tools. Such a framework requires a database model that is appropriate for complex multi-dimensional scientific data. Scientific databases require query and retrieval functions that are fundamentally different from those of traditional databases. In a traditional database, the database designer defines the relationships that exists among the data, creates and compiles the schema, and then populates the database with data. The situation in a scientific database application is exactly reversed — the database is first populated by the data, and then the scientist attempts to discover what relationships exist in that data. As the scientific investigation continues, the schema must be able to incorporate new metadata identifying newly discovered structure in the data.

We propose a high-level view of a scientific database environment based on the concept of "model refinement". Any database (or subset thereof) is treated as representing an abstract model of some phenomena. A retrieval request from the database is interpreted as representing a refinement of the model. Model refinement often aims at identifying subsets of the data that are not relevant to the particular data analysis task and removing these subsets from further analysis. The newly created data is *derived* from the original data. Derived information can also be acquired by processing the database to determine "summary" information such as statistical properties or periodicity information from a fast Fourier transform, etc. Such information contributes to the model refinement and must become part of the database.

In fact, the determination of appropriate retrieval requests is intimately tied to the analysis functions to be applied to the data, and vice versa. It is also clear that it must be possible to present any retrieval from a scientific database to the user in a visual format that somehow incorporates "all" of the data retrieved.

2 A Paradigm for Interdisciplinary Scientific Research

A primary goal of the research is the development of structures and procedures which facilitate interdisciplinary investigations conducted by scientists from one or more scientific disciplines or specialities aided by one or more mathematical and/or statistical analysts. In a typical scientific investigation the individual steps outlined below form the components of a cyclic process in which trial and

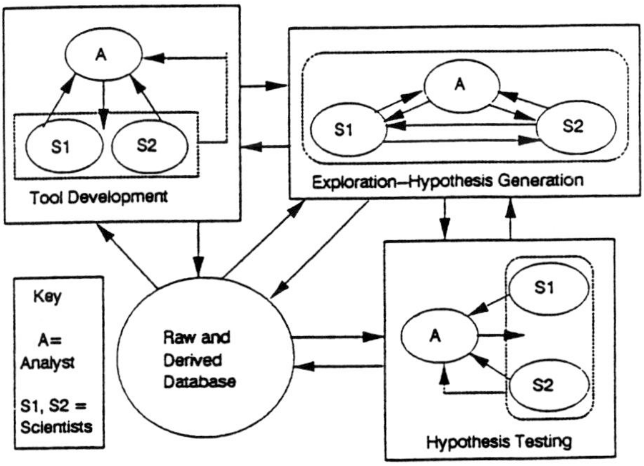

Fig. 1. Paradigm for interdisciplinary scientific research.

error (and serendipity) play important roles. This process is summarized in Fig. 1 in which "A" denotes the analysis components of the interdisciplinary team whose other members are the scientists "Si". The process of investigation we envision consists of cycles formed from one or more of the following components:

- Analytical Tool Development
- Data Validation, Exploratory Analysis, and Hypothesis Formulation
- Hypothesis Testing
- Summary and Presentation of Scientific Results
- Analytical Tool Development.

The analyst develops new analytic techniques and/or utilizes existing techniques focused on particular needs of the scientific team. In the tool development process the analyst must be able to work freely in an environment consisting of *high-level* tools for:

a. accessing the *scientific database* under investigation;

b. mathematical analysis (such as provided by *Matlab* or *Gauss*);

c. *statistical analysis* (such as provided by *Systat* or *SAS*);

d. *data visualization and presentation*;

e. implementing *specialized analysis functions*;

f. interactively defining the *analysis process* that integrates components of all the tools into a problem-specific solution.

Data Validation, Exploratory Analysis, and Hypothesis Formulation

Individually or jointly the analyst and scientists explore segments of the database for the purpose of:

a. assessing the validity of the database through the testing of known relationships among data components and the detection of extreme "outliers" indicating possible errors in the data;

b. developing summary measures (such as means, variances, covariances);

c. exploring the feasibility of existing hypotheses;

d. experimenting with analysis and visualization tools in order to show (potential) relationships within the data; and/or

e. generating new hypotheses on the basis of observed relationships within the data.

Hypothesis Testing

The analyst and scientists formulate specific hypotheses and the analyst develops or uses appropriate statistical and analytical techniques in their testing.

Summary and Presentation of Scientific Results

Appropriate members of the interdisciplinary team use subsets of derived data to summarize the scientific implications of the investigation and prepare them for publication and presentation to the scientific community. Generally, the most effective presentation techniques are exactly those that led to either the development of the hypothesis or its verification or both.

3 Nature of Scientific Data

A scientific database is a collection of data which is the subject of scientific investigation. Data to support scientific research is often multidisciplinary, multisource (public and private, local and global), taken from both general and specific collection programs. Increasingly, it is essential that this diverse data from disparate sources be interrelated. In meteorological applications, for example, air mass information relates to terrain information; in hydrology, runoff data is related to rainfall, terrain, soil, and man-made structures such as streets, storm sewers, cities, dams and irrigation projects.

3.1 Lattice-Oriented Data

For this investigation, we consider that multidimensional data are amassed from observations at points in a lattice (although with certain sampling techniques, values may represent local averages). Each observation includes parameters which identify the location within the lattice and parameters under study. Typically temporal and spatial attributes explicitly identify the sample although this is not always the case (e.g. for geological cores, sampling is time-independent and time of deposition is a derived attribute). Many examples of scientific disciplines fit this paradigm including meteorology, hydrology, geology, reservoir analysis, groundwater analysis, oceanography, environmental science, astronomy and space physics.

3.2 Relationships Among Data

Data values may or may not have embedded interactions. These interactions can be among attributes, among lattice points or both. Geological core sample data may have no (known or relevant) interactions. In astronomical data, there is known interaction between lattice points (e.g., gravity between points of concentration of mass) which must be incorporated into explanations of phenomena observed. Pollution data in fluids may involve interdependencies both among parameters and among lattice points. For example, concentration of nutrients at time t_1 and location l_1 may affect organism growth at t_2 and l_2. An interaction may be measured directly, computed from directly measured attributes, or computed from derived attributes (as with astronomical data, where gravity depends on mass and distance). These interactions may be linear or non-linear. In fact, the nature of the interaction may be affected by other factors, hopefully revealed in the data. For example, a reaction rate may be exponentially related to substance concentration within a certain range of temperature; outside that range, the behavior may be completely different.

Interrelationships in scientific data are fundamentally more complex and interdependent than relationships in most other database applications. In other applications, relationships are known (or assumed) in advance and expressed directly as metadata (e.g., as in an Entity Relationship diagram or other schema design tool) before data are added. Schema design, implementation, and operation are separate and sequential steps.

In contrast, a scientific database system often must manage data before its entire semantic structure is known. Some interrelationships among data parameters may be known in advance (i.e., some patterns in data can be attributed to known phenomena within the problem domain), but many others are not known;

the objective of the research is to discover them. Exploratory analyses on the data are used to discover additional semantic structure, which corresponds to new understanding. One common approach is to remove patterns caused by known phenomena which may mask the phenomena under study, search for other remaining patterns, develop a mathematical/statistical model for those remaining patterns, then use that model to discover possible phenomena which may explain the data. This is usually an iterative process. Relationships are discovered, added, and refined in stages as a scientific model which illuminates underlying physical phenomena is developed. The database software must be able to add derived and user specified interrelationships as metadata to a schema already populated with existing data.

3.3 Data and Models of Data

Scientific applications require that the database system manage both data and models of the data. As data represents an abstraction or model of the real world, further interpretation and derivation can introduce a model of the data (and the real world) which is even more abstract, hopefully eliminating irrelevancy and retaining significance. Such models can take the form of statistical summarizations, cluster analysis, contour analysis, interpolation, normalization, smoothing, units conversion, change in representation, spectral and time series analysis, image analysis, finite element modeling, curve/surface fitting, and registration/integration of data from multiple sources. Models can be represented as data (as in statistical summarizations), graphs, images (visualizations), formulas depicting parameter relationships, and/or rules. The database system must update data and metadata to enable derived data models to be immediately available for further derivation and refinement. A derived model is metadata because it contains (application dependent) semantics about the data; it is also data in that it must be available as input to further analyses.

3.4 Meaning of Updates

In scientific investigation, change to data and data models is endemic to the research process. As the investigation progresses, new understanding may necessitate revisions based on new and better evaluation techniques. This situation may come from developments outside the local investigation (e.g., an improved radiocarbon dating technique may cause recomputation of deposition times in geological core samples; this could result in the need to re-evaluate derived models). Whenever a change is made, the database system must be able to identify data models which need to be updated. Because investigators may want to retain and work with older versions as well as the most up-to-date ones, the database system must offer version control.

3.5 Related Database Developments

Database systems have been widely applied and used in many applications. A number of schema models have been proposed and studied. Semantic schema models (Hull and King [20]) attempt to express rich semantic relationships among data but they have not been integrated into database systems with the exception of limited attempts for the Entity Relationship model (Chen [8]). The hierarchical, network, relational and lately the object schema models have received extensive attention (Chamberlin [7]; Kim [27]; Taylor and Frank [39]; Tsichritizis and Lochovsky [41]). While implementations exist, these models lack comparable semantic richness. Database systems development has been driven primarily by commercial record oriented applications. However, database systems developed for commercial applications often do not meet the needs of other areas. Technical areas in particular require the ability to specify long transactions, represent large chunks of information, maintain complex qualitative and quantitative interrelationships among data, and retain multiple versions of data (Katz, Chang and Bhateja [25]; Kim and Lochovsky [28]). Although improvements in database technology have occurred in some fields, such as office administration and engineering (Encarnacao and Lockemann [12]; Jones and Martin [24]; Ozoyogly et al. [33]), database support for scientific research applications has not been adequately studied.

Although there is obvious overlap, the needs of scientific database applications have fundamental differences from those in engineering. Engineering databases represent carefully controlled worlds where the environment is designed and engineered for a specific predetermined purpose. Interrelationships among entities in that world are relatively well understood in advance. Irrelevancy has been engineered out to enhance productivity. Scientific environments are not pre-engineered; the data represents the state(s) of a (partially) unknown world from which the scientist wants to deduce behavior. The goal in scientific database applications is to incorporate essentially unstructured information where what is extraneous is not immediately apparent. Through analysis and visualization techniques, the scientist hopes to separate the significant from the irrelevant, and thereby to discover the unknown. Schema evolution techniques are essential in the highly interactive discovery support environment we envision (Banerjee et al. [1]).

For many years, programmers have hand coded individual file processing solutions to support scientists and their aides; and this trend persists today. Early scientific database activity involved integration of specific analysis tools with file and later data management (Johansson and Shilling [23]; Montgomery et al. [32]; Westlake and Kleinschmidt [44]). The need to incorporate accuracy information was recognized in (Sparr and Hann [38]; Sparr [36]). The need for database systems to handle units conversion automatically in a sophisticated way was recognized in (Sparr [37]).

Use of scientific database systems has frequently been directed toward governmental regulation rather than scientific discovery. Attention to database support for research activities has lagged. Statistical database systems are the most common type of scientific database system (Bragg [5]; Michalewicz [30]). These often lack generality to facilitate seamless integration of more comprehensive analytical tools; metadata is limited to description of how the data interfaces to the specific statistical package. Relatively little attention has been paid to development of a generic schema model for scientific databases.

More versatile tools, such as spreadsheets and comprehensive analysis packages, are standalone products not currently integrated into data management environments. Investigations into advanced tools such as scientific data visualization typically (and rightly) focus on the analysis techniques rather than the data management aspects which are considered peripheral. Recently, some efforts have been initiated to develop Work common data interchange standards such as Common Data Format (CDF) from NASA (Gough et al. [16]; Treinish and Gough [40]) and Hierarchical Data Format (HDF) from NCSA (Haber [18]; NCSA [14]).

4 Nature of Scientific Data Visualization

4.1 Traditional Visualization Techniques

Traditional scientific visualization techniques range from simple two dimensional techniques such as line graphs, bar charts, and scatterplots to three dimensional displays using realistic image synthesis techniques, such as ray tracing (Glassner [15]), sophisticated lighting models (Hall [19]), volume rendering (Kaufman [26]), etc. Animation of time-dependent data is also extremely effective for providing insight into such data.

All these techniques have proven to be particularly effective in presenting data for which a natural two or three-dimensional representation exists, such as in medical imaging, computational fluid dynamics, and finite element analysis. Even in these cases, however, the dimensionality of the data that can be presented is relatively low. For example, a three-dimensional view of a fluid flow simulation utilizes synthetic light sources and shading variations to represent the surface of the fluid more realistically. An additional parameter, such as the temperature at each point on the surface could be represented by the color at that point. Some additional information, such as flow direction, speed, and velocity, could be represented (at a lower resolution) by arrows or ribbons on the surface. Such techniques are absolutely critical components of a scientific visualization environment and need to be available to the scientist in a standardized form.

4.2 Multi-Dimensional Data Visualization

On the other hand, how might we display very high-dimensional data, where we have 10 or more data parameters at every point in our sample space? What if the important relationships embedded in the data are only evident when most or all of these data parameters interact in certain ways? This problem of *multi-dimensional* data visualization and interpretation has begun to be addressed in recent years by a variety of novel and intriguing approaches.

A number of researchers have experimented with *iconographic* displays in which each multi-dimensional sample point is represented by an icon which changes shape, color, etc. based on the parameters at the data point (Pickett and Grinstein [34]; Grinstein, Pickett and Williams [17]; Beddow [2]; LeBlanc, Ward and Wittels [29]; etc.). When the sample points are relatively dense, the iconographic display presents texture patterns that vary according to the nature of the data. The Appendix includes several examples of iconographic images generated from the same Landsat satellite data. The proper selection of icon and parameter mapping can produce visual patterns that represent important multi-dimensional relationships among the data parameters. The individual icons themselves are not intended to be discernible, but the overall texture pattern that results can show areas in which the combinations of the parameters appear to be "interestingänd thereby can be used to focus the application of other analysis techniques.

Other approaches may be particularly useful for data with sparse multi-dimensional sample points. Inselberg (Inselberg and Dimsdale [22]) has proposed an intriguing representation, called *parallel coordinates*, in which each point in n-dimensional space is mapped to a line in two-dimensional space. The patterns created by mapping a set of n-dimensional points may give some insight into the nature of that data. Mihalisin (Mihalsin et al. [31]) describes a hierarchical nesting of plots which can show interesting properties of high dimensional data. In n-Vision, Feiner (Feiner and Beshers [13]) uses another form of nesting of dimensionality allowing a user to interactively select a point on a surface at which a new graph is displayed with three more data dimensions representing the next level surface.

4.3 Systems for Scientific Data Visualization

Although analysis and visualization packages tailored to particular disciplines have been available for a number years, it has only been very recently that interactive *general purpose* analysis and visualization packages have been developed. Some, such as IDL (IDL [21]), require the user to write simple interactive programs to specify the analysis and visualization steps. PV-WAVE from Precision Visuals provides extensive built-in functionality that the user accesses by menu selection, while also supporting user-defined programming. More flexibility is provided in *flow-based* environments such as AVS (Upson et al. [43]; AVS [10]) from Stardent, apE (Dyer [11]) from the Ohio Supercomputing Center, and our own Flexvis system (Bergeron and Grinstein [4]; Behari [3]; Calder [6]). In these systems the user defines a visualization pipeline or graph composed of independent processes connected via arcs representing data flow. The scientist-user can create arbitrary visualization specifications from a set of pre-defined tools, and can add new tools to the set. Such systems all provide the user with the ability to enter and change analysis parameters interactively. In addition, the Flexvis

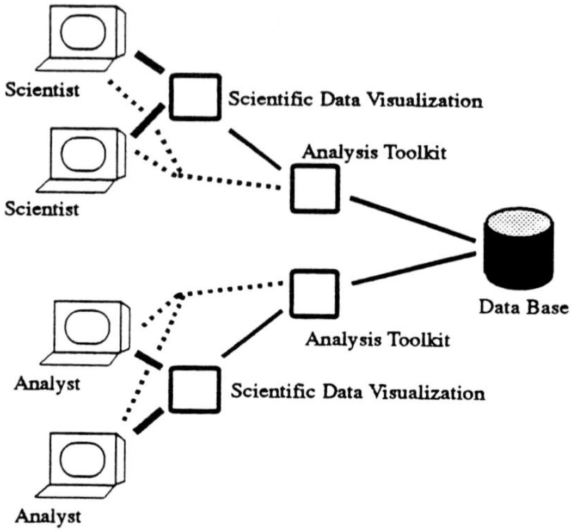

Fig. 2. Computational environment for scientific research.

system also supports the specification of an *interaction graph* so that a user can define specialized interaction processing that occurs *during* execution of the flow graph (Calder [6]). These flow-based environments are the most promising approach to the integration of analysis and visualization since they allow the scientist to *experiment* with both analysis and visualization techniques at the same time.

5 Toward a Target Scientific Database System

Our target database system for scientific research is illustrated in Fig. 2. Relevant data are extracted from the repository(ies), reduced if necessary and entered into an integrated database with appropriate metadata. Known relationships among data are included but the conduct of the research may uncover additional relationships which are also added to the database.

To be effective, our design must incorporate three principles.

1. Metadata – The syntactic and semantic information required for this database system includes syntactic and semantic information about data models, queries, and analytical tools. Queries are expressed as process graphs. This form permits invocation of compound sequences of sophisticated analytical tools and also serves as a derivation record for data models.

2. New Schema Model – Current schema models are inadequate to represent a comprehensive conceptual structure for the complexities of scientific databases.

3. Schema Evolution – the database is in a state of continuing evolution as new data models are derived. Derived models are data in that they are subject

to further analysis. They are also metadata in that they convey application dependent semantics about the data and can be used to influence further processing (e.g. detection of outliers). A schema must evolve to fully integrate new data models and their complex patterns of relationships.

5.1 Terminology

The term *repository data* refers to data as captured for a computer database system. Transformation from original raw data can include verification, sampling (in the sense of signal sampling), summarizing, digitizing, normalizing, or the attachment of uncertainty or accuracy information. Often, this is performed by the collecting agency before publication and distribution. Repository data is in that form, least removed from the raw data, that is available to the investigators. We assume that repository data contains only actual observations and is the base data for the research.

Derived data results from a transformation of existing data, such as the application of analysis techniques. Derivation can also include the addition of preliminary structure and relationships. Derivations are abstractions or models of the input data (and the world). Derived data can be in several forms: data sets (as in statistical summarizations), graphs, images (visualizations), and formulas or rules depicting interparameter relationships. We use *data set* to mean a subset of repository data or derived data treated as a unit during analysis. The internal form used to represent a data set (e.g., relation or object) is determined by the schema model.

The term *parameter* refers to a measurement taken as part of a data observation which describes the state of the real world (such as the temperature of an air sample). Thus, an *attribute* is the corresponding database property (i.e., the name and value of a characteristic of a database entity representing a data observation). An attribute is the database representation of a parameter which portrays the state of the corresponding real world condition.

We use the term *model* in several ways. A *scientific model* refers to phenomena which control or explain real world behavior. A *data model* refers to an abstraction of (a portion of) a scientific database. Data models contain derived data. Normally they result from the application of one or more analytical techniques although in some cases, users may define them directly. Data models are created to discover scientific models. Our usage of the term data model contrasts with the normal meaning of the term in database literature where it refers to a structure for organization of metadata.

A *schema model* is a structure for organizing metadata in a database system. This corresponds to the term data model in the database community. Common schema models are relational, object and network. A *schema* is an instance of a schema model which contains metadata to describe a specific database.

Metadata is schema information which provides syntactic and semantic description of database contents. This includes internal structure and location, application independent semantic information, and application dependent semantic information. Metadata are also needed to describe the capabilities and requirements of analysis tools.

5.2 Metadata

Metadata must describe data models, the derivation of data models, and the capabilities and interfaces of analytical procedures. The schema model must be able to represent data models of various forms: data sets, images, graphs, formulas, and rules. Data sets can be represented conventionally except that provision for attaching accuracy and units information to either specific observations, selected parameters, data sets, or entire lattices must be incorporated. Another result form is graphic; examples are graphs and visual data images. A third form is equations or formulas expressing interrelationships among data parameters.

In our environment queries derive data models and are defined and represented in the database as graphs connecting input nodes (data sources) through sequences of internal nodes (visualization or analytical processes) to output nodes (results) via arcs (data paths). For each data model, the database system retains the query graph defining the derivation ancestry. Resultant data models are available as input data sources for subsequent queries. The query interface supports abstraction of processes by permitting the creation of a single symbol to represent an entire subgraph of processes.

Metadata about procedures must include a view (in the schema sense) which defines the interface to data and allows the database system to map data values to program variables. The interface must be sophisticated enough to map multiple instances onto array variables in the analysis tool, not normally part of extant embedded database systems. Examples of metadata information about procedures include:

parameters needed with accuracy and sampling requirements

parameters created and/or changed and the effect on accuracy

preconditions and postconditions

Preconditions and postconditions represented in metadata enable the data management system to detect attempts to make inappropriate use of analysis tools. Examples of checks include fixed/variable sampling grid or sampling too infrequently in temporal and/or spatial domains for stability criteria. A key part of the metadata for each analytical technique is the representation of the impact the technique has on accuracy.

5.3 Schema Model

We believe a new schema model is required for scientific applications because existing schema models are not suitable for representing the complex metadata needed.

The value orientation of the relational model does not readily handle changes in key attributes (e.g. to incorporate a better estimate of the location of a specific sample in the lattice). Furthermore, the relational model does not readily support long transactions, representation of information in large chunks, nor dynamic evolution of data relationships.

By contrast, the object database approach can build interrelationships among data instances at runtime. The object model can express abstraction/specialization and aggregation/composition relationships among data in a way that allows inheritance of general characteristics to specific instances. Methods (procedures) can be attached to objects (individual observations or groupings of observations) although it is not clear what role encapsulation should play in a database for research and discovery. Object identity solves the problem of updates to key attributes. The object model also addresses aspects of version control and schema evolution. Among extant database technologies, the object database approach seems to come closest to what is needed for our scientific data model. However, the object data model lacks the capability for rich semantics present in semantic data models (Hull and King [20]); we are presently unwilling to make this sacrifice. Available object database systems are not yet mature enough to use directly at this time. Their effectiveness has not been demonstrated in dynamic integrated environments such as we envision for collaborative scientific research with the extreme complexity and fluidity of data interrelationships.

5.4 Schema Evolution

The system must create a separate identity for (some) newly derived results but maintain relationships which give derivation ancestry. During the exploratory stage of research, the database system may need to retain multiple versions of data sets reflecting alternative derivations. The metadata must be updated to enable the immediate use of derived models to be flexibly and seamlessly used in the derivation of new models.

5.5 Bridge to Knowledge Based Systems Technology

Formula models, whether defined or derived, are of further interest. A formula which expresses a functional relationship among data parameters closely corresponds to the notion of a rule in an expert system. The ability to derive and represent relationships as rules and subsequently use them as query inputs presents a natural opportunity to integrate reasoning and knowledge base systems technology into the scientific investigation. Our approach supports the evolution toward knowledge based techniques; processes can represent procedural information in many forms including formulas (rules) as well as flow graphs.

The ultimate goal is a knowledge system which can infer refinements to data models using data combined with preliminary models (inferences). Eventually, as techniques are developed to express the semantics of complex mathematical analysis routines in more explicit form, knowledge base system techniques can be applied to the scientific discovery process. This is beyond the scope of this work, although our approach is consistent with that direction.

5.6 Project Overview

Three interrelated themes are addressed by this research:

1. a visualization-based model of a scientific database;
2. support for new and existing tools that aid the cooperation between the scientist and analyst;
3. the development of effective data visualization tools.

Integrated Testbed Environment for Exploratory Data Analysis

Our testbed environment uses as many existing tools as possible. This approach maximizes the functionality available at the expense of some degradation to the user interface. Since our principal goal is aimed at validating the *approach*, we feel this is an appropriate trade-off. The environment consists of three principal components:

1. a process flow-based visualization system,
2. interfaces to existing analysis tools, and a
3. database interface to implement the scientific database model that we are proposing.

Flow-based Visualization System

The user interface to our environment is provided by a process flow-based scientific visualization system, Flexvis (Behari [3]; Calder [6]). Because other similar systems, such as AVS-2 (AVS [10]) and apE (Dyer [11]), are becoming more widely available in the scientific and computing community, it is desirable and feasible to minimize the reliance of the remainder of the system on the visualization front-end.

Interfaces to Analysis Tools

Scientists currently have a wide range of general purpose and specialized analysis packages available. It is clearly not feasible to re-implement all these systems in order to fully integrate all such functionality into a single monolithic system. Furthermore, there are considerable advantages to allowing the scientist to utilize analysis tools that he or she already understands and trusts. Our current goal, therefore, is to provide a framework that allows us to integrate existing tools into our environment. Anything is potentially a candidate for integration. The limitation to this approach, however, is that existing software can usually only act as a *data filter* to transform data from one form into another. This is particularly constraining when the "analysis" is a complex simulation that could benefit from some interaction with the user. It is preferable to *integrate* such a simulation into the process graph, allowing the user to control it interactively. That is, as the simulation proceeds, the user gets immediate feedback of partial results and uses that to alter the simulation process itself. This, however, requires recoding of the simulation process.

Database Interface

The entire scientific database available to a user consists of the original (essentially unstructured) data along with derived data models and associated metadata that represent the *discovered* structure in the data. This information can be maintained using a conventional relational database. Although there are characteristics of object-oriented databases that make them attractive as the underlying tool for building our database, their performance, reliability, and availability are still rather limited. On the other hand, the actual choice of database tool (even with such disparate models as relational and object-oriented) is not nearly as important as the concepts and the methodology that we have adopted as the model for a scientific database environment. Consequently, we have designed the database interface for minimal reliance on the particular choice of database tool.

Refinement of the SDB Model

A major goal of the research is to refine our schema model for a scientific database and evaluate the effectiveness of this model in fostering cooperation among scientists and analysts. The facilities of the scientific database interface should provide analysis and visualization tools with a unified integrated access to original repository data, all saved derived data, including statistical and mathematical models of the data (or parts of it), and user specified hypothesized models of the data. The principal problems to be addressed in this area include:

1. developing an efficient and usable mechanism for representing the *process graph* associated with derived data;

2. developing an interface for defining/accessing the metadata from visualization and analysis tools;

3. addressing the critical (and very difficult) issue of representing the reliability of data and for incorporating that in the analysis tool interfaces;

4. developing mechanisms by which users can *browse* through the known information about a database (i.e., the structure that has so far been induced or deduced about the data);

5. evaluating the advisability/feasibility of utilizing both traditional and object databases as the integrating database medium.

Scientific Data Visualization

The human visual perception system represents an extremely powerful tool for providing insight about complex data. The data visualization tools in our environment are an important vehicle for providing both the scientist and the analyst with information for directing the analytical process. In particular, we believe that the visual presentation itself serves as a very important form of communication between scientists of different disciplines and between the scientist and the analyst. Specific issues to be addressed include:

1. integrating traditional and novel visualization techniques with desired analysis functionality;
2. experimenting with iconographic display techniques to evaluate their effectiveness for helping to direct the analysis process;
3. supporting dynamic displays of results from time series, as well as for transforming other data dimensions to the time domain for the purpose of data presentation;
4. evaluating data visualization as a communication medium among scientists and analysts.

Multidisciplinary Collaboration

A major goal of the SDB environment is to foster cooperation among scientists and between the scientist and the analyst. The database and visualization interfaces must be tailored to the needs of both. Consequently, the facilities within the environment and the design of the interface must be developed with the collaboration of both.

6 Conclusion

We believe that our model for a scientific database represents an innovative approach to integrating analysis and visualization in a context that will foster multidisciplinary collaboration. The representation of the semantics of the analysis process in terms of a flow graph provides an effective basis for further expansion into a knowledge base system.

References

[1] J. Banerjee, et al. (1987). *Semantics and Implementation of Schema Evolution in Object-Oriented Databases.* In *Proc. ACM SIGMOD 1987 Annual Conference, San Francisco, CA, SIGMOD Record*, volume 16(3).

[2] J. Beddow (1990). *Shape Coding of Multidimensional Data on a Microcomputer Display.* In *Proc. Visualization '90, San Francisco*, pp. 238–246. IEEE Computer Society Press.

[3] A. Behari (1990). *Flexvis - A Flow-based Visualization System.* In *Computer Science Technical Report 90-71.* University of New Hampshire.

[4] R.D. Bergeron, G.G. Grinstein (1989). *A Reference Model for Scientific Visualization.* In *Proc. Eurographics '89*, pp. 393–399. North-Holland Publ. Co.

[5] A.W. Bragg (1981). *Data Manipulation Languages for Statistical Databases – The Statistical Analysis System (SAS).* In *Proceedings Workshop on Statistical Database Management, Menlo Park, California.*

[6] B.H. Calder (1991). *An Interactive Scientific Visualization Application Development Environment.* In *Computer Science Technical Report 91-09.* University of New Hampshire.

[7] D.D. Chamberlin (1976). *Relational Data-Base Management Systems. ACM Computing Surveys*, 8(1):43–66.

[8] P. Chen (1976). *The Entity Relationship Mode — Toward a Unified View of Data. ACM Trans. Database Systems*, 1(1):9–36.

[9] E. Cohen, Jr. R.A. Hay (1981). *Why are Commercial Database Management Systems Rarely Used for Research Data?* In *Proceedings Workshop on Statistical Database Management, Menlo Park, California.*

[10] Stardent Computer Corp. (1989). *Application Visualization System Developer's Guide.* Stardent Computer Corp., Newton MA.

[11] D.S. Dyer (1990). *A Dataflow Kit for Visualization. IEEE Computer Graphics & Appl.*, 10(4):60–69.

[12] J.L. Encarnacao, P.C. Lockemann (Eds.), editors (1987). *Engineering Databases.* Springer-Verlag.

[13] S. Feiner, C. Beshers (1990). *Visualizing n-dimensional Virtual Worlds with n-Vision. Computer Graphics*, 24(2):37–38.

[14] National Center for Supercomputer Applications (NCSA) (1989). *DataScope Reference Manual.* NSCA, University of Illinois, Champaign-Urbana, IL.

[15] A. Glassner, editor (1989). *An Introduction to Ray Tracing.* Academic Press.

[16] M. Gough (1988). *CDF Implementer's Guide.* National Space Science Data Center, NASA Goddard Space Flight Center, Greenbelt, MD.

[17] G.G. Grinstein, R.M. Pickett, M. Williams (1989). *EXVIS: An Exploratory Visualization Environment.* In *Proc. Graphics Interface '89*, pp. 254–261. CIPS, Toronto.

[18] R. Haber (1989). *Scientific Visualization and the RIVERS Project at the National Center for Supercomputing Applications. IEEE Computer*, 22(8):84–88.

[19] R. Hall (1989). *Illumination and Color in Computer Generated Imagery.* Springer-Verlag, New York.

[20] R. Hull, R. King (1987). *Semantic Database Modeling: Survey, Applications, and Research Issues. ACM Computing Surveys*, 19(3):201–260.

[21] Research Systems Inc. (1989). *Introduction to IDL.* Research Systems Inc., Denver, Colorado.

[22] A. Inselberg, B. Dimsdale (1990). *Parallel Coordinates: A Tool for Visualizing Multi-Dimensional Geometry.* In *Proc. Visualization '90, San Francisco*, pp. 361–378. IEEE Computer Society Press.

[23] J.H. Johansson, J.D. Shilling (1981). *Toward the Development of an Integrated Economic Data Base at the World Bank.* In *Proceedings Workshop on Statistical Database Management, Menlo Park, California.*

[24] F.H. Jones, L. Martin (1987). *The AutoCAD Database Book.* Ventana Press, Chapel Hill, NC.

[25] R.H. Katz, E. Chang, R. Bhateja (1986). *Version Modeling Concepts for Computer-Aided Design Databases.* In *Proc. ACM SIGMOD 1987 Annual Conference, Washington, D.C, SIGMOD Record*, volume 15(2).

[26] A. Kaufman, editor (1990). *Volume Visualization.* IEEE Computer Society Press.

[27] W. Kim (1990). *Introduction to Object-Oriented Databases.* MIT Press, Cambridge, MA.

[28] W. Kim, F.H. Lochovsky (1989). *Object Oriented Concepts, Databases, and Applications.* ACM Press, New York, NY.

[29] J. LeBlanc, M.O. Ward, N. Wittels (1990). *Exploring N-Dimensional Databases.* In *Proc. Visualization '90, San Francisco*, pp. 230–237. IEEE Computer Society Press.

[30] Z. Michalewicz (1990). *Current Research in Statistical and Scientific Databases: the V SSDBM. IEEE Data Engineering*, 13(3).

[31] T. Mihalisin, E. Gawlinski, J. Timlin, J. Schwegler (1990). *Visualizing a Scalar Field on an N-dimensional Lattice.* In *Proc. Visualization '90, San Francisco, Oct. 1990*, pp. 255–262. IEEE Computer Society Press.

[32] R.H. Montgomery, et al. (1981). *A Reservoir Water Quality Database System.* In *Proceedings Workshop on Statistical Database Management, Menlo Park, California.*

[33] G. Ozoyogly, et al. (1990). *A Scientific DBMS for Programmable Logic Controllers. IEEE Data Engineering*, 13(3).

[34] R.M. Pickett, G.G. Grinstein (1988). *Iconographic Displays for Visualizing Multidimensional Data.* In *Proc. IEEE Conf. on Systems, Man, and Cybernetics, Beijing.*

[35] S.M. Pizer (1981). *Intensity Mappings to Linearize Display Devices. Computet Graphics and Image Processing*, 17(3):262–268.

[36] T.M. Sparr (1981). *Units and Accuracy in Statistical Databases.* In *Proceedings Workshop on Statistical Database Management, Menlo Park, California.*

[37] T.M. Sparr (1987). *A Units Expert for Engineering Databases.* In *Proceedings IEEE Computer Society Dallas Chapter Conference on New Directions in Database and Knowledge Management Systems, Dallas, Texas.*

[38] T.M. Sparr, R.W. Hann Jr. (1970). *A Water Quality Storage and Retrieval System for Regional Application.* In *Proceedings of National Symposium on Data and Instrumentation for Water Quality Management, Madison, Wisconsin*, pp. 215–220.

[39] R.W. Taylor, R.L. Frank (1976). *CODASYL Data-Base Management Systems. ACM Computing Surveys*, 8(1):67–103.

[40] L. Treinish, M. Gough (1987). *A Software Package for the Data-independent Management of Multidimensional Data. EOS Transactions of the American Geophysical Union*, 68(28).

[41] D.C. Tsichritizis, F.H. Lochovsky (1976). *Hierarchical Data-Base Management. ACM Computing Surveys*, 8(1):105–123.

[42] J.D. Tukey (1977). *Exploratory Data Analysis.* Addison-Wesley Publ. Co., Reading, MA.

[43] C. Upson, et al. (1989). *The Application Visualization System: A Computational Environment for Scientific Visualization. IEEE Computer Graphics & Appl.*, 9(4):30–42.

[44] A.J. Westlake, I. Kleinschmidt (1990). *The Implementation of Area and Membership Retrievals in Point Geography using SQL. IEEE Data Engineering*, 13(3).

Appendix: Examples of Iconographic Data Displays

The color illustrations of this chapter at the end of the book show examples of iconographic data displays. They simultaneously present satellite data containing 5 different wavelength measurements from the Eastern end of the Great Lakes region.

Volume Synthesis Principles

Arie E. Kaufman

There is a recent increase in the use of discrete voxel representation for a variety of geometry-based applications. These include CAD, simulation, and scientific visualization, as well as those that intermix geometric objects with 3D sampled or computed datasets. In these applications the inherently continuous 3D geometric scene is sampled employing voxelization (3D scan-conversion) algorithms, which generate a 3D raster of voxels. The voxelized objects have to conform to some 3D discrete topological requirements such as connectivity and absence of tunnels. During the voxelization process, also termed the volume synthesis process, each voxel is assigned precomputed numeric values that represent some measurable view-independent properties of a tiny cube of the real or simulated object. These values are then readily accessible for speeding up the rendering or the discrete ray tracing process. The voxelization algorithms are the 3D counterparts of the 2D scan-conversion algorithms, and the 3D raster generated by them is the 3D counterpart of the conventional 2D raster.

1 Background

Volume visualization is a method of interpreting volumetric datasets by providing visual insight into them. The objective is to peer inside volumetric objects in order to see and steer the unseen. It encompasses an array of techniques for extracting meaningful information primarily from sampled or simulated volumetric datasets and displaying it in a visual manner [36]. The volumetric dataset is commonly represented as a regular grid of voxels, where all voxels are identical cubes aligned with the axes and where the dataset is often stored as a full 3D raster of voxels, called the *volume buffer*. Alternatively, arbitrary irregular grids, either rectilinear, structured (curvilinear), or unstructured, are employed (e.g., [66]). *Volume synthesis*, which is the subject of this paper, employs a regular volume buffer for object representation, and it focuses on a set of techniques for representing and synthesizing voxelized synthetic objects. Exploiting a volume buffer for synthetic models representation, manipulation, and rendering, as well as for sampled or simulated datasets, is an emerging technology, referred to as *volume graphics* [12].

Figure 1 portrays the dataflow and the taxonomy for volume visualization and volume graphics. Volume graphics is limited to the stages of volume visualization marked with solid arrows in Fig. 1. One source of volume data, which is the primary input for volume visualization, is empirical, generated by either a sampling device or a computer model that creates discrete sampling points of

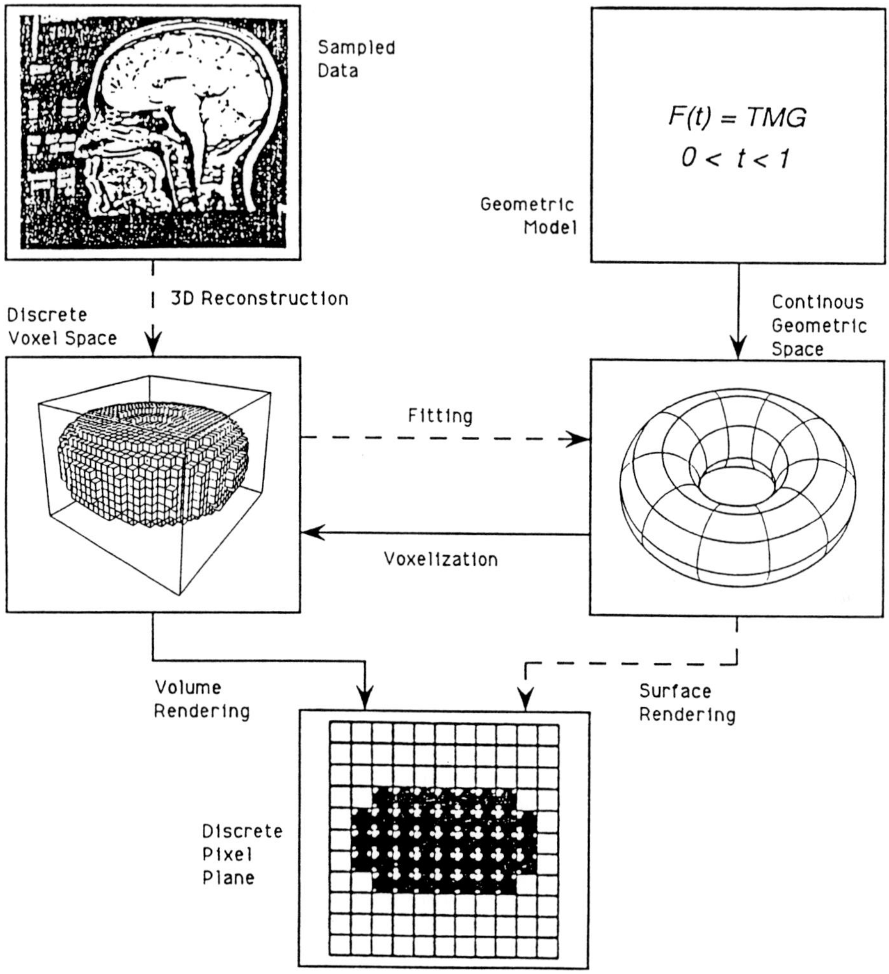

Fig. 1. Dataflow and taxonomy for volume visualization and volume graphics. The latter is limited to the stages marked with solid arrows.

a real or simulated object or phenomenon. This empirical dataset is 3D reconstructed into a 3D volumetric dataset of voxels by interpolating between the data points.

Another source of volume data, which is a primary input for volume synthesis, is a 3D continuous geometric model, commonly represented as a display list. Such a geometric model is *3D scan-converted* (*voxelized*) into a set of voxels that "best" approximates the synthetic model within the discrete voxel space [10, 34, 35, 40, 51]. This process is referred to also as *volume synthesis*. Once converted, the digital synthetic model may be intermixed with the empirical dataset to form a hybrid voxel model [24, 41, 47]. The motivation for volume synthesis and the fundamentals of voxelization are the foci of this paper.

In order to visualize the volume dataset, the volume primitives can be directly projected into 2D pixel space. This process, which is termed *volume rendering*, involves both the viewing and the shading of the volume image and can be accomplished by forward projection (e.g., [17]), ray casting (e.g., [45]), or by a discrete ray tracing [75].

Alternatively, volume data can be first converted into surface primitives in a process called *fitting* (e.g., [9]). Then, the surface primitives (e.g., polygon mesh, contours) are rendered to the screen by a conventional process of *surface rendering*. In volume graphics, which exploits a volume buffer as a representation, synthesis, and rendering medium, fitting followed by surface rendering is thus inappropriate (cf. [21, 70]).

2 Motivation

The swift advances in hardware – primarily faster, cheaper, and denser memories – have been transforming revolutionary approaches in computer graphics into reality. One typical example is the revolution of raster graphics that took place in the seventies. Another approach, which has a similar potential and is currently shaping up, is volume graphics. Below, we introduce the characteristics of volume graphics and discuss its advantages and weaknesses.

The display of graphics in the sixties and seventies was based on continuous vector drawing devices and on object-based approach to scene representation, manipulation, and display. The alternative approach, termed *raster graphics*, has been predominant in the eighties and today. Figure 2 contrasts vector graphics with raster graphics.

The object-based approach had also been adapted for 3D graphics. This approach in 3D, termed *surface graphics*, combines raster technology for the display, and object-based approach for the representation, manipulation and rendering of 3D scenes. This method is supported by powerful geometry engines that have flourished in the past decade, making surface graphics the state-of-the-art in 3D graphics.

	Vector Graphics	Raster Graphics
Scene complexity	sensitive	insensitive
Object complexity	sensitive	insensitive
Block operations	difficult	trivial
Sampled data	no	yes
Interior information	almost impossible	yes
Scan-conversion	embedded in drawing	decoupled from drawing
Space	variable	large but constant
Aliasing	nonexistent	frequent
Transformation	continuous	discrete
Geometric information	exact	approximate
	Surface Graphics	Volume Graphics

Fig. 2. A comparison between vector graphics and raster graphics and between surface graphics and volume graphics

Surface graphics strikingly resembles vector graphics in many ways (see Fig. 2). In surface graphics, any change to the scene, viewing parameters, or rendering parameters requires re-processing of the list of primitives through the image drawing pipeline. Surface graphics presents merely the surfaces of 3D filled objects, and similar to the limitations of vector graphics, rendering the interior of these 3D objects is almost impossible.

Volume graphics is an emerging alternative to surface graphics. Volume graphics also employs a 2D frame buffer for display purposes, but unlike surface graphics, it employs a volume buffer for 3D representation, manipulation, and rendering. The 3D scene is digitized at the beginning of the image generation pipeline, in a volume synthesis stage, and the resulting 3D discrete form is used repeatedly as a database of the 3D scene for manipulation and rendering purposes.

The problems associated with the volume buffer representation for synthetic models, such as the discreteness, memory size, processing time, and loss of geometric representation, echo similar problems encountered when raster graphics emerged as an alternative technology to vector graphics. Discreteness, as in raster graphics, can be handled with anti-aliasing, smoothing, and supersampling techniques. Memory size and processing time cease to be an issue as memory and hardware systems advance beyond our imagination. The loss of geometric information might be problematic, for example, for transformations, measurements, and normal calculation. It can be alleviated either by storing such information with each voxel or by keeping it in an auxiliary structure. This auxiliary information might also be useful for hit verification by discrete rays in discrete ray tracing (see Sect. 4), and for re-voxelizing the scene in case of a change in the scene itself or the environment.

Since the 3D scene is digitized early in the pipeline, digitization is decoupled from rendering. Furthermore, all objects are converted into one uniform meta-object – the voxel. Each voxel is atomic and represents the information about one single object that pierces that voxel. Volume graphics offers benefits similar to surface graphics, with several advantages due to these decoupling, uniformity, and atomicity features. The advantages of volume graphics are discussed below.

Due to the use of a volume buffer, the rendering is actually insensitive to the complexity of the scene, and its performance depends mainly on the constant resolution of the volume buffer and not on the number of objects in the scene. It is, therefore, especially attractive for rendering scenes consisting of a large number of objects (e.g., a large polygon mesh generated by a polyhedral smoothing algorithm [16, 52, 72]). The rendering process is also insensitive to the complexity of the objects, and is thus especially appealing for objects that are hard to render, such as curved surfaces, fractals, and constructive solid models.

Another type of complexity involves objects that are enhanced with texture mapping, photo mapping, or solid texturing. Unlike conventional techniques, in volume graphics texture mapping is performed once during the voxelization stage and the texture color is stored in each voxel. In fact, in volume graphics, the scene is created (voxelized) once for multiple viewpoints and lighting conditions, while surface graphics must repeatedly scan-convert the scene after every change in the viewing parameters. In anticipation of repeated access to the volume buffer, all view-independent attributes can be precomputed during the voxelization stage, stored with the voxel, and be readily accessible for speeding up the rendering, making the rendering independent of the observer's viewpoint.

In addition, the volume graphics approach naturally and directly supports sampled and simulated datasets, as well as intermixing thereof with geometric objects [41]. Another central feature of volumetric representation is that, unlike surface representation, it contains the inner structures of the objects, which can be revealed and explored with the appropriate manipulation and rendering techniques.

Furthermore, the volume buffer discrete representation lends itself to block operations such as *voxblt* [37], which are the 3D counterparts of the *bitblt* operations on windows. Moreover, the discrete representation is suitable for spatial set operations, which can be performed on a voxel-by-voxel basis during the voxelization stage. This property can be exploited for pre-generating constructive solid geometry (CSG) voxelized models. This further provides for spatial matte masking operations, such as merging two volumes, extracting cut planes, or removing regions [17]. The spatial presortedness of the voxels supports incremental processing (e.g., transformations) that is based on inter- and intra-scan-line and scan-plane coherency. The presortedness also lends itself to grouping and aggregating neighboring voxels. For example, in a flight simulator (such as in the Hughes Aircraft Company RealScene® flight simulator) voxels below the terrain are represented as tall cuboids extending from sea level to the terrain height. Similarly, voxels can be aggregated into super-voxels in an octree-like or pyramid-like hierarchy [46, 49, 62], when a less detailed model is required, such

Fig. 3. A scene from Hughes Aircraft Company RealScene® flight simulator, which simulates a flight over voxel-represented terrain enhanced with satellite photo mapping and synthetic raised objects such as the fifteen buildings in front of an airport terminal building. The terrain and building have been voxelized and texture-mapped during the voxelization phase.

as when an aircraft ascends during a flight simulation, thus making rendering independent of object scale (see Fig. 3). Finally, the simplicity and repetitiveness of the representation lend themselves to hardware realizations [38].

Naturally, the use of the voxel-based representation appeared first in applications involving sampled 3D datasets in which the data is in volumetric form. These applications still provide the major driving force for volume visualization. Some examples of these applications are medical imaging (e.g., computed tomography, magnetic resonance imaging [69]), biology (e.g., volume microscopy [39]), geoscience (e.g., seismic measurements [73]), industry (e.g., inspection [43]), earth science [29], and molecular systems [24].

Although the voxel-based representation seems to be more natural for empirical imagery, the advantages of this representation, as discussed in Sect. 2, are also attracting traditional surface-based applications (cf. [48]). Some examples are the rendering of fractals [27, 54], solid textures [57, 58], fur [32], gasses [18], plant growth [25], and complex models [64]. The most convincing example is the Hughes RealScene flight simulator, which exploits very successfully the voxel representation for terrain data, "raised" objects, and moving objects, in an application which traditionally has been identified with extremely high-performance surface graphics.

Fig. 4. Newell's teapot over a textured floor, mirrored in a disk mirror. The teapot, floor, and mirror have been voxelized and pre-textured during the voxelization phase. Discrete ray-tracing time for this 256 × 83 × 9 resolution image is about 31 seconds, running on a 20 MIPS machine.

Furthermore, in many applications, like radiation therapy planning and surgical planning, the sampled data need to be visualized along with synthetic objects that may not be available in digital form, such as isodose surfaces and radiation beams [47], osteotomy surfaces, surgical cuts, prosthetic devices, grafts [65], injection needles, and scalpels [41]. These objects can be digitized and intermixed with the sampled organ in the same volume buffer and then rendered together [41]. This has been also demonstrated in mixing an MRI dataset of a head and a confocal microscope dataset of a neuron with geometric objects representing oblique cuts, mirrors, and back drop and floor for casting shadows [76]. Additional examples of intermixing are: molecular graphics (e.g., overlaying stick figure on an electron density map), computational fluid dynamics (e.g., projectile density clouds with embedded flow ribbons and mirrors [63]), modeling (e.g., stream rising from a glass [18]), and many others [22, 60, 71].

3 Voxelization Algorithms

The process of generating from a continuous geometric model a set of voxels in 3D discrete space that closely approximates the continuous model is called *voxelization*. As this process mimics the scan-conversion process that pixelizes 2D geometric objects, it is referred to also as *3D scan-conversion* algorithms. A large body of 2D scan-conversion algorithms, which are fundamental to 2D raster systems, has been published. The objects that have been studied are: straight lines [3, 5, 67, 68], curves [3, 15, 31, 61] and, more specifically, circles and circular arcs [6, 30, 50, 68] and ellipses and parabolae [59, 2]. Scan-conversion algorithms for 2D polygons [20, 53, 55] and planar regions bounded by curves [28, 56, 74] have also been investigated.

In 2D scan-conversion, geometric objects are scan-converted directly into a frame buffer just before they are displayed and thus aesthetic and aliasing reduction considerations are commonly integrated into the conversion process. The voxelization process, on the other hand, is decoupled from the rendering process. It generates a view-independent 3D data-base, which represents the objects within a volume buffer.

In the past, digitization of solids was performed by spatial enumeration algorithms that employ point or cell classification methods in an exhaustive fashion, or, preferably, by recursive subdivision [44]. However, subdivision techniques for model decomposition into rectangular subspaces are computationally expensive. The literature of 3D digitization by scan-conversion is relatively small. Danielsson [15] generalized his 2D incremental curve algorithm to 3D, where the 3D curve is defined by the intersection of two surfaces. Mokrzycki [51] has elaborated on Danielsson's method and has yielded a similar algorithm. Kaufman and Shimony [40] pioneered in introducing 3D scan-conversion algorithms for voxelizing 3D lines, polygons, polyhedra, parametric curves and patches, 3D circles, spheres, cylinders, and cones. Algorithms for traversing 3D rays have been developed for ray tracing using uniform subdivision [4, 8, 23, 64]. Kaufman has also presented an efficient algorithm for voxelizing polygons using an integer-based decision mechanism embedded within a scan-line filling algorithm [33, 35]; efficient algorithms for voxelizing cubic parametric curves, bicubic parametric surface patches, and tricubic parametric volumes using an integer-based forward differencing technique [34]; and algorithms for voxelizing quadric objects, such as cylinders, spheres, and cones using methods that sweep a discrete circle/line along a discrete circle/line [10]. All these algorithms have been developed without considering the theoretical framework of 3D discrete topology [42], and the voxelized objects have not been fully categorized by their connectivity, absence of tunnels, or absence of cavities, and are not necessarily accurate or minimal. This is a subject of current research [11].

Another subject of current research is non-binary voxelization. The voxelization algorithms developed so far generate a binary voxelization of the object, namely, each voxel is either in the object or part of the background. A binary voxelized object exhibits aliasing in three spatial coordinates, and a technique for reducing aliasing in 3D is desirable. Many researchers have incorporated the concept of sampling theory into 2D scan-conversion algorithms for generating anti-aliased gray-level 2D primitives [1, 7, 13, 14, 19, 26]. The goal is to extend the concept of sampling theory into 3D voxelization algorithms. Unlike 2D anti-aliasing algorithms whose sole purpose is image quality, 3D anti-aliasing algorithms focus on the generation of a gray-level fuzzy database, which provides the renderer with a more precise model of the scene. One approach is to enhance the neighboring information, especially near the continuous surface, by assigning to the neighboring voxels appropriate gray-scale values, representing, for example, partial volume coverage or material mixture, instead of binary values. Such an algorithm generates a voxelized object with a cloud of gray-level voxels around it. We call this process *non-binary* or *fuzzy* voxelization.

4 Discrete Ray Tracing

A ray tracing approach that employs a volume buffer, called *discrete ray tracing*, *volumetric ray tracing*, or *3D raster ray tracing* (RRT) [76, 75] is a vivid example of the power of volume graphics. Unlike existing ray tracing methods, which use geometric representation for the 3D scene, RRT employs a volume buffer for representing the 3D scene. The volume buffer maintains in every voxel all view-independent attributes of the object piercing the volume unit represented by this voxel.

RRT operates in two phases: a preprocessing voxelization phase and a discrete ray tracing phase. In the voxelization phase the geometric model is digitized, using voxelization algorithms, which convert the continuous representation of the model into a discrete representation and stores it in the volume buffer. For datasets that are already digitized, as in 3D medical imaging and 3D computational visualization, the discretization step is, of course, unnecessary. In the second phase a discrete variation of the conventional recursive ray tracer is employed. Unlike conventional algorithms, in which analytical rays are intersected with the object list in order to find the closest intersection, in RRT 3D discrete rays (which are essentially voxelized lines) are traversed through the 3D raster in order to find the first surface voxel. Encountering a non-transparent voxel indicates a ray-surface hit. The attributes stored in the voxel are then readily available for spawning offspring rays and computing the illumination at that point.

In conventional ray tracing, computation time grows with the number of objects and performance is greatly influenced by the type of objects comprising the scene; intersection calculation between a ray and a parametric surface is significantly more complex than intersecting the ray with a sphere or a polygon. In contrast, RRT completely eliminates the computationally expensive ray-object intersections calculation, and instead relies solely on a fast discrete ray traversal mechanism and a single, simple type of object – the voxel. Consequently, RRT is in effect independent of the number of objects in the scene or the object complexity or type. RRT performance, however, is sensitive to the resolution of the volume buffer. Therefore, for a given resolution, ray tracing time is nearly constant and can even decrease as the number of objects in the scene increases, as less stepping is necessary before an object is encountered.

Any change in the view-dependent parameters requires a traditional ray tracer to execute a full fledged rendering, which consists of recomputing many view-independent attributes such as surface normal, texture color, and light source visibility and illumination. In contrast, RRT precomputes the view-independent attributes during the voxelization phase and stores them within each voxel. These attributes are readily accessible for multiple rendering of the fixed scene, in which viewing, lighting, or shading parameters change. RRT is a typical example of a volume graphics technique demonstrating the attractive features intrinsic to volume graphics, described in Sect. 2. See example in Fig. 4.

5 Concluding Note

The close resemblances between surface graphics and vector graphics and between volume graphics and raster graphics, exhibited in Fig. 2, suggest that volume graphics has the potential to become an alternative to traditional surface graphics the same way raster graphics superseded vector graphics in the seventies. The progress to date in volume graphics, the associated empirical and intermixing applications, computer hardware, and memory systems, combined with the search for faster interactive visualization and the desire to reveal the inner structures of volumetric objects, guarantee that computer graphics will experience a paradigm shift towards volume graphics.

Acknowledgment

This project has been supported by the National Science Foundation under grants MIP-8805130 and IRI-9008109, and grants from Hughes Aircraft Company and Hewlett Packard Company. Figure 3 is courtesy of Hughes Aircraft Company. This paper is based on joint work with Dany Cohen and Roni Yagel [12, 75, 76].

References

[1] G. Abram, L. Westover, T. Whitted (1985). *Efficient Alias-Free Rendering Using Bit-Masks and Look-Up Tables. Computer Graphics*, 19(3):53–59.

[2] J.R. Van Aken (1984). *An Efficient Ellipse-Drawing Algorithm. IEEE Computer Graphics & Appl.*, 4(9):24–35.

[3] J.R. Van Aken, M. Novak (1985). *Curve-Drawing Algorithms for Raster Displays. ACM Transactions on Graphics*, 4(2):147–169.

[4] J. Amanatides, A. Woo (1987). *A Fast Voxel Traversal Algorithm for Ray Tracing.* In *Proceedings of EUROGRAPHICS '87, Amsterdam, The Netherlands*, pp. 3–9.

[5] J.E. Bresenham (1965). *Algorithm for Computer Control of a Digital Plotter. IBM Systems Journal*, 4(1):25–30.

[6] J.E. Bresenham (1977). *A Linear Algorithm for Incremental Digital Display of Circular Arcs. Commun. of the ACM*, 20(2):100–106.

[7] L. Carpenter (1984). *The A-Buffer, an Antialiased Hidden Surface Method. Computer Graphics*, 18(3):103–109.

[8] J.G. Cleary, G. Wyvill (1988). *Analysis of an Algorithm for Fast Ray Tracing using Uniform Space Subdivision. The Visual Computer*, 4:65–83.

[9] H.E. Cline, W.E. Lorensen, S. Ludke, C.R. Crawford, B.C. Teeter (1988). *Two Algorithms for the Three-Dimensional Construction of Tomograms. Medical Physics*, 15(3):320–327.

[10] D. Cohen, A. Kaufman (1990). *Scan Conversion Algorithms for Linear and Quadratic Objects.* In A. Kaufman, editor, *Volume Visualization*, pp. 280–301. IEEE Computer Society Press, Los Alamitos, CA.

[11] D. Cohen, A. Kaufman (1991). *Fundamentals of Surface Voxelization.* Tech. Rep. 91.06.09, Computer Science, SUNY at Stony Brook, June.

[12] D. Cohen, A. Kaufman, R. Yage (1991). *Volumetric Graphics.* Tech. Rep. 91.01.30, Computer Science, SUNY at Stony Brook.

[13] F.C. Crow (1977). *Shadow Algorithms for Computer Graphics. Computer Graphics*, 11(2):242–248.

[14] F.C. Crow, M.W. Howard (1981). *A Frame Buffer System with Enhanced Functionality. Computer Graphics*, 15(3):63–69.

[15] P.E. Danielsson (1970). *Incremental Curve Generation. IEEE Transactions on Computers*, C-19:783–793.

[16] D.W.H. Doo, M.A. Sabin (1978). *Behaviour of Recursive Subdivision Surfaces Near Extraordinary Points. CAD Journal*, 10(6):356–360.

[17] R.A. Drebin, L. Carpenter, P. Hanrahan (1988). *Volume Rendering. Computer Graphics*, 22(4):64–75.

[18] D.S. Ebert, R.E. Parent (1990). *Rendering and Animation of Gaseous Phenomena by Combining Fast Volume and Scanline A-buffer Techniques. Computer Graphics*,, 24(4):367–376.

[19] E. Fiume, A. Fournier, L. Rudolph (1983). *A Parallel Scan Conversion Algorithm with Anti-Aliasing for a General Purpose Ultracomputer. Computer Graphics*, 17(3):141–150.

[20] J.D. Foley, A. van Dam, S.K. Feiner, J.F. Hughes (1990). *Computer Graphics – Principles and Practice, (2nd ed.).* Addison-Wesley, Reading, MA.

[21] K.A. Frenkel (1989). *Volume Rendering. Communications of the ACM*, 32(4):426–435.

[22] M. Frühauf (1991). *Combining Volume Rendering with Line and Surface Rendering.* In *Proceedings EUROGRAPHICS Conference, Vienna, Austria, September*, pp. 21–32.

[23] A. Fujimoto, T. Tanata, K. Iwata (1986). *ARTS: Accelerated Ray-Tracing System. IEEE Computer Graphics & Appl.*, 6(4):16–26.

[24] D.S. Goodsell, S. Mian, A.J. Olson (1989). *Rendering of Volumetric Data in Molecular Systems. Journal of Molecular Graphics*, pp. 41–47.

[25] N. Greene (1989). *Voxel Space Automata: Modeling with Stochastic Growth Processes in Voxel Space. Computer Graphics*, 23(3):175–184.

[26] S. Gupta, R.F. Sproull (1981). *Filtering Edges for Gray-Scale Displays. Computer Graphics*, 15(3):1–5.

[27] J.C. Hart, D.J. Sandin, L.H. Kauffman (1989). *Ray Tracing Deterministic 3-D Fractals. Computer Graphics*, 23(3):289–296.

[28] R.D. Hersch (1986). *Descriptive Contour Fill of Partly Degenerated Shapes. IEEE Computer Graphics & Appl.*, 6(7):61–70.

[29] W. Hibbard, D. Santek (1989). *Visualizing Large Data Sets in the Earth Sciences. IEEE Computer*, 22(8):53–57.

[30] B.K.P. Horn (1976). *Circle Generators for Display Devices. Computer Graphics and Image Processing*, 5:280–288.

[31] B.W. Jordan, W.J. Lennon, B.D. Holm (1973). *An Improved Algorithm for the Generation of Nonparametric Curves. IEEE Transactions on Computers*, C-22(12):1052–1060.

[32] J.T. Kajiya, T.L. Kay (1989). *Rendering Fur with Three Dimensional Textures. Computer Graphics*, 23(3):271–280.

[33] A. Kaufman (1987). *An Algorithm for 3D Scan-Conversion of Polygons.* In *Proc. EUROGRAPHICS'87, Amsterdam, Netherlands, August*, pp. 197–208.

[34] A. Kaufman (1987). *Efficient Algorithms for 3D Scan-Conversion of Parametric Curves, Surfaces, and Volumes. Computer Graphics*, 21(4):171–179.

[35] A. Kaufman (1988). *Efficient Algorithms for 3D Scan-Converting Polygons. Computers & Graphics*, 12(2):213–219.

[36] A. Kaufman (1990). *Volume Visualization.* IEEE Computer Society Press Tutorial, Los Alamitos, CA.

[37] A. Kaufman (1991). *The voxblt Engine: A Voxel Frame Buffer Processor.* In A.A. M. Kuijk, W. Strasser, editors, *Advances in Graphics Hardware III.* Springer-Verlag, Berlin.

[38] A. Kaufman, R. Bakalash, D. Cohen, R. Yagel (1990). *Architectures for Volume Rendering - A Survey. IEEE Engineering in Medicine and Biology*, 9(4):18–23.

[39] A. Kaufman, R.Yagel, R. Bakalash, I. Spector (1990). *Volume Visualization in Cell Biology.* In *Proceedings Visualization '90, San Francisco, CA, October*, pp. 160–167.

[40] A. Kaufman, E. Shimony (1986). *3D Scan-Conversion Algorithms for Voxel-Based Graphics.* In *Proc. ACM Workshop on Interactive 3D Graphics, Chapel Hill, NC, October*, pp. 45–76.

[41] A. Kaufman, R. Yagel, D. Cohen (1990). *Intermixing Surface and Volume Rendering.* In K. H. Hoehne, H. Fuchs, S. M. Pizer, editors, *3D Imaging in Medicine: Algorithms, Systems, Applications*, pp. 217–227.

[42] T.Y. Kong, A. Rosenfeld (1989). *Digital Topology: Introduction and Survey. Computer Vision, Graphics and Image Processing*, 48(3):357–393.

[43] R.P. Kruger, T.M. Cannon (April 1978). *The Application of Computerized Tomography, Boundary Detection, and Shaded Graphics Reconstruction to Industrial Inspection. Materials Evaluation*, 36:75–80.

[44] Y.T. Lee, A.A.G. Requicha (1982). *Algorithms for Computing the Volume and Other Integral Properties of Solids: I-Known Methods and Open Issues; II-A Family of Algorithms Based on Representation Conversion and Cellular Approximation. Communications of the ACM*, 25(9):635–650.

[45] M. Levoy (1988). *Display of Surfaces from Volume Data. IEEE Computer Graphics & Appl.*, 8(5):29–37.

[46] M. Levoy (1990). *Efficient Ray Tracing of Volume Data. ACM Transactions on Graphics*, 9(3):245–261.

[47] M. Levoy (1990). *A Hybrid Ray Tracer for Rendering Polygon and Volume Data. IEEE Computer Graphics & Appl.*, 10(3):33–40.

[48] M. Levoy (1991). *Photorealistic Volume Rendering in Scientific Visualization.* In *SIGGRAPH Course Notes*, volume 27, pp. 2–1 – 2–13.

[49] M. Levoy, R. Whitaker (1990). *Gaze-Directed Volume Rendering. Computer Graphics*, 24(2):217–223.

[50] M.D. McIlroy (1983). *Best Approximate Circles on Integer Grids. ACM Transactions on Graphics*, 2(4):237–263.

[51] W. Mokrzycki (1988). *Algorithms of Discretization of Algebraic Spatial Curves on Homogeneous Cubical Grids. Computers & Graphics*, 12(3/4):477–487.

[52] A.H. Nasri (1987). *Polyhedron Subdivision Methods for Free-Form Surfaces. ACM Transactions on Graphics*, 6(1):29–73.

[53] W.M. Newman, R.F. Sproull (1979). *Principles of Interactive Computer Graphics, (2nd ed.).* McGraw-Hill, New York.

[54] V.A. Norton (1982). *Generation and Rendering of Geometric Fractals in 3-D. Computer Graphics*, 16(3):61–67.

[55] T. Pavlidis (1982). *Algorithms for Graphics and Image Processing.* Computer Science Press, Rockville, MD.

[56] T. Pavlidis, C.J. Van Wyk (1985). *An Automatic Beautifier for Drawing and Illustrations. Computer Graphics*, 19(3):225–234.

[57] D.R. Peachey (1985). *Solid Texturing of Complex Surfaces. Computer Graphics*, 19(3):279–286.

[58] K. Perlin, E.M. Hoffert (1989). *Hypertexture. Computer Graphics*, 23(3):253–262.

[59] M.L.V. Pitteway (1967). *Algorithm for Drawing Ellipses or Hyperbolae with a Digital Plotter. Computer Journal*, 10(3):282–289.

[60] K. Roang, M. Kangasabai (1991). *Volume Visualization of Fluid Dynamics Simulations in Complex Geometries.* In *Second EUROGRAPHICS Workshop on Visualization in Scientific Computing, Delft, Netherlands, April.*

[61] J. Roberge (1985). *A Data Reduction Algorithm for Planar Curves. Computer Vision, Graphics, and Image Processing*, 29:168–195.

[62] H. Samet (1989). *Applications of Spatial Data Structures.* Addison Wesley, Reading, MA.

[63] P. Shirley, H. Neeman (1989). *Volume Visualization at the Center for Supercomputing Research and Development.* In *Proceedings of the Chapel Hill Workshop on Volume Visualization, Chapel Hill, NC, May*, pp. 17–20.

[64] J.M. Snyder, A.H. Barr (1987). *Ray Tracing Complex Models Containing Surface Tessellations. Computer Graphics*, 21(4):119–128.

[65] L. Sobierajski, D. Cohen, A. Kaufman, R. Yagel, D. Acker (1990 (Submitted)). *Trimmed Voxel Lists for Interactive Surgical Planning.* Tech. Rep. 90.05.22, SUNY Stony Brook, May.

[66] D. Speray, S. Kennon (1990). *Volume Probes: Interactive Data Exploration on Arbitrary Grids. Computer Graphics*, 24(5):5–12.

[67] R.F. Sproull (1982). *Using Program Transformations to Derive Line-Drawing Algorithms. ACM Transactions on Graphics*, 1(4):257–273.

[68] Y. Suenaga, T. Kamae, T. Kobayashi (1979). *High-Speed Algorithm for the Generation of Straight Lines and Circular Arcs. IEEE Transactions on Computers*, TC-28(10):728–736.

[69] U. Tiede, K.H. Hoehne, M. Bomans, A. Pommert, M. Riemer, G. Wiebecke (1990). *Investigation of Medical 3D-Rendering Algorithms. IEEE Computer Graphics & Appl.*, 10(3):41–53.

[70] J.K. Udupa, H.M. Hung (1990). *Surface Versus Volume Rendering: A Comparative Assessment.* In *Proceedings of the First Conference on Visualization in Biomedical Computing, Atlanta, GA, May*, pp. 83–91.

[71] T. van Walsum, A.J.S. Hin, J. Versloot, F.H. Post (1991). *Efficient Hybrid Rendering of Volume Data and Polygons.* In *Second EUROGRAPHICS Workshop on Visualization in Scientific Computing, Delft, Netherlands, April.*

[72] Y. Wang, D. Cohen, A. Kaufman (1991). *Generating a Smooth Voxel-Based Surface from an Irregular Polygon Mesh.* Tech. Rep. 91.02.15, Computer Science, SUNY at Stony Brook.

[73] R.H. Wolfe, C.N. Liu (1988). *Interactive Visualization of 3D Seismic Data: A Volumetric Method. IEEE Computer Graphics & Appl.*, 8(7):24–30.

[74] C.J. Van Wyk (1984). *Clipping to the Boundary of a Circular-Arc Polygon. Computer Vision, Graphics, and Image Processing*, 25:383–392.

[75] R. Yagel, D. Cohen, A. Kaufman (1992). *Discrete Ray Tracing. To appear in IEEE Computer Graphics & Appl.*

[76] R. Yagel, A. Kaufman, Q. Zhang (1991). *Realistic Volume Imaging.* In *Proceedings Visualization '90 San Diego, CA, October*, pp. 226–231.

Surface Interpolation from Cross Sections

Heinrich Müller, Arnold Klingert

An essential technique of visualizing the inner structure of a body is tomography. Tomographic methods represent the body by a sequence of cross sectional slices. For the observer, however, it is often rather difficult to imagine the true three-dimensional shape from the slices. For this reason, numerous methods for reconstructing a three-dimensional representation from a given stack of slices were suggested in the past. We give a unifying survey of such methods and sketch several new approaches not published yet. The emphasis is on techniques which can be also applied to larger distances between the cross sections and in the case of distorted slices. The methods usually yield a surface representation which can be used beyond visualization for purposes of geometric modeling, simulation, and manufacturing.

1 Introduction

The problem dealt with in the following is

Surface from cross sectional contours (SFCSC).

Input. A sequence of parallel planes in space consisting of a collection of disjoint two-dimensional cells.

Output. A collection of spatial cells whose cross sections along the given planes induce the cells of the planes.

The spatial cells are assumed to be *solid*. A cell is solid if its boundary consists of two-dimensional points only. This means points which have an environment topologically equivalent to a circular disk in the plane. For instance, point **p** in Fig. 1 (left) is not two-dimensional in this sense. It is also assumed that the intersection of a spatial cell with a given plane is two-dimensional. This means that the cells labeled a and c in the right part of Fig. 1 are not correct, while cell b is. Cell a has just a single point in common with the lower plane, while cell c intersects the lower plane in a line segment only. Cell b has a complete contour in common with upper and the lower plane and thus is admissible.

The SFCSC problem has found considerable interest related with acquiring data by techniques of tomography. Particularly known techniques of tomography are X-ray tomography, NMR tomography, and ultra-sound tomography which are widely used in medicine. In microscopy, the slicing of objects for analysis is

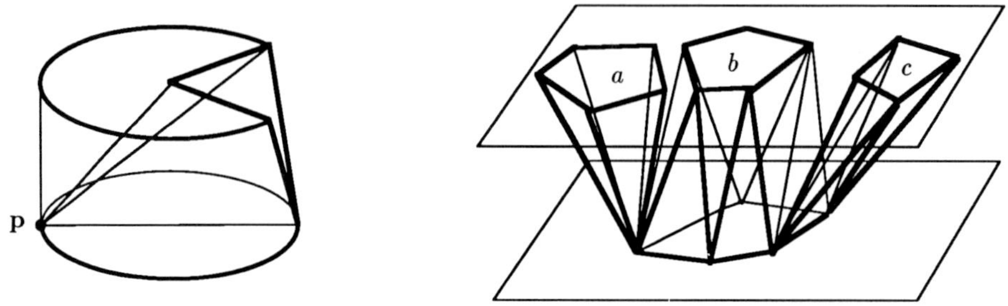

Fig. 1. Forbidden interpolations from contours.

wide-spread, too. But also non-microscopic objects can be analyzed by cutting them or grinding them off piecewise, like it is for example done in anatomy or geology.

An example of the usefulness of tomography, but also of its difficulties, is the following approach of analyzing the inner structure of the liver [58]. The histological structure of liver tissue can be considered as a conglomeration of liver lobules. With a microscope the lobules can be recognized in cross sectional slices. Figure 2 shows a cross section which represents a region of approximately

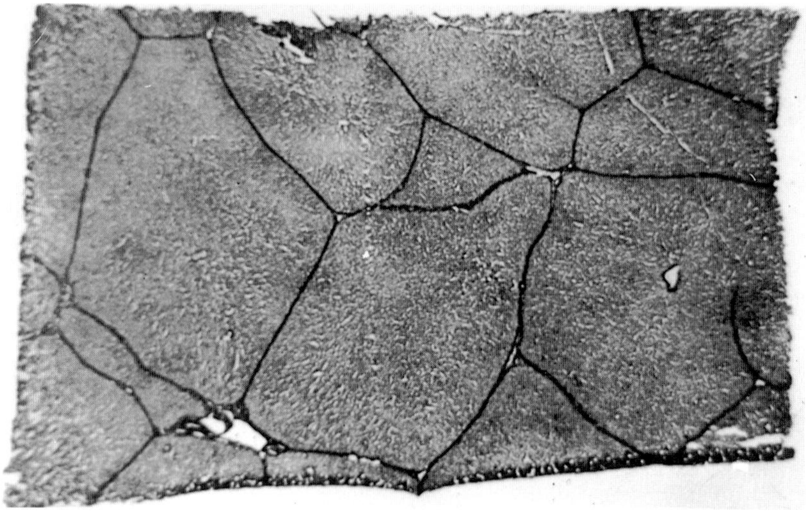

Fig. 2. Sector of a cross sectional slice of liver tissue.

6 mm×4 mm. It shows a faceted structure. The facets represent slices of the liver lobules. In the inner of the facets, the central vein can sometimes be realized. A question of anatomy controversially discussed is how the spatial structure of the lobules does look like.

The answer to this question can be approached by solving the SFCSC problem. The input is a sequence of two-dimensional polygonal cell decompositions which is derived from the digitized cross sections by segmentation (Fig. 3). The result of spatial reconstruction is shown in Fig. 4. The liver lobules are displayed

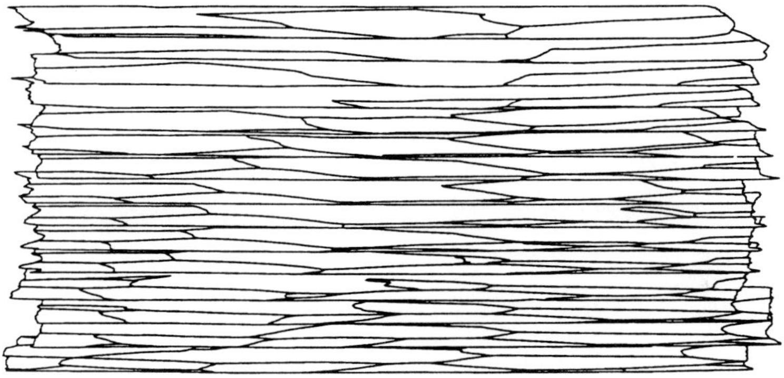

Fig. 3. A stack of planar cell decompositions obtained from a sequence of cross sectional slices of liver tissue.

with different shrinking factors in order to elucidate their spatial arrangement. The cells are shrunk with respect to their centers, with factors 0.95, 0.75 and 0.50, respectively.

In other applications, the cell decompositions will usually be less complex than in this example. In the most elementary case the cell decomposition is given by exactly one closed curve. The curve divides the plane into two parts, the bounded interior and the unbounded exterior.

The techniques of tomography differ in several aspects. One of them is the size of distances available between slices. Small distances may be excluded by technical, in particular mechanical limitations. But also the time requirements for generating cross sections or the encumbrance of the material, for instance by X-rays, exclude small distances of slices. A further problem apart from slice distance are distortions by mechanical influences or by varying positions and deformations of living subjects. The zig-zag structure in direction of the spatial axis which can be clearly recognized in Fig. 4 is caused by different distortions and displacements of the slices.

Usually the slices are loaded into the computer as raster images. In the raster images the contours of the shape to be reconstructed have to be detected. Given small distances between cuts which are additionally assumed geometrically correct, segmentation as well as the calculation of the surface can be performed directly in 3D space based on the rastered structure (Rhodes [55], Artzy, Frieder, Herman [4], Hermann, Webster [29], Gordon, Udupa [24], Udupa, Ajjanagudde [69], Lorensen, Cline [42]). If nothing else than a pictorial representation is required, direct volume rendering methods can also be applied in this case, cf. other chapters in this book.

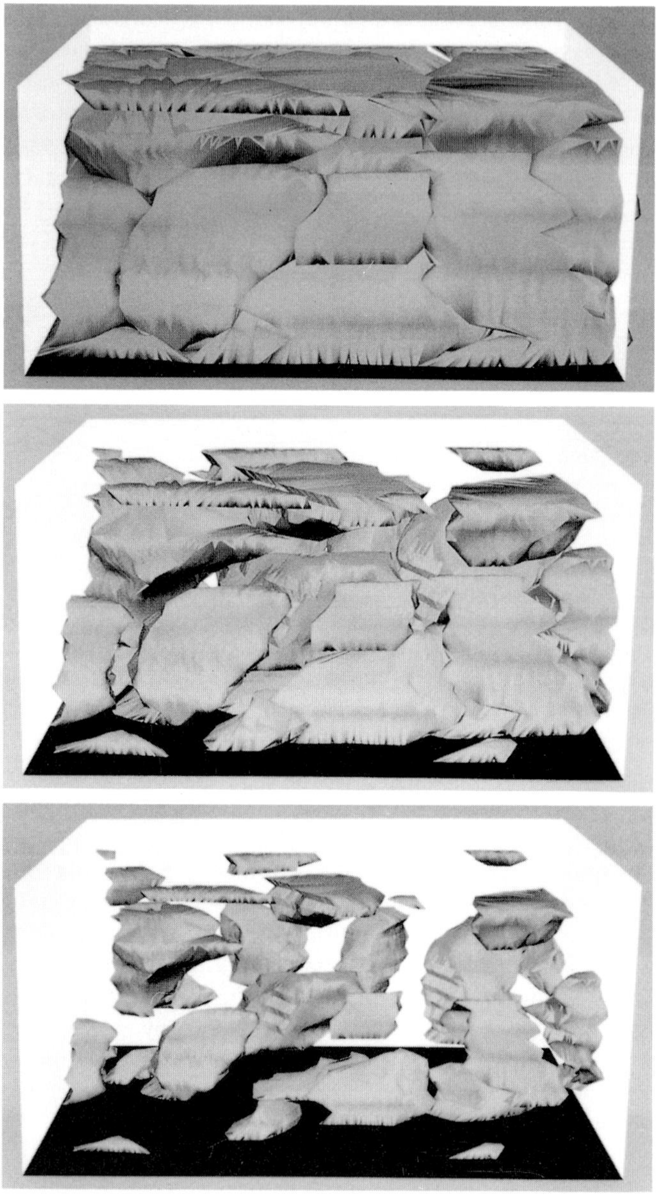

Fig. 4. Visualization of the interpolated three-dimensional cell decomposition with cells shrunk by factors of 95%, 75% and 50% of the original size.

In the following we will be concerned with methods which are suitable for larger distances between cuts, too. The material presented is organized along the two main subproblems which must be solved by these methods. These subproblems are *reconstruction of topology* which will be treated in Sect. 2, and *reconstruction of geometry* which is subject of Sect. 3–9. The problems arising during data acquisition, like non-exact slice geometry or transformation of raster images into a suitable geometric description are the contents of Sect. 10. In Sect. 11 interactive techniques for solving the SFCSC problem are discussed.

2 Topological Reconstruction

Topological reconstruction means the following:

Reconstruction of topology.

Input. The same as for the SFCSC problem.

Output. An assignment of the contours form slice to slice in a way that there exists a solid with the property that the assigned contours are continuously transferred into one another when moving a plane in parallel to the given sectional planes through the solid.

The problem of topological reconstruction is of particular difficulty if there are several contours on each plane and if it is known that the corresponding solid can have branchings.

2.1 Assignment Graphs and Nesting Trees

The assignment of contours can be formally described by an *assignment graph*. The vertices of the assignment graph correspond to contours. Its edges connect vertices which belong to contours of consecutive slices forming a connected subsolid. An assignment graph is called *valid* if there exists a solid which induces the assignment described by that graph. Vaild assignment graphs are related to so-called *Reeb graphs* [54] which were used by Shinagawa and Kunii [63] for surface interpolation from cross sections.

For a sequence of contours usually more than one valid assignment graph does exist. Figure 5 shows a stack of contours and different assignment graphs. On the other hand not all graphs assigning contours between slices are in fact valid. In Fig. 5 also a non-valid assignment graph is displayed. For this assignment the inner and the outer boundary interchange from slice to slice. This can only happen if the solid has points on its boundary which are not two-dimensional.

Figure 6 shows that a planar slice in general is composed of nested contours. Nested contours occur if the sliced solid has holes, like for instance a torus. A valid assignment of contours has to preserve the nesting. An assignment graph with this property is called *topologically valid.* Differently expressed, an

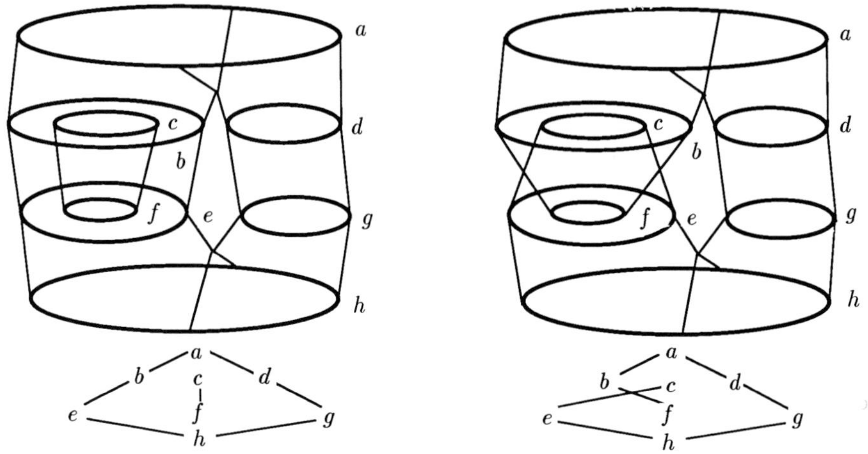

Fig. 5. A stack of contours with a feasible and a non-feasible assignment graph.

assignment graph is topologically valid if there exists any solid and any sequence of cutting planes so that the nesting of contours for the intersections of this solid with these cutting planes coincide with those of the given planes.

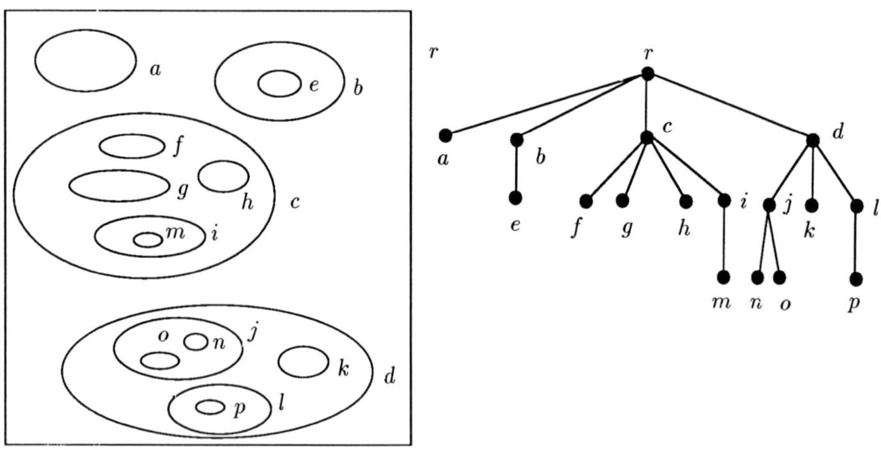

Fig. 6. A slice of nested contours and the corresponding tree of nesting.

The nesting of contours can be formally described by a *nesting tree* (Fig. 6). Each vertex of the nesting tree corresponds to a contour. The root of the nesting tree corresponds to a virtual outer contour which envelops all other contours. The successors of a vertex of the nesting tree correspond to the contours immediately contained in the contour corresponding to that vertex.

2.2 Enumeration of Assignment Graphs

As we have seen, the solution of the problem of topological reconstruction is not unique in general. Each valid assignment graph represents a solution. The following algorithm for enumerating topologically valid assignment graphs is based on criteria resulting from the comparison of the nesting trees of two consecutive slices. They can be derived by analyzing the events happening on a cutting plane which is moved continuously through a solid. Observing its intersection with the surface of the solid, we note that the number of contours changes exactly if a point on the surface is reached at which the cutting plane becomes tangential. These points are called *singular* with respect to the given direction of movement. There are two types of singular points, *extremal points* and *saddle points*. Extremal points are characterized by the fact that the surface lies totally on one side of the tangential cutting plane. At a saddle point the tangential cutting plane is intersected by the surface. In mathematics, the analysis of surfaces in this manner is subject of the so-called *Morse theory* (Milnor [44], Morse, Cairns [45]). Morse theory has found applications with surface classification in medicine and computer vision (Shinagawa, Kunii, Kergosien [64], Kergosien [36]).

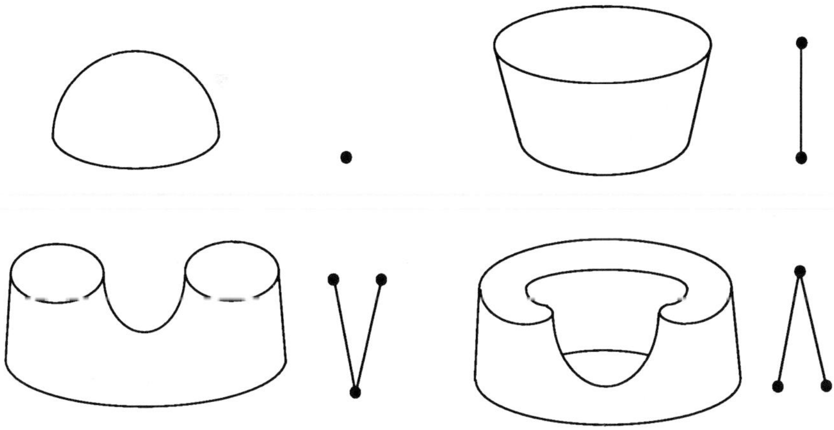

Fig. 7. The four cases of elementary branchings: the extremal case (top left), the cylindric case (top right), the disconnected branching (bottom left) and the connected branching (bottom right).

The algorithm of enumeration only considers assignment graphs with elementary branchings. Formally, a *branching* is a connected component of a subgraph of the assignment graph induced by the vertices of two consecutive planes. A branching is called *elementary* if the connected components are single vertices, single edges or are star-shaped of degree 2. The latter means that one vertex is connected by two edges with two vertices on the opposite plane. We introduce this restriction in order to simplify the discussion. Now four types of elementary branchings can be distinguished: the *extremal case*, the *cylindric case*, the

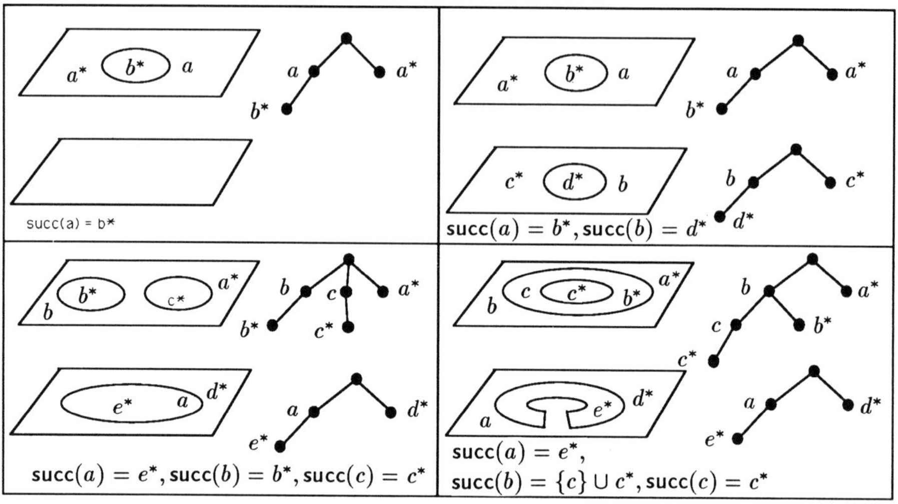

Fig. 8. The four cases of recursion in the enumeration algorithm. Letters with * denote sets of contours, the others stand for single contours.

disconnected branching, and the *connected branching*. Figure 7 sketches their geometric incarnations and the corresponding connected components of the assignment graph.

Briefly sketched, the enumeration algorithm processes a pair of nesting trees recursively, starting at their roots. A step of recursion starts with two sets of contours, one for each of the two neighboring planes under consideration. First, all possible sets of branchings between these contours are determined. Then, for each of these branchings, the contours lying on the next nesting level with respect to the contours involved in the branching are detected. According to the types of branching, four cases of contour detection have to be distinguished. They are compiled in Fig. 8. With the detected contours the algorithm is called recursively.

Unfortunately, this procedure only works if connected saddle point components are excluded. The reason is that connected saddle point components (cspc) behave differently from the cylindric components (cc) and the disconnected saddle point components (dspc). They may "liberate" more deeply nested contours and allow to establish assignments with higher level contours. Figure 9 explains this situation. Although contour f is on level 4 of its nesting tree it can be assigned to contour b which is on level 2 of its tree. The reason is that a hole is opening between the two planes which is shown in the geometric sketch on the right side of the figure. From a global view this means that both cylinders are on the same topological level, whereas from the local view of the second plane their base contours g and f are on different levels.

This problem can be solved by the following modified strategy. In the first step of the recursion a set of cspc is chosen. Then the given set of contours is extended by the contours liberated by them. This procedure is iterated through

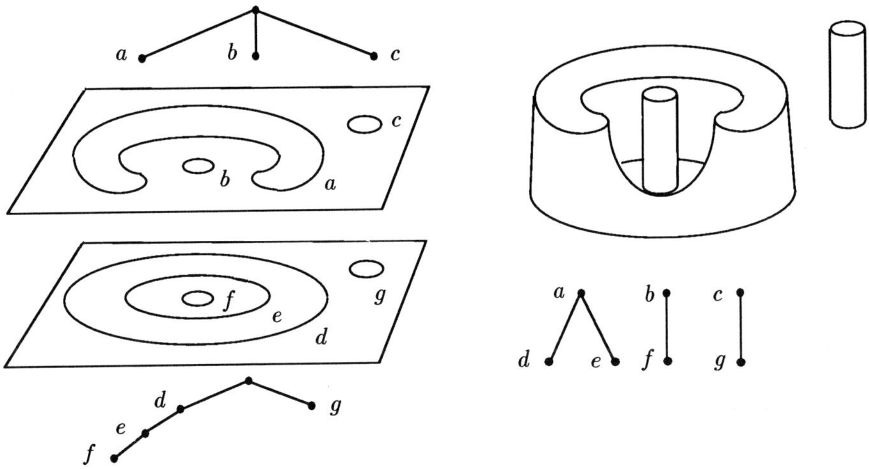

Fig. 9. Liberation of contours by connected saddle point components. From a global view, the two cylinders on the right lie on the same level of nesting whereas their base contours g and f lie on different nesting levels.

all levels of the nesting trees. Now, the assignment on the resulting set of contours is extended by all combinations of assignments inducing components of the remaining three types only. These components are treated recursively as before, with the modification just described.

In the following semi-formal sketch of the algorithm the nesting trees are traversed using an operation *succ(.)*. The call *succ(a)*, a a contour, delivers the set of successors of the node belonging to a in the corresponding nesting tree. The main part of the algorithm is the recursive subalgorithm *assign(.)* which just performs the steps described above. For the connected components the formal notation $G = (V_1 \cup V_2, E)$ of a bipartite graph is used. V_1 and V_2 are the two vertex sets, here corresponding to the contours on planes one and two. E is the set of edges connecting vertices of V_1 with vertices of V_2. Comments are embraced in (* *).

Algorithm Enumeration of Assignments.

Input. The roots r, r' of the nesting trees of two consecutive slices.

Output. The set A^* of possible assignment graphs for the given nesting trees.

Procedure.
subalgorithm assign(W,W',A,A^*);
W: a given set of contours of the first plane;
W': a given set of contours of the second plane;
A: a given assignment graph with cspc only;
A^*: the resulting set of assignment graphs induced by A;

```
  ΔW: set of contours of the first plane liberated by a cspc;
  ΔW': set of contours of the second plane liberated by a cspc;

BEGIN
  ΔW := ∅; ΔW' := ∅;
  A* := ∅;
  FOR all assignment graphs B augmenting A w.r.t.
    W and W' with cspc only DO
    (* liberation of contours on this level, cf. Fig. 8d: *)
  BEGIN
    FOR all cspc C = ({a} ∪ {b,c}, {(a,b),(a,c)}) in B − A DO
      ΔW := ΔW ∪ succ(c);
    FOR all cspc C = ({a,b} ∪ {c}, {(a,c),(b,c)}) in B − A DO
      ΔW' := ΔW' ∪ succ(a);
    IF ΔW ∪ ΔW' ≠ ∅ THEN
    BEGIN (* liberation of contours on deeper levels: *)
      assign(ΔW, ΔW', B, B*);
      A* := A* ∪ B*
    END
    ELSE (* no new contours liberated: *)
      FOR all assignment graphs D augmenting B by
        connected components others than of type cspc DO
      BEGIN (* recursion for the other types of components *)
        FOR all connected components C of D − B DO
          CASE C OF
            1a: (* extremal point component ({a} ∪ ∅,∅}), Fig. 8a *)
              assign(succ(a),∅,∅, C*);
            1b: (* extremal point component (∅ ∪ {a},∅}) *)
              assign(∅,succ(a),∅,C*);
            2: (* cylindric component ({a} ∪ {b}, {(a,b)}), Fig. 8b *)
              assign(succ(a),succ(b),∅,C*);
            3a: (* disconnected saddle point component, Fig. 8c *)
                ({a,b} ∪ {c}, {(a,c),(b,c)}) *)
              assign(succ(a) ∪ succ(b),succ(c),∅, C*);
            3b: (* disconnected saddle point component, Fig. 8c *)
                ({a} ∪ {b,c}, {(a,b),(a,c)}) *)
              assign(succ(a),succ(b) ∪ succ(c),∅, C*);
          END(* CASE *);
        (* report the assignment graphs: *)
        A* := A* ∪ {{D} ∪ ⋃_{C cc of D} E(C) : E(C) ∈ C*}
        (* E(C) denotes an arbitrary graph induced by C in C* *)
      END
  END
END;
```

main algorithm.
BEGIN
  assign(root(r), root(r'), $\emptyset$, A^*);
  report(A^*)
END.

The large number of assignment graphs which can be delivered by this algorithm can be restricted somewhat if a "sampling density" of the given planes is assumed which leads to a topologically correct interpolation. This means that for a solid from which the contours on the given cutting planes are obtained, the nesting tree of the contours on an arbitrary cutting plane lying between two of the given planes is the same as that of one of these planes. The consequence is that nested singular points like those displayed between the upper and the lower plane on the lower right side of Fig. 10 are excluded. A topologically sufficient

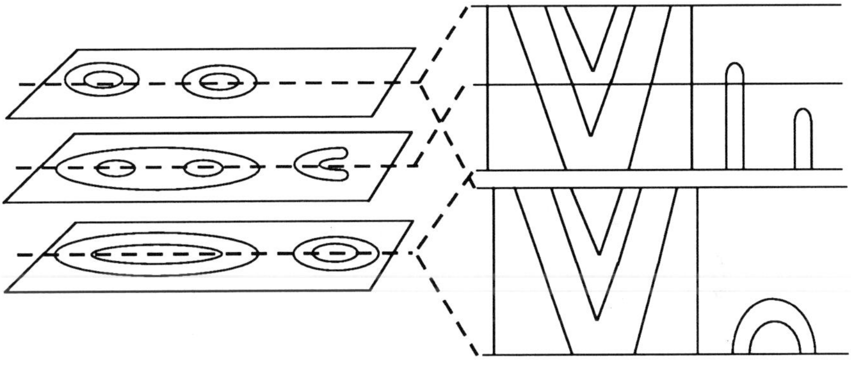

Fig. 10. Sampling with cutting planes. The result depends on the configuration of the intermediate plane which has to belong to the sampling planes in the case of topologically sufficient sampling. The upper slice on the right side shows a correct interpolation whereas in the lower one the intermediate plane is not considered which results in a topologically incorrect result.

sampling density helps to exclude ambiguities like that one displayed on the upper right side of the figure which is a legal alternative if only the upper and the lower planes are seen. In case of sufficient sampling, the enumeration algorithm is only allowed to enumerate assignments which have geometric incarnations fulfilling the property of topologically sufficient sampling with respect to the given planes.

In practice the number of topologically valid assignment graphs can also be restricted using properties known from the area of application. For example, it might be known that the solid is connected, that it does not have holes or that it does not have branchings. These topological properties can be immediately considered in the enumeration algorithm.

Another type of restriction may be based on simple *geometric criteria* which have to be fulfilled by mutually assigned contours. They may be classified in criteria of *similarity* and criteria of *mutual location.* Examples of such criteria are given in the next two sections. They can be introduced into the enumeration algorithm by testing for each assignment whether it satisfies the criterion. If it does not the assignment is skipped.

2.3 Similarity of Contours

A quite natural geometric criterion is that the shapes of assigned contours have to be similar. In the field of pattern recognition a number of features were suggested whose coincidence indicate the similarity of shapes – although they are usually not sufficient (Gonzalez, Wintz [23]). Examples are the *perimeter* U of the contour, the *area* F of the surrounded region, and the *size invariant form factor* $S := U^2/4\pi F$ derived from them. Other features are the *central moments* $\mu_{p,q}$ of order p, q which are defined by

$$\mu_{p,q} := \sum_{i=0}^{n}\sum_{j=0}^{n}(x_i - x)^p(y_j - y)^q, \; x = \sum_{i=0}^{n} x_i/n, \; y = \sum_{j=0}^{n} y_j/n,$$

for $n + 1$ points (x_i, y_i).

A feature describing the shape of a contour more precisely is the *average of the k-curvature.* The *k-curvature*, $k \in I\!N$, of a polygonal chain $\mathbf{p}_0, \ldots, \mathbf{p}_n$ at a vertex $\mathbf{p}_i$, $i = 1, \ldots, n-1$, is defined as the angle between the vectors $\mathbf{p}_{i-k}\mathbf{p}_i$ and $\mathbf{p}_i\mathbf{p}_{i+k}$. A more realistic measure should take the lengths of the edges into consideration, for example by taking the inverse of the radius of the circle defined by the three consecutive vertices $\mathbf{p}_{i-1}$,$\mathbf{p}_i$, and $\mathbf{p}_{i+1}$.

An entire family of feature values results from the Fourier representation of the contour. Let be $\mathbf{k}(t)$, $t \in [0, 1]$, a closed polygonal chain with vertices $\mathbf{p}_k \in I\!R^2$, $k = 0, \ldots, m$, $\mathbf{p}_0 = \mathbf{p}_m$, which is piecewise linearly parameterized by assigning the parameter value

$$t_k := \sum_{j=1}^{k} \Delta t_j/T, \; \Delta t_j := \|\mathbf{p}_j - \mathbf{p}_{j-1}\|, \; T := \sum_{j=1}^{m} \Delta t_j,$$

to point $\mathbf{p}_k$. Then (cf. Kuhl, Giardina [38])

$$\mathbf{k}(t) = \mathbf{a}_0/2 + \sum_{j=1}^{\infty}(\mathbf{a}_j \cos(2\pi jt) + \mathbf{b}_j \sin(2\pi jt))$$

with

$$\mathbf{a}_0 = \frac{1}{T}\sum_{p=1}^{m}(\mathbf{p}_p + \mathbf{p}_{p-1})(t_p - t_{p-1}),$$

$$\mathbf{a}_k = \frac{T}{2k^2\pi^2}\sum_{p=1}^{m}\frac{\mathbf{p}_p - \mathbf{p}_{p-1}}{t_p - t_{p-1}}(\cos(2\pi kt_p) - \cos(2\pi kt_{p-1})), \; k > 0,$$

$$\mathbf{b}_k = \frac{T}{2k^2\pi^2} \sum_{p=1}^{m} \frac{\mathbf{p}_p - \mathbf{p}_{p-1}}{t_p - t_{p-1}} (\sin(2\pi k t_p) - \sin(2\pi k t_{p-1})), \; k > 0.$$

This representation $\mathbf{k}(t)$ is called the *Fourier representation* of the closed polygonal chain.

In the Fourier representation, the first $\mathbf{a}_k$, $\mathbf{b}_k$ up to a fixed k_0 can be taken as feature descriptors. They represent an approximation of the given polygonal chain by the corresponding terms of the Fourier representation. Figure 11 shows the behavior of approximation with an increasing number of terms. Already for a few terms an approximation of the given shape can be recognized.

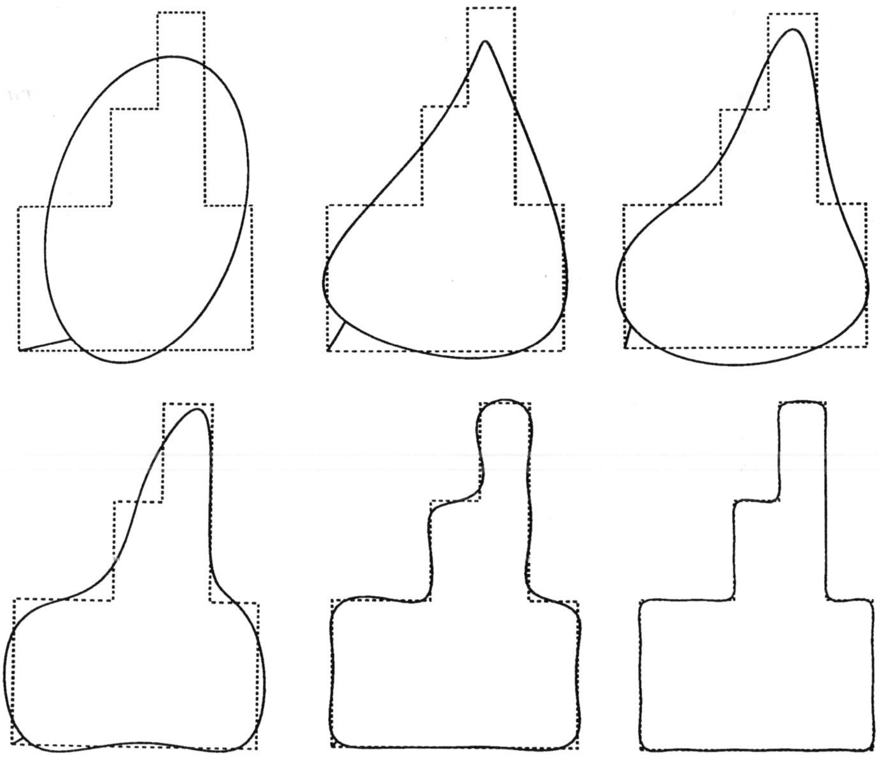

Fig. 11. Approximations with cut-offs of the Fourier representation of a closed polygonal chain with 1, 2, 3, 5, 10 and 20 terms of the sum.

The features listed above are useful since they all are *invariants of motion*. This means that they have the same values when the contour is shifted or rotated.

2.4 Mutual Location of Contours

A reason for considering geometric criteria based on the mutual location of contours is that the formulation of the SFCSC problem implies that the interpolated surface does not penetrate itself. For the assignment graph this means that its connected components must have non-penetrating geometric incarnations. Although the property of penetration depends on the type of geometric interpolation applied, the chance of penetration is less if contours which lie close together are mutually assigned.

A simple criterion is to assign two contours to each other if their centers are close together. A more sophisticated definition of the distance of two contours c_i^j and c_k^{j+1} on levels j and $j+1$ is

$$w(c_i^j, c_k^{j+1}) = \frac{d_j \int \int_{B_i^j} dS}{\int \int_{B_i^j} g(\mathbf{x}, B_k^{j+1}) dS}.$$

d_j denotes the vertical distance between the slices and B_k^{j+1} is the area of the region inside the contour c_k^{j+1}. The function g measures the minimum Euclidean distance among points $\mathbf{x} \in B_i^j$ to any point inside the other contour. A simple practical approximation of the value w is

$$\bar{w}(c_i^j, c_k^{j+1}) = \frac{n_i^j n_k^{j+1} d_j}{\sum_{m=0}^{n_i^j} \sum_{l=0}^{n_k^{j+1}} d(\mathbf{x}_{i,m}^j, \mathbf{x}_{k,l}^{j+1})}.$$

Here the Euclidean distance d of the vertices $\mathbf{x}_{i,l}^j$, $l = 1, \ldots, n_i^j$, is taken.

Another simple criterion is the relative mutual overlapping of two contours,

$$\tilde{w}(c_i^j, c_k^{j+1}) = \frac{d_j}{1 - B_i^j / B_{j+1}^k}.$$

Christiansen, Sederberg [9] suggest to assign contours if the bounding rectangles formed by their extremal coordinates overlap, which is easier to calculate.

The horizontal distance as well as mutual overlapping are not symmetric in general, that is

$$\tilde{w}(c_i^j, c_k^{j+1}) \neq \tilde{w}(c_k^{j+1}, c_i^j),$$

and therefore are no metrics.

A number of other weights which can be used as criteria can be derived from these values. The reliability of a weight is satisfactory if w/d_j grows unlimited for a correct assignment whenever d_j gets close to zero. For w and $\tilde{w}$ this property is given whereas $\bar{w}/d_j$ is limited. In Shinagawa, Kunii [62] the graph for some human organs is automatically created by using $\bar{w}$ only.

3 Geometric Reconstruction

The task of geometric reconstruction is to derive the surface of a solid whose slices coincide with the given contours and which corresponds with a given assignment graph. The surface can be seen as a piecewise composition of surface segments

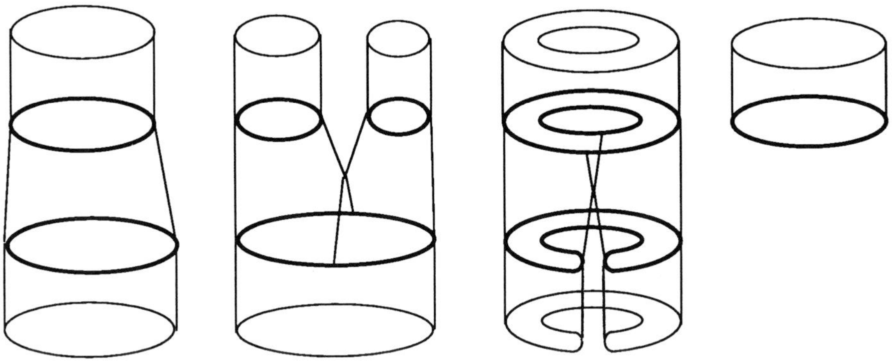

Fig. 12. The cases of the contour interpolation problem.

which are interpolated from mutually assigned contours in two neighboring cutting planes. For the restricted assignment graphs of the previous section the following cases can be distinguished:

Reconstruction of geometry.

Input.

Cylindric interpolation.
One simply closed curve in two parallel planes which are mutually assigned.

Saddle point interpolation with disconnected saddle.
Two disjoint nested contours in one plane and one contour in the other. The two contours are assigned to the single contour.

Saddle point interpolation with connected saddle.
Two disjoint contours in one plane which are not nested, and one contour in the other plane. The two contours are assigned to the single contour.

Extremal point interpolation.
One contour in the first plane to which no contour in the second plane is assigned.

Output. A surface which bounds a solid when extended cylindrically on both sides, so that the slices by the given planes show the given contours (Fig. 12).

The surface segments have to satisfy the additional boundary condition that they do not penetrate themselves and that they do not penetrate any other segment.

For geometric reconstruction two main types of approaches exist: the *surface oriented approaches* to which the next three sections are devoted, and the *volume oriented approaches* which are outlined in Sect. 7–9.

4 Triangulated Surfaces

A wide-spread approach of surface oriented interpolation is to construct surfaces composed of triangles. In this case it is assumed that the given contours are closed polygonal chains.

The triangles are usually constructed by connecting the vertices of the polygonal chains by line segments in a suitable manner. However, sometimes it makes sense to introduce additional points. Additional points are also called *Steiner points.* An example is saddle point interpolation where the saddle point is usually expected to lie between the two given planes.

Before we go into the details of triangulation we will briefly discuss some methods of *point reduction.* The reason for point reduction is to reduce the number of triangles which depends directly on the number of vertices of the given polygonal chains. In order to keep the number of triangles low, a point reducing step often precedes triangulation in practice.

4.1 Point Reduction

The aim of point reduction is to replace a given polygonal chain by one with less vertices which does not deviate too much from the original one. Shantz [60] has proposed a reduction based on the k-curvature of the polygonal chain. The new chain is initialized with the first vertex of the given chain. Then the k-curvatures are summed along the chain until a given threshold $maxcurvature$ is exceeded. The vertex of the chain at which this happens is added to the new chain. The sum is set back to 0 and the process is continued analogously.

The values k and $maxcurvature$ are suitably chosen for the given class of polygonal curves. An optimal choice can be achieved by dynamically measuring the deviation between the new and the old chain. The values of the parameters are changed if the deviation becomes too large. For possible measures cf. Sect. 4.6.2.

Point reduction may cause a self-intersecting new curve. Self-intersection can be avoided by additionally checking in the algorithm sketched above whether the new edge to the vertex that is supposed to be added next intersects the new chain. If this happens, its predecessor (which has passed this test) is added to the chain instead. For polygonal chains with many vertices the intersection test with all edges of the already calculated edges can be avoided by sorting these edges with respect to the x-coordinates of their vertices. Only those edges have to be tested against the new edge which are overlapping the new edge in x-projection. These edges can be efficiently determined with the help of the sorted list.

4.2 Cylindric Triangulation

In this section we will only treat cylindric triangulations without Steiner points since they are sufficiently flexible for practice. A more precise definition of this type of triangulation is

Cylindric triangulation.

Input. Two disjoint polygonal chains $\mathbf{p}_0, \ldots, \mathbf{p}_m$ and $\mathbf{q}_0, \ldots, \mathbf{q}_n$ in space, $\mathbf{p}_0 = \mathbf{p}_m$, $\mathbf{q}_0 = \mathbf{q}_n$.

Output. A set of triangles with vertices $\mathbf{p}_i$, $\mathbf{q}_j$, so that

1. each triangle has vertices on both polygonal chains,
2. each edge of the polygonal chains occurs on exactly one triangle,
3. each edge $\mathbf{p}_i\mathbf{q}_j$ of a triangle belongs to exactly two triangles.

The number of cylindric triangulations is considerable for larger values m and n. There are $N(m,n) = n \cdot \binom{m+n-1}{m-1}$ different cylindric triangulations. In order to show this we first calculate the number $N'(m,n)$ of triangulations containing the triangle $\mathbf{p}_0\mathbf{p}_1\mathbf{q}_j$ for some fixed j. It is not difficult to see that $N'(m,n)$ satisfies the recurrence $N'(m,n) = N'(m-1,n) + N'(m,n-1)$, $N'(m,0) = 1$, $N'(1,n) = 1$. This recurrence is solved by $N'(m,n) = \binom{m+n-1}{m}$. Since we have n possibilities of triangles $\mathbf{p}_0\mathbf{p}_1\mathbf{q}_j$, $N(m,n) = n \cdot N'(m,n)$.

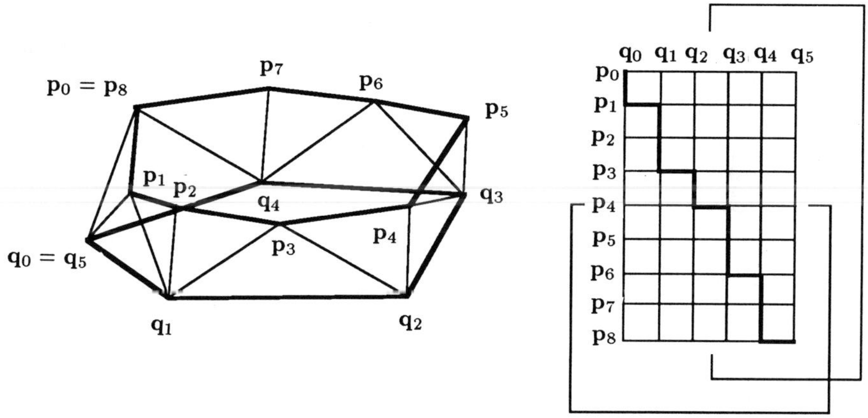

Fig. 13. A cylindric triangulation and the corresponding path in the toroidal graph.

By Keppel [35] a description of contour triangulations by directed graphs was introduced. His *toroidal graph* can be graphically represented by a two-dimensional grid (Fig. 13). It has $m \cdot n$ nodes $\mathbf{n}_{i,j}$, $i = 0, \ldots, m$, $j = 0, \ldots, n$, $\mathbf{n}_{0,j} = \mathbf{n}_{m.j}$, $\mathbf{n}_{i,0} = \mathbf{n}_{i,n}$. Each node $\mathbf{n}_{i,j}$ corresponds to a line segment $\mathbf{p}_i\mathbf{q}_j$ between the two given polygonal chains. The arcs of the toroidal graph have the form $\mathbf{n}_{i,j}\mathbf{n}_{i,j+1}$ and $\mathbf{n}_{i,j}\mathbf{n}_{i+1,j}$. Arcs of the first kind represent triangles $\mathbf{q}_j\mathbf{q}_{j+1}\mathbf{p}_i$, whereas those of the second kind correspond to triangles $\mathbf{p}_i\mathbf{p}_{i+1}\mathbf{q}_j$. A cylindric triangulation is expressed by a directed closed path which visits each row and column at least once. Each such path corresponds to a cylindric triangulation and vice versa.

4.3 Saddle Point Triangulation

In contrast to the cylindrical case the introduction of Steiner points may be helpful in order to model saddle points. If no Steiner points are allowed the saddle must be modeled on one of the given planes which at the first glance does not coincide with the requirements of practice. However, if the cutting planes lie close together, the shapes of the contours in the region around the expected saddle point may be so simple that the true shape and that one interpolated without Steiner points may not differ severely. Also, the trouble with Steiner points is that it is not clear where they should be geometrically located. The decision may be as arbitrary as the triangulation without Steiner points. Summarizing, there are some reasons to discuss saddle point interpolation without Steiner points, too.

Modeling saddle point on one of the given cutting planes for triangulations without Steiner points can be achieved by one of the following two alternatives:

Contour merging.

Input. Cf. contour interpolation.

Output.
The two contours on the same plane are transferred into one by introducing two nonintersecting line segments (Fig. 14). In the disconnected case both line segments lie totally in the exterior of both contours. In the connected case the line segments are located in the ring between the inner and the outer contour. A special case is that both edges are identical (Shantz [60]). In the disconnected case the resulting surface consists of a cylindric triangulation and a plane polygon. In the connected case the surface is formed by two cylindric triangulations and a plane polygon which has to be triangulated.

Contour splitting.

Input. Cf. contour interpolation.

Output.
The single contour is divided by two nonintersecting line segments into two contours (Fig. 15). In the connected case both line segments lie totally in the exterior of the contour, in the disconnected case in its inner. Both line segments are introduced so that the resulting contours are nonintersecting. In the connected case, the resulting surface consists of a cylindric triangulation and a plane polygon. In the disconnected case the surface is composed of two cylindric triangulations and a plane polygon. Again the polygon may be triangulated.

Contour merging and contour splitting can be seen as operations on the assignment graph. Contour merging transfers two vertices of the graph into one and can be applied only on vertices with exactly one incident edge connecting

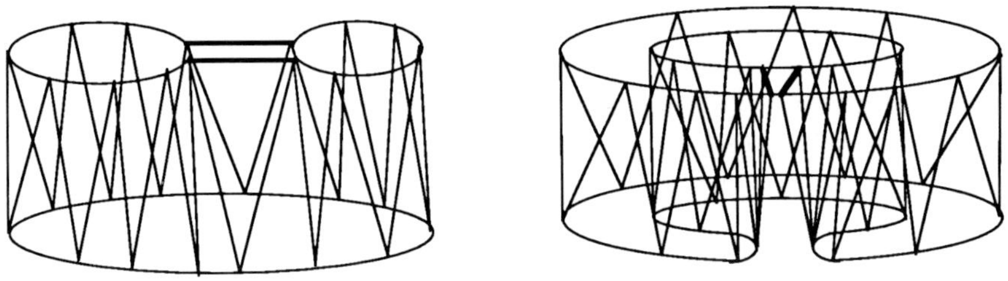

Fig. 14. Contour merging for triangulating saddles without additional nodes, in the disconnected and the connected case.

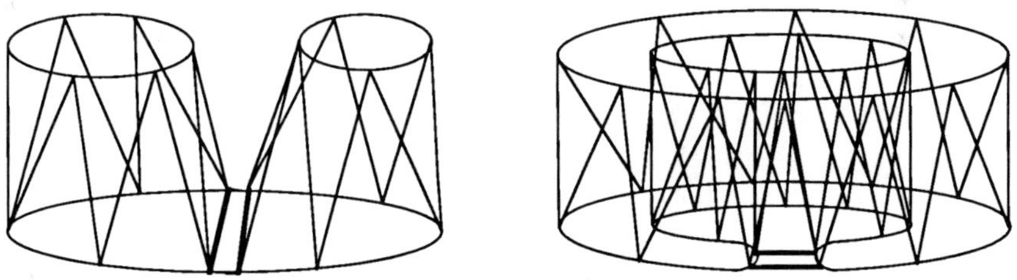

Fig. 15. Contour splitting for triangulating saddles without additional nodes, in the disconnected and the connected case.

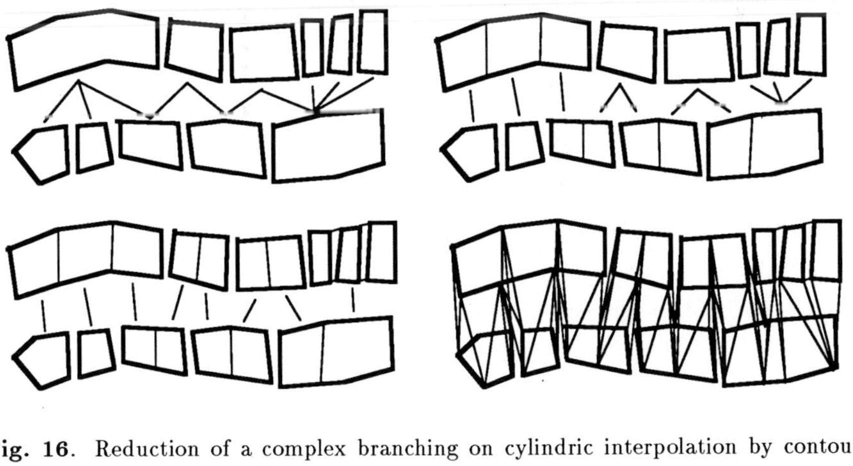

Fig. 16. Reduction of a complex branching on cylindric interpolation by contours splitting and contour merging.

each of them to the same vertex on the opposite plane. Contour splitting splits a vertex of the assignment graph into two vertices and can only be applied on a vertex with more than one outgoing edges. The outgoing edges are distributed among both resulting vertices, each obtaining at least one of them. Reduction on the cylindric case means to apply a sequence of merging and splitting operations so that the final graph consists of isolated edges only. These edges connect contours which have to be interpolated cylindrically.

The splitting and merging rules can be also applied on more complex branchings than those characterized by star-shaped connected components of degree 2 of the assignment graph. Figure 16 shows an example. A discussion of reduction on cylindric interpolation is given by Ekoule, Peyrin, Odet [14] in a more general context of interpolation.

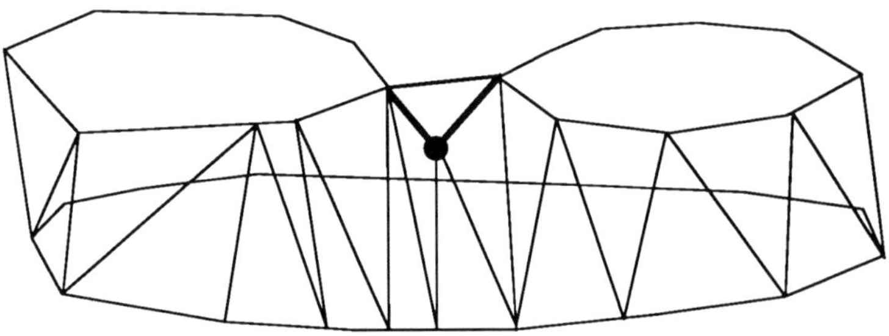

Fig. 17. Modeling a saddle point by subdividing an interconnecting line segment of two contours.

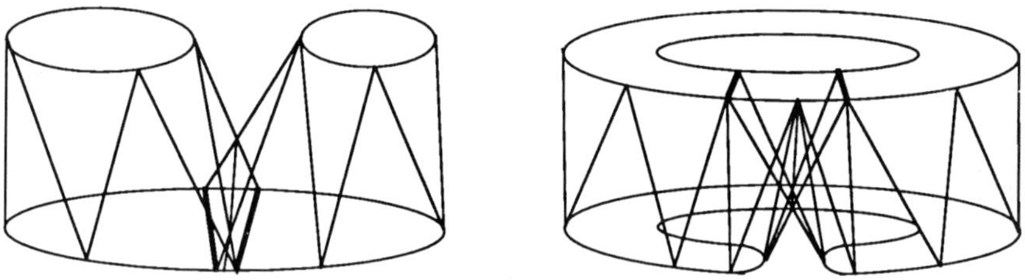

Fig. 18. Penetrating cylinders for triangulating saddles with additional nodes, in the disconnected and the connected case.

One approach of triangulating with Steiner points is to insert Steiner points for modeling the line of intersection. Christiansen, Sederberg [9] apply contour merging in the disconnected case by inserting one line segment between the closest points of the two contours. The center of the line segment is used as a Steiner point. In the resulting cylindrical interpolation the Steiner point is moved into the middle between both planes (Fig. 17).

Another possibility of saddle triangulation with Steiner points is to give up the disjointness of the cylinders in the contour splitting approach:

Penetrating cylinders.

Input. Cf. the disconnected case of contour interpolation.

Output.
The interpolating surface is constructed by allowing intersecting contours in the disconnected case of contour splitting (Fig. 18). The surface is obtained by intersecting both cylindric surfaces and eliminating those parts of one cylinder that are encountered in the other.

Algorithms for polyhedron intersection can be found in [43].

4.4 Extremal Points

Extremal points appear if on one of the neighboring cutting planes no continuing contour is assigned to a given contour. The only possibility without adding Steiner points is to triangulate the inner of the polygon. This triangulation closes the surface as it is required at an extremal point.

4.5 Penetrations

The formulation of the triangulation problem demands triangulated surfaces which are free of penetrations, at least if penetrations are not used for introducing Steiner points. A triangulation is free of penetrations if there are no two triangles which intersect. Two triangles with vertices **a**, **b**, **c** respectively **d**, **e**, **f** intersect if they are not co-planar and have a common inner point. The latter can be expressed by

$$\mathbf{a} + \mu_1(\mathbf{b} - \mathbf{a}) + \nu_1(\mathbf{c} - \mathbf{a}) = \mathbf{d} + \mu_2(\mathbf{e} - \mathbf{d}) + \nu_2(\mathbf{f} - \mathbf{d}),$$

$$0 \leq \mu_i,\ 0 \leq \nu_i,\ \mu_i + \nu_i \leq 1,\ i - 1, 2,$$

where at least one of the inequalities $\mu_i + \nu_i \leq 1$, $i = 1, 2$, must be satisfied strongly. By solving the system of equations, the variables μ_1, ν_1 and μ_2 can be expressed dependent on ν_2,

$$\mu_1 = (a_1 + \nu_2 a_2)/d,\ \nu_1 = (b_1 + \nu_2 b_2)/d,\ \mu_2 = (c_1 + \nu_2 c_2)/d$$

with

$$\begin{array}{rclrcl} a_1 &=& |\mathbf{d}-\mathbf{a}\ \ \mathbf{c}-\mathbf{a}\ \ \mathbf{d}-\mathbf{e}|, & a_2 &=& |\mathbf{f}-\mathbf{d}\ \ \mathbf{c}-\mathbf{a}\ \ \mathbf{d}-\mathbf{e}|, \\ b_1 &=& |\mathbf{b}-\mathbf{a}\ \ \mathbf{d}-\mathbf{a}\ \ \mathbf{d}-\mathbf{e}|, & b_2 &=& |\mathbf{b}-\mathbf{a}\ \ \mathbf{f}-\mathbf{d}\ \ \mathbf{d}-\mathbf{e}|, \\ c_1 &=& |\mathbf{b}-\mathbf{a}\ \ \mathbf{c}-\mathbf{a}\ \ \mathbf{d}-\mathbf{a}|, & c_2 &=& |\mathbf{b}-\mathbf{a}\ \ \mathbf{c}-\mathbf{a}\ \ \mathbf{f}-\mathbf{d}|, \\ d &=& |\mathbf{b}-\mathbf{a}\ \ \mathbf{c}-\mathbf{a}\ \ \mathbf{d}-\mathbf{e}|. & & & \end{array}$$

$|.|$ denotes the determinant of the embraced matrix. The inequalities for μ_i, ν_i, $i = 1, 2$, thus lead to the inequalities

$$\begin{array}{rclrcl} 0 &\leq& (a_1 + \nu_2 a_2)/d, & 0 &\leq& (b_1 + \nu_2 b_2)/d, \\ 0 &\leq& (c_1 + \nu_2 c_2)/d, & 0 &\leq& \nu_2, \\ 1 &\geq& (a_1 + b_1 + \nu_2(a_2 + b_2))/d, & 1 &\geq& (c_1 + \nu_2(c_2 + 1))/d, \end{array}$$

which must be simultaneously satisfied by ν_2. They can be brought into a form $a \leq \nu_2$ or $a \geq \nu_2$, respectively, in which the existence of a solution can be easily decided.

In an efficient implementation a test on overlapping coordinates of the triangles will precede this calculation. A pairwise test of all triangles can often be avoided by storing the coordinate intervals of all triangles sorted according to at least one coordinate. Then those triangles can be quickly found out whose coordinate intervals overlap that of a new one, for example by binary search. A more profound alternative is e.g. the interval tree data structure (Mehlhorn [43]).

A global approach for speeding up the intersection test in the context of enumerating all nonintersecting triangulations is storing those pairs of triangles which were already checked, together with the corresponding result, "intersecting" or "non-intersecting". The penetration test is only performed if the pair is not found in the data structure. A possible implementation might be a hash structure. The indices of the vertices of the triangles can be used as key.

In the liver example of section 1 there are adjacent cells with touching contours. If two cells have touching contours on consecutive planes, it may happen that they induce touching surfaces of the reconstructed 3D cells. In order to get penetration-free surfaces the triangulation of the touching parts of the two surfaces must be the same. However, there is still a considerable degree of freedom for valid triangulations.

4.6 Quality of Triangulations

Among the large number of triangulations only a considerably reduced subset will usually be acceptable for an area of application. For characterization of acceptable triangulations different criteria were proposed. They can be distinguished in *measure criteria* and *shape criteria*.

4.6.1 Measure Criteria

In the case of cylindrical triangulations, measure criteria can often be expressed formally using the toroidal graph. For this purpose, weights $w(\mathbf{e})$ and $w(\mathbf{n})$, respectively, are assigned to its edges $\mathbf{e}$ and nodes $\mathbf{n}$ (Ganapathy, Dennehy [21]). Typical measure criteria of this sort are

minimum area of the total surface (Fuchs, Kedem, Uselton [20]):
$\min_{\mathbf{c}\ \text{path}} \sum_{\mathbf{e} \in \mathbf{c}} w(\mathbf{e})$ with $w(\mathbf{e})$ = area of triangle definded by edge $\mathbf{e}$;

minimum largest area of a triangle:
$\min_{\mathbf{c}\ \text{path}} \max_{\mathbf{e} \in \mathbf{c}} w(\mathbf{e})$ with $w(\mathbf{e})$ = area of the triangle induced by edge $\mathbf{e}$;

minimum total length of the edges of the triangulation:
$\min_{\mathbf{c}\ \text{path}} \sum_{\mathbf{n} \in \mathbf{c}} w(\mathbf{n})$ with $w(\mathbf{n})$ = length of edge defined by node $\mathbf{n}$;

minimum longest edge:
$\min_{\mathbf{c}\ \text{path}} \max_{\mathbf{n} \in \mathbf{c}} w(\mathbf{n})$ with $w(\mathbf{n})$ = length of the edge defined by node $\mathbf{n}$;

minimum largest angle of a triangle:
$\min_{\mathbf{c}\ \text{path}} \max_{\mathbf{e} \in \mathbf{c}} w(\mathbf{e})$ with
$w(\mathbf{e})$ = largest angle of the triangle defined by edge $\mathbf{e}$;

maximum smallest angle of a triangle:
$\max_{\mathbf{c}\ \text{path}} \min_{\mathbf{e} \in \mathbf{c}} w(\mathbf{e})$ with
$w(\mathbf{e})$ = smallest angle of the triangle defined by edge $\mathbf{e}$;

maximum smallest slope of a triangle w.r.t. the contour planes:
$\max_{\mathbf{c}\ \text{path}} \min_{\mathbf{e} \in \mathbf{c}} w(\mathbf{e})$ with $w(\mathbf{e})$ = slope of the triangle belonging to edge $\mathbf{e}$.

These optimization criteria lead to versions of the classical problem to find a shortest path between two given nodes of a graph. Fuchs, Kedem, Uselton [20] have developed an algorithm finding an optimum closed path in an $m \times n$ toroidal graph within time $O(mn \log m)$.

Triangulations based on other weight criteria can have less efficient algorithms of solution. In the worst case, even exhaustive search may become necessary. Kaneda, Harada, Nakamae, Yasuda, Sato [32] for example have proposed to connect each vertex on a contour with its nearest neighbor on the other contour. However, it is not always possible to obtain a triangulation containing all these interconnecting line segments as edges. The alternative is to consider those triangulations containing a maximum number of such edges. No algorithm more efficient than exhaustive search seems to be known yet. Shinagawa, Kunii, Nomura, Okuno, Hara [65] give an approximate solution for the maximum number of shortest edges.

An alternative to criteria of optimization is to define special classes of triangulations algorithmically. An example of an algorithmic framework in this context is the *greedy approach*. Starting with an initial node in the toroidal graph, the greedy algorithm of Christiansen, Sederberg [9] successively constructs a path by adding that neighbor of the currently last node of the path which has the smallest weight. The weight used is the length of the connecting line segment represented by the node. Ganapathy, Dennehy [21] choose the next node in such a way that the absolute difference between the lengths of the pieces processed yet and belonging to the upper respectively to the lower contour is minimal. Cook, Batnitsky [10] suggest to construct the triangles in such a way that their orientation is as close as possible to the orientation of the line joining the centroids of the two contours.

A last type of criteria we mention here is oriented at considerations of volumes. Keppel [35], for example, maximizes the volume bounded by the triangulated surface. Tönnies [68] uses the so-called *difference volume*. The difference volume is obtained by first projecting one contour region c_1 onto the plane of its neighbor c_2 in the stack by a parallel projection. This leads to two, possibly

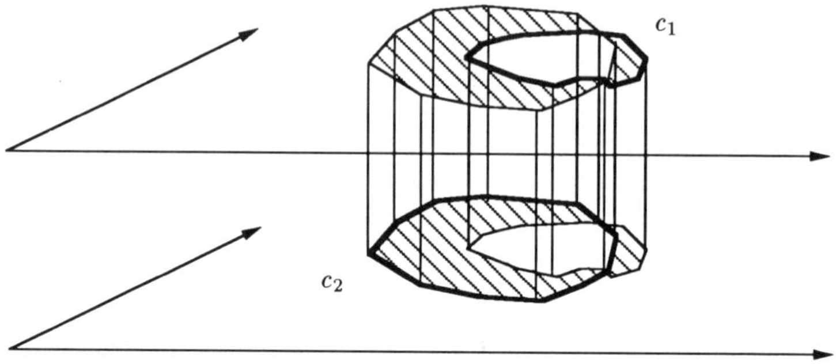

Fig. 19. The difference area/volume for two closed contours.

intersecting, contour regions (2D cells) on one plane. The normalized symmetric difference of these regions defined by

$$B = (c_1^- - c_2^o) \cup (c_2^- - c_1^o)$$

yields in general a number of 2D cells. A^- is the *closure* of a set A, A^o its open kernel. These independent cells are called the *difference area* of c_1 and c_2. They are interconnected by points (0D cells) or chains (1D cells). If we expand the cells of B to a third dimension by the vertical distance between c_1 and c_2, we obtain a volume that bridges the gap in the contour stack (Fig. 19). This volume is called *difference volume.*

The reason for considering the difference volume is that the more a given triangulation is inside this region the more it is intuitively close to a "minimal" surface between the given contours which an elastic skin tiedly wrapped around the contours would deliver. For identical contours the difference volume degenerates to a surface which is the natural cylindric interpolation of the contours. This motivates the usefulness of this criterion at least for very similar contours.

If the projected contours are intersecting, a triangulation which lies totally in the difference volume can be computed by a sweep plane algorithm. A plane is moved according to a fixed direction over the contours with the goal to find events of interest. In fact, the motion is simulated by jumping from one event of interest to the next. Events of interests are vertices of the contours and intersection points of contour edges in the parallel projection mentioned above. The sweep plane located at two consecutive events of interest slices the contours and the difference volume in quite elementary pieces. By connecting parts of the contour that lie inside such a slice an interpolating triangulation is build up successively. If the edges of a one-slice-interpolation are determined to be inside the one-slice-difference volume, and if the resulting surface is closed the global result is a valid triangulation within the global difference volume. Details can be found in [37].

Evidently, a triangulation thus obtained will usually be one with Steiner points. Some of the triangles together with their new points can be eliminated

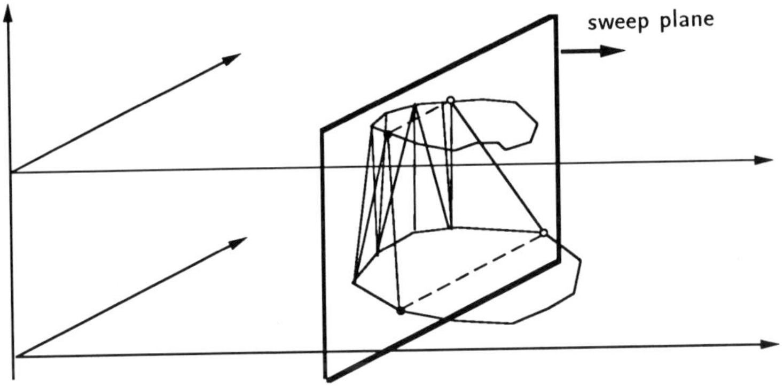

Fig. 20. The principle of the sweep plane triangulation.

if e.g. the triangles are coplanar. Therefore the number of points needs not to increase significantly in practice. For further reduction of triangles and for the aesthetic look of the result it might be useful to allow the triangulation to leave the difference volume up to a certain extend [37].

4.6.2 Shape Criteria and Deformation

Shape criteria take into consideration the shape properties of the given polygonal chains. An example is shape analysis based on curvature. Curvature allows to identify changes between left and right windings, and regions of extremal curvature. The shape of the polygonal chains is transferred onto the shape of the triangulation by inserting edges between both contours which connect points with the same property in suitable order. By the new edges, the given stripe is subdivided into substripes which are triangulated according to some further criterion. A similar idea was followed by Chang, Chen, Ho [8]. Ekoule, Peyrin, Odet [14] carry out a "decomposition of concavities" using iterated convex hulls of subchains.

We will now present a more general approach for the application of shape similarity on the problem of triangulation. The idea is to interpret the problem of interpolating polygonal chains as a problem of deformation. Suppose that a triangulation is already constructed for two polygonal contours lying in two parallel planes. When moving a cutting plane in parallel to the given planes from the plane of the first chain in direction to the plane of the second chain, the images of intersection define an animation showing a deformation of the first chain into the second. Thus the triangulation induces a deformation between the two polygonal chains. We will now proceed the other way round, namely we will construct triangulations using (other) techniques of deformation.

In doing this, we do not insist that a chain is exactly mapped onto the other, an approximation will do. The deformation is applied on one chain, the other remains unchanged. The deformed chain and the other unchanged one are then triangulated by a simple weight criterion. From this triangulation, a triangulation of the original chains is derived by inserting an edge between two vertices if and only if the corresponding vertices are connected by an edge in the triangulation of the deformed situation. It is important to note that the resulting triangulation might be self-intersecting even if the original one was not. Thus a check for self-intersection has to be performed and possibly a process of re-triangulation has to be started.

In the following we will present an exemplary class of deformations which has proven itself to be quite useful. It is based on the observation that, in the real world, deformation can be reached by thinking of the image as being drawn on an elastic skin. The elastic skin is deformed by moving some of its points $\mathbf{p}_i$ in a new position $\mathbf{q}_i$. This causes a deformation of the image too. This idea is formalized as follows:

Deformation Problem.

Input. n pairs $(\mathbf{p}_i, \mathbf{q}_i)$ of control points, $\mathbf{p}_i, \mathbf{q}_i \in \mathbb{R}^2$, $i = 1, \ldots, n$.

Output. An at least continuous function $\mathbf{f} : \mathbb{R}^2 \to \mathbb{R}^2$ with $\mathbf{f}(\mathbf{p}_i) = \mathbf{q}_i$.

Besides continuity, other properties of the function $\mathbf{f}$ are desirable which concern the visual pleasantness and which are hard to describe in an exact manner. These properties will become recognizable in the following discussion of concrete propositions.

The control points required by this approach can be determined by shape criteria. For example, points of distinguished curvature as described above can be used. These points can also be chosen using a Fourier approximation. Fourier approximations preserve the coarse shape and eliminate fine structures which are often not required. Searching for a suitable deformation will start with only a few control points and will continue by successively adding suitable new control points. The latter are chosen based on a comparison of the deformed shape with the shape of destination.

For *comparison of shapes* a number of different criteria can be used to measure the resemblance or distance between two given shapes or contours. Some of them, already mentioned in Sect. 2.3, are common in pattern recognition and concentrate on shape similarity. Others have a more mathematical background and are not supposed to be invariant of rotation or translation. Among those are the Hausdorff-distance and the Fréchet-metric for closed polygonal chains.

Using the Euclidean-distance d in $\mathbb{R}^2$, the *Hausdorff distance* δ_H of two curves a, b is defined by

$$\delta_H(a,b) := max(\sup_{p\in a}(\inf_{q\in b}(d(p,q))), \sup_{p\in b}(\inf_{q\in a}(d(p,q)))).$$

The Hausdorff distance distance has a pleasant simplicity and can be calculated efficiently for polygonal chains. In [2] an algorithm is given that takes time

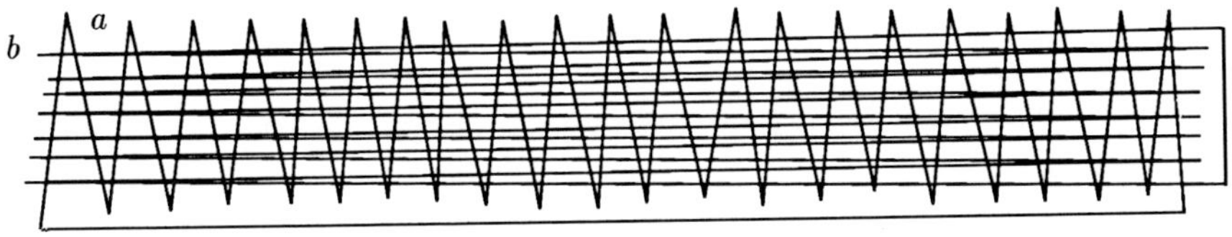

Fig. 21. Two contours with small Hausdorff-distance and high dissimilarity.

$O((n+m)\log(n+m))$, where n, m are the numbers of vertices of the two polygons involved.

A major drawback of the Hausdorff distance, however, is the fact that the distance can become relatively small without being close to matching (Fig. 21).

The *Fréchet-distance* overcomes this problem by considering paths (parameterizations) on the contours or curves instead of set of points. Given two closed curves c_1, c_2 the Fréchet-distance is defined by [49], [19]

$$\delta_{FC}(c_1, c_2) := \inf_{v_1, v_2} \inf_{\alpha_1, \alpha_2} \sup_{t \in [0,1]} d(\alpha_1(t), \alpha_2(t))$$

where for $i = 1, 2$, v_i is any point on c_i and α_i varies over all monotone parameterizations $\alpha_i : [0,1] \rightarrow \mathbb{R}^2$ of c_i with $\alpha_i(0) = \alpha_i(1) = v_i$. For polygonal chains the time requirement to determine δ_{FC} is limited by $O(nm \log^3(nm))$ [3]. For convex contours δ_H and δ_{FC} are identical.

Another measure is the *difference area* between two closed contours which is defined as the area of the symmetric difference of the regions enclosed by the contours. An approximation of the differential area can be obtained by thinking one of the two contours perpendicularly moved into a parallel plane. The two contours are then triangulated cylindrically according to the "minimum sum of areas" measure where "area" means the area of the perpendicular projection of the triangles.

Independent of the measure chosen, the deformation is accepted if its value lies below a given threshold.

The problem of deformation in this form can be understood as a problem of scattered data interpolation. The question is to find a function $\mathbf{f}$ which performs a mapping between the given pairs of control points $(\mathbf{p}_i, \mathbf{q}_i)$. An interpolation of this sort is known as *scattered data interpolation* since the control points can lie arbitrarily and need not be arranged e.g. on a grid. Many suggestions of interpolating functions for scattered data were made in literature. Usually the scattered data problem is tackled in the following form:

Scattered Data Interpolation Problem.

Input. n data points $(\mathbf{x}_i, y_i)$, $\mathbf{x}_i \in \mathbb{R}^d$, $y_i \in \mathbb{R}$, $i = 1, \ldots, n$.

Output. An at least continuous function $f : \mathbb{R}^d \rightarrow \mathbb{R}$ interpolating the given data points, i.e. $f(\mathbf{x}_i) = y_i$, $i = 1, \ldots, n$.

A deformation is obtained by applying a scattered data method in the latter sense on each component $f_j : I\!R^d \to I\!R$, $j = 1, 2$, of $\mathbf{f}$ separately. The data points to be used are $(\mathbf{p}_i, q_{i,j})$, $q_{i,j}$ the j-th component of $\mathbf{q}_i$, $j = 1, 2$.

A classical example of a scattered data interpolation method is that by Shepard [61]. Shepard uses a weighted average of the data values at the data points, with weights dependent on the distance of the observed point from the given data points,

$$f(\mathbf{x}) \;=\; \sum_{i=1}^{n} w_i(\mathbf{x}) y_i.$$

$w_i : I\!R^n \to I\!R$ is the weight function and $y_i \in I\!R$ the data value at the data point $\mathbf{x}_i \in I\!R^d$. The weight functions satisfy the conditions

$$\begin{aligned} w_i(\mathbf{x}) &\geq 0, \quad i = 1, \ldots, n, \quad \sum_{i=1}^{n} w_i(\mathbf{x}) = 1, \quad \text{and} \\ w_i(\mathbf{x}_j) &= \delta_{ij}, \quad \text{i.e.} \quad w_i(\mathbf{x}_i) = 1 \quad \text{and} \quad w_i(\mathbf{x}_j) = 0, \quad j \neq i, \; i, j = 1, \ldots, n. \end{aligned}$$

These conditions guarantee the property of interpolation.

Shepard uses the weight function

$$\begin{aligned} w_i(\mathbf{x}) &= \frac{\sigma_i(\mathbf{x})}{\sum_{j=1}^{n} \sigma_j(\mathbf{x})}, \text{ with} \\ \sigma_i(\mathbf{x}) &= \frac{1}{(d_i(\mathbf{x}))^{\mu}}, \quad d_i(\mathbf{x}) \text{ the distance between } \mathbf{x} \text{ and } \mathbf{x}_i. \end{aligned}$$

The smoothness is determined by the exponent μ. $\mu > 1$ assures the continuity of the first derivative. The value disappears at the data points.

The Shepard interpolation in this form shows an usually unfavorable behavior when applied on the problem of deformation. Better results are achieved with local interpolants (Gordon, Wixom [25], Franke, Nielson [18], Lancaster, Salkauskas, [39], Farwig [16]), locally bounded weight functions (Franke, Nielson [18]), and radial basis functions (Hardy [28], Duchon [11], Eck [13]). Experiments show that in the case of small deformations good results can be achieved with many of these approaches (Ruprecht, Müller, [56]). For larger deformations, however, undesired foldovers can occur in the deformed picture. These are caused by the fact that, contrary to the scattered data problem, the image space is mapped onto itself. The influence of the control points on a transformed point should therefore be dependent not only on the initial relative position but also on the trajectory along which a control point is moved to its new location.

Moving the control points can be simulated by decomposing the deformation into several smaller deformations which are executed sequentially. When choosing a linear trajectory, the deformation with pairs of control points $(\mathbf{p}_i, \mathbf{q}_i)$ is linearly subdivided into m deformations $(\mathbf{p}_{i,k}, \mathbf{q}_{i,k})$, $\mathbf{q}_{i,k} = \mathbf{p}_{i,k+1}$, $k = 0, \ldots, m-1$, with

$$\mathbf{p}_{i,k} = \mathbf{x}_i + \frac{k}{m} \cdot (\mathbf{q}_i - \mathbf{p}_i), \quad k = 0, ..., m. \tag{7.1}$$

Figure 22 shows the behavior of these methods when applied to a grid.

4.7 Intermediate Layers

Another application of deformation is to introduce additional layers between the given planes in cases of higher dissimilarity. Introducing additional layers is one approach to achieve triangulations with additional points. Suppose our intention is to introduce the additonal layer on level α, $0 \le \alpha \le 1$. Level 0 is supposed to be the first plane, level 1 to be the second, and the new layer divides the distance between the given planes according to the ratio $\alpha : (1-\alpha)$. Then the contour on level 0 is deformed in direction to the contour on level 1 by the method of piecewise deformation sketched above so that one of the intermediate contours, c_0, belongs to the points on the trajectories with parameter α. Analogously, the contour on level 1 is deformed in direction to that on level 0, with an intermediate contour, c_1, at the trajectory points with parameter $(1-\alpha)$. The contours c_0 and c_1 are then interpolated by any approach. Since they can be expected to be similar, one of the measure heuristics should do. If c_0 is seen to be on level 0 and c_1 is seen on level 1, the final intermediate contour on level α is obtained by cutting the resulting surface in parallel to the planes of c_0 and c_1 on level α.

5 Pyramidal Extrapolation

Pyramidal extrapolation yields polyhedrons with additional vertices. Intuitively, the approach is to shrink the given contour continuously and to move it at the same time perpendicular to the plane spanned by it. In the end the contour is a degenerated closed polygonal chain which surrounds an empty inner region. The resulting surface is defined as the set of points touched by the contour during the process of shrinking (Fig. 23). The resulting surface may also be represented by a finite sequence of closed polygonal chains with a decreasing number of vertices. Each vertex and each edge of a polygon has a corresponding vertex and edge on the previous one of the sequence. An edge and its corresponding one are in parallel. Corresponding vertices are connected by line segments. The facets lying between two consecutive polgons are plane. The shape of the resulting surfaces can be influenced more flexible by letting the edges shrink with different but constant speeds.

This construction can be immediately applied to the case of extremal points. Furthermore, it helps to reduce the case of saddle points to the cylindric case. In the disconnected situation the two contours which have to grow together are developed by cylindrical extrapolation into direction of the other plane. The process is terminated when both extrapolations touch for the first time. The contact point is taken as the saddle point. In this location both contours are merged into one contour which touches itself at the saddle point. The surface between the new contour and the opposite single contour is obtained by cylindrical interpolation. The connected case is treated analoguously.

Cylindric interpolation by pyramidal extrapolation can be achieved by extrapolating both contours in direction of the opposite plane. If the extrapolation is done in a way that the intersection of the extrapolations with the opposite

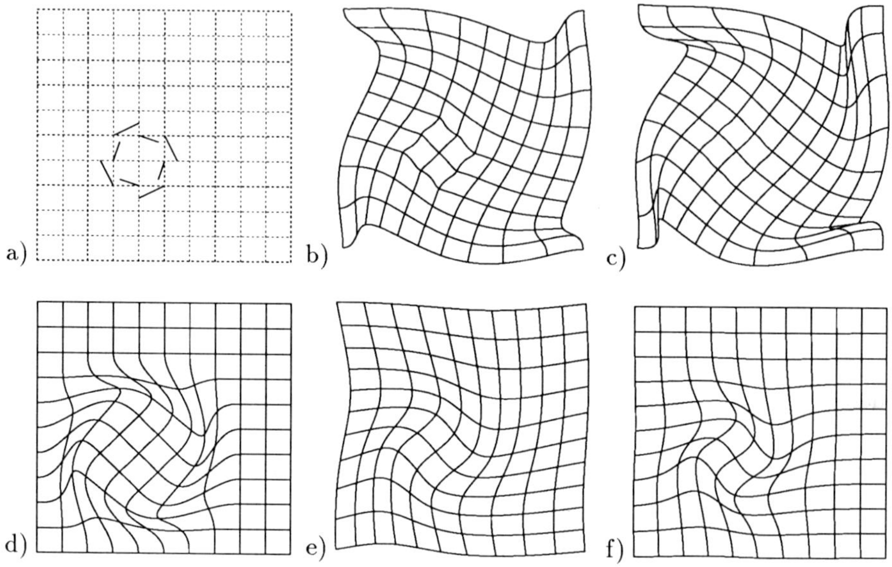

Fig. 22. Deformation of a square grid. An embedded 2×2-square is rotated by 45° within an square grid with fixed corners: a) Original grid with displacements; b) Shepard interpolation with local interpolation ($\mu = 2$); c) local Shepard interpolation ($\mu = 3$); d) Shepard interpolation with weight functions of limited influence; e) Hardy's multiquadrics ($\mu = 1$); f) Hardy's multiquadric with limited influence.

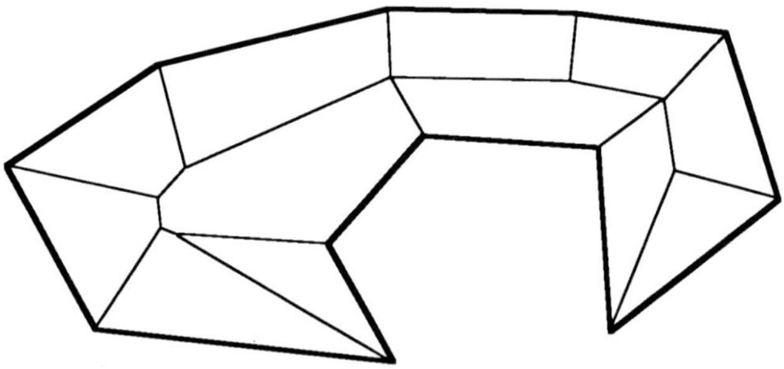

Fig. 23. Contour shrinking for modeling extremal points.

plane lies totally within the contour of this plane, the union of the two polyhedrons obtained by extrapolation is taken as interpolating polyhedron. If the intersection of the extrapolations with the opposite plane contains the contour on this plane completely, the union of the extrapolating polyhedrons is taken for interpolation.

6 Smooth Surface Interpolation

If it is known that the solid to be reconstructed has a smooth surface the construction of interpolating surfaces of a higher degree of smoothness than simple continuity is necessary. Surfaces of higher smoothness can be derived from triangulated surfaces by methods known from the area of computer aided geometric design. Examples are the techniques by Barnhill, Birkhoff, Gordon [5] and Nielson [50] which were generalized and modified in quite different ways, cf. e.g. Nielson [51] or the survey by Abramowski, Müller [1]. For a larger number of triangles the direct application of these methods is expensive. In this case it should be combined with a strategy of patch reduction. Chang, Chen, Ho [8] connect the given contours with quadrilaterals instead of triangles and use so-called Hermite patches for smoothing.

A more simple approach is *lofting* (Faux, Pratt [17], Farin [15]). Here a surface is filled in between two given curves given in parameter representation, by connecting points with the same parameter value by a curve of a fixed type, for example a line segment. The smoothness of the surface depends on the smoothness of the connecting curve. An example is the work of Sunguroff, Greenberg [66].

If both given contours are polygonal chains, they first are approximated by smooth curves [66, 8]. For choosing the parametrization a triangulation previously performed may be useful. The parametrization is chosen so that vertices on both contours connected by an edge of the triangulation have similar parameter values.

In topology, the existence of a continuous family of curves between two given curves is known as *homotopy*. From this view, Shingawa, Kunii, Nomura, Okuno, Hara [65] unify different approaches of smooth interpolation. One of them is a translation of the toroidal graph to the continuous case. Kehtaranavaz [33] takes into consideration the local shape characteristics of the contours.

Other approaches of smooth contour interpolations are based on physical models. Tello, Mann, Rowell [67] transform the given contours into the Fourier description sketched in Sect. 2.3 and apply a Fourier filter along the third coordinate. Lin, Liang, Chen [41], Lin, Chen [40] propose a dynamic elastic interpolation technique between contours. Weiss [71] represents the formed surface as a network of springs and minimizes its energy. Sander, Zucker [57], Xu, Lu [72], Kehtarnavaz, De Figueiredo [34], Chang, Chen, Ho [8] use curvature arguments for characterizing the smoothness of a surface.

In order to get a globally smooth surface during the interpolation of the surface between two neighboring planes the contours in one or more additional planes in the environment of both sides have to be considered. For the lofting approach, for example, it must be guaranteed that the connecting curves are continued smoothly along several cuts. In Sect. 9 we will present an approach based on *implicit surface representation* allowing a globally smooth interpolation of contours without patch compositing, which also works in complex situations like saddle points. Less traditional methods of smoothing polygonal networks which can be globally applied, too, are *techniques of subdivision* (e.g. Dyn, Levin, Gregory [12], Abramowski, Müller [1]). A sequence of finer networks converging to a smooth surface is constructed according to a given algorithm of subdivision.

7 Volume Oriented Reconstruction

In the previous sections techniques were presented which had the goal to derive a description of an interpolating solid by its surface. Now we will focus on methods which compose the solid by spatial cells. This eliminates difficulties which are typical for surface oriented approaches. Examples of such difficulties are to guarantee that the resulting surface in fact is the boundary of a solid, for example that it does not penetrate itself. From the cell decomposition a surface can be reconstructed if necessary by collecting those parts of the boundaries of the cells which are not common for two neighboring cells. Typical shapes of spatial cells are tetrahedrons and orthogonal parallelepipeds.

In the following two approaches are described: *interpolation by spatial Delaunay triangulations* and *interpolation by grids*. Both strategies simultaneously yield solutions for the contour assignment problem and for the geometric interpolation problem if they are applied on two consecutive planes containing several contours. The heuristics applied implicitly is to connect neighboring contours. This is useful in many applications. Other situations may be modified by deformation so that these heuristics can be useful, too. On the other hand, both approaches can also be applied locally on the elementary cases of cylindric, saddle point and extremal point interpolation. In this case, any other strategy of global assignment can precede. However, joining the resulting pieces continuously or smoothly may cause difficulties which have to be solved dependent on the approach.

8 Interpolation by Spatial Delaunay Triangulations

The interpolation by Delaunay triangulation (Boissonnat [6]) is performed between consecutive pairs of cutting planes. The contours are given by simply closed polygonal chains which do not intersect but which may be nested. The arrangement of the vertices of the polygonal chains is assumed to be so that the inner of the solid represented by them lies on the right hand side when the contour is followed according to the arrangement (Fig. 24)

The core of the approach is a decomposition of a rectangular bounding box of the contours by tetrahedrons. The triangulation also contains tetrahedrons not

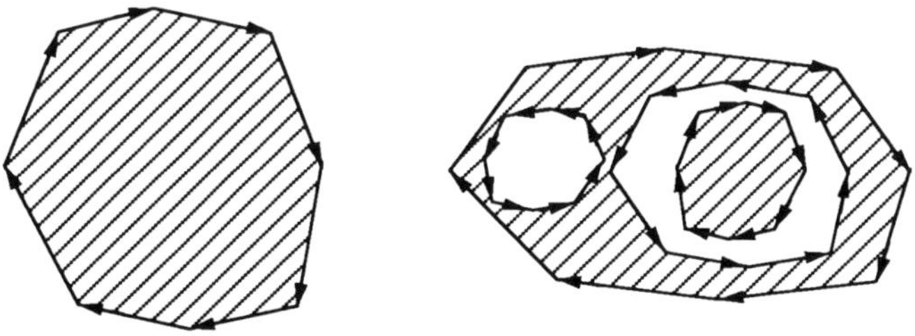

Fig. 24. Inner and outer side of polygonal chains.

belonging to the object to be constructed. They are eliminated in a subsequent step. The result is a representation of the interpolated polyhedron by a set of disjoint tetrahedrons. If only the surface of the interpolated solid is of interest, the triangles which belong to the surface have finally to be collected. The triangulation carried out is a *Delaunay triangulation.* The Delaunay triangulation leads to balanced compact tetrahedrons that tend to interconnect neighbored points. A triangulation of a finite set of points in the plane is a decomposition of its convex hull into triangles by connecting its points by line segments. A triangulation of a finite set of points in the plane is called a *Delaunay triangulation* if the smallest surrounding circle of each triangle does not contain any points of the given point set in its interior. Similarly, a triangulation of a point set in 3D space is a decomposition of the convex hull of the points into tetrahedrons induced by interconnecting line segments between the given points. A 3D triangulation is a Delaunay triangulation if the smallest surrounding sphere of each tetrahedron does not contain any of the given points in its interior. A triangle or a tetrahedron having this property is called a *Delaunay triangle* or a *Delaunay tetrahedron*, respectively. A *Delaunay edge* is a line segment connecting two given points that possess a surrounding circle (sphere) not containing any points in its interior.

A 2D Delaunay triangulation can be constructed incrementally by successively inserting new points in an already existing triangulation [27].

Up to now, the Delaunay triangulation was described for a point set only. What might happen is that edges of the given triangulation do not belong to the Delaunay triangulation – this happens if they are not Delaunay edges. Fig. 25 shows an example. However, this property is required in the following. This dilemma is solved by subdividing these edges in sufficiently short sub-edges. The new points can efficiently be added to the triangulation using the incremental algorithm [27].

For the solution of our interpolation problem a spatial Delaunay triangulation of the points in the two given planes is required that contains the edges of the contours in both planes. In principle, an incremental algorithm similar to the planar case is possible in space [7, 70]. However, in our special situation of $2\frac{1}{2}$*D-Delaunay triangulation* a more simple possibility does exist. It is based on the following observation. Suppose a set of points is given distributed on two

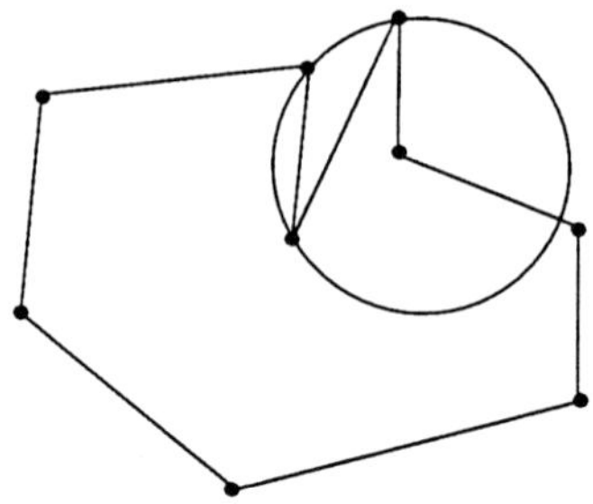

Fig. 25. Polygon edges which are not Delaunay edges.

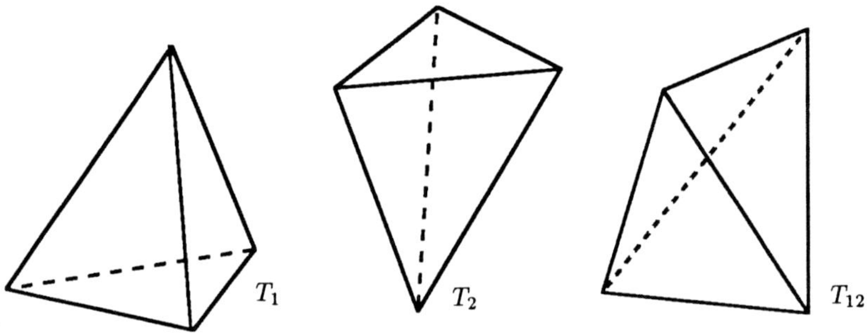

Fig. 26. The three types of tetrahedrons in a spatial Delaunay triangulation.

parallel planes E_i, $i \in \{1,2\}$, in space. Let D_i be the Delaunay triangulation on the planes E_i. Let further be DT the spatial Delaunay triangulation belonging to the given polygons. Then $DT \cap E_i = DT_i$, $i \in \{1,2\}$, that is the restriction of the spatial triangulation onto the planes E_i, coincide with the plane Delaunay triangulations on these planes. Furthermore, each tetrahedron of DT belongs exactly to one of the following types:

Type T_i is defined by a triangle in DT_i and by the point closest to the center of its smallest surrounding circle in the other plane, $i \in \{1,2\}$.

Type T_{12} is defined by two edges $e_i \in DT_i$, $i \in \{1,2\}$, whose projected Voronoi edges are intersecting.

Fig. 26 shows these three types of polyhedrons.

Using this property, the tetrahedrons of the spatial Delaunay triangulation can be calculated easily. The tetrahedrons of type T_1 are determined by connecting the triangles in the lower triangulation by a point in the upper triangulation according to the rule above. The tetrahedrons of type T_2 are obtained by switching the role of the upper and the lower triangulation. Determining the tetrahedrons of type T_{12} is more difficult. Their characterization above uses the Voronoi diagram belonging to a Delaunay triangulation. The *Voronoi diagram*

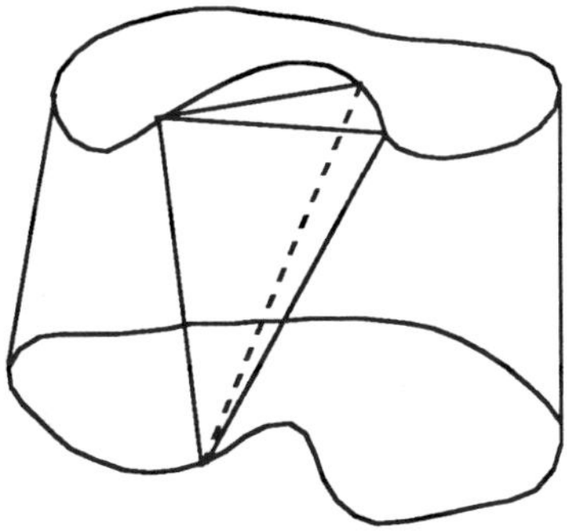

Fig. 27. The elimination of tetrahedrons with edges outside the contours.

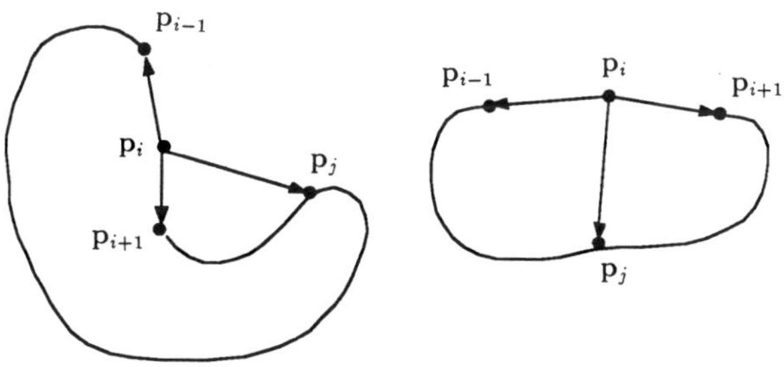

Fig. 28. Classification of edges with respect to a contour.

(cf. e.g. [43]) is obtained from the line segments perpendicular to the mid-points of the edges of the Delaunay triangulation

The next step is to decide which of the tetrahedrons generated belong to the triangulated solid to be reconstructed and which of them do not. This elimination of the outer tetrahedrons uses the orientation of the given contours (Fig. 27). Since the points of a contour are arranged so that the inner of the expected solid is to the right, it is easy to detect whether a Delaunay edge in one of the two planes is inside, outside, or on a contour. The edge $\mathbf{p}_i\mathbf{p}_j$ is on the contour if $\mathbf{p}_i\mathbf{p}_j = \mathbf{p}_i\mathbf{p}_{i+1}$ or $\mathbf{p}_i\mathbf{p}_j = \mathbf{p}_i\mathbf{p}_{i-1}$. It lies outside the contour if the edges $\mathbf{p}_i\mathbf{p}_{i-1}$, $\mathbf{p}_i\mathbf{p}_j$, and $\mathbf{p}_i\mathbf{p}_{i+1}$ are arranged clockwise. Otherwise, $\mathbf{p}_i\mathbf{p}_j$ is inside the contour (Fig. 28).

The set of tetrahedrons remaining after eliminating all those tetrahedrons having at least one edge outside the contours in the sense above is called a *simple Delaunay triangulation.* The simple Delaunay triangulation does not solve the interpolation problem completely. The reason are possibly existing non-solid connections. A *non-solid connection* is a set of adjacent tetrahedrons which touch one of the two planes at most one-dimensionally (Fig. 29). A *solid Delaunay triangulation* is obtained be eliminating all nonsolid interconnected tetrahedrons.

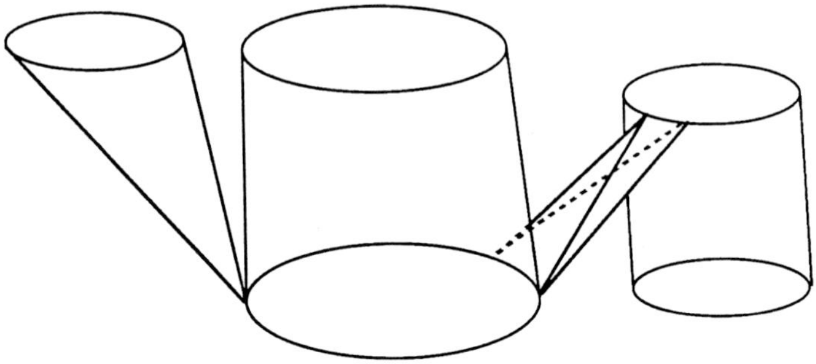

Fig. 29. Non-solid connections.

They can be found by following chains of non-solid tetrahedrons. The details are omitted here.

With the implementation of geometric algorithms of this sort, it is important to consider that the mathematical real numbers are approximated by floating point numbers [26, 30, 31, 52, 59]. Floating point numbers may cause numerical inaccuracies in the 2D triangulation and in the calculation of intersection points during $2\frac{1}{2}$D triangulation which may lead to topological inconsistencies in the data structures. These problems can be avoided by carrying out the calculation as close as possible to the input data. In the algorithm presented here it is possible to use the input data directly without too much overhead. The major operations of the algorithm are testing whether a point lies in a circle given by three points, testing the orientation of a triangle, and testing whether a point lies in a circle induced by three Voronoi points.

The Delaunay interpolation will only yield satisfiable results in situations where the contours to be mutually assigned are displaced only slightly and do not enclose rather thin regions. By combination with the deformation techniques of Sect. 4.6.2 the spectrum of useful applications can be extended:

Algorithm Delaunay Interpolation Under Deformation.

Input. Contours in two parallel planes.

Output. An interpolation of the given contours, if possible.

Procedure.

BEGIN

deform the first plane so that its contours coincide well
with those of the second plane;
perform the 2 $\frac{1}{2}$D-Delauany triangulation with the deformed
first plane and the original second plane;
deform the triangulation by replacing the coordinates of the points
of the deformed contours by their original ones;

IF the deformed triangulation does not have twisted tetrahedrons
THEN
determine the surface of the solid by tetrahedron elimination
ELSE the algorithm does not yield a result
END.

Deformation of the $2\frac{1}{2}$D-Delaunay triangulation yields a triangulation of the original contours which is not a Delaunay triangulation in general. If contour edges were subdivided by additional vertices during the process of Delaunay triangulation, their images on the original contours are obtained by subdividing the original edges with the same ratio as they do on the transformed edges. Contour deformation is performed so that it is a one-to-one mapping in that region of the plane containing the given contours. Thus the deformation of Delaunay triangulation on the first plane induces a triangulation of the original contours. Let be the vertices of the triangular faces of a tetrahedron arranged so that the normal, calculated by the cross product of vectors, is directed outwards. If the normal vectors of the deformed tetrahedron calculated from the same arrangement of the vertices are not all directed outwards the tetrahedron is called *twisted*. For tetrahedrons of types T_1 or T_2 this happens if one of its vertices is crossing its opposite edge during deformation. Tetrahedrons of type T_{12} become twisted if the angle between the two opposite edges changes its sign.

If the algorithm does not yield a result, insertion of intermediate levels as described in Sect. 4.7 may help to achieve a more favorable situation.

9 Reconstruction with Spatial Grids

The geometric shape of the cells of a Delaunay triangulation depends on the given data. An alternative is to choose initially the structure of the cell decomposition independent of the given data, and to inspect only those cells in more detail which are traversed by the boundary of the solid. For this purpose, the contours of the given planes are enveloped by a rectangular parallelepiped. This bounding box is chosen so that the contours lie completely in its inner. The bounding box is subdivided by an axis-parallel grid in a way that the given planes are also planes of the grid. For each vertex of the grid its membership to the solid is determined by classifying it against the contours in the given planes. According to this classification, a separating surface is introduced between the inner and the outer points with the marching cubes method [42].

In order to yield a good approximation of the given contours the grid should be chosen so that each rectangle of the grids induced on the given planes is traversed by at most one contour segment. If the original data are raster images this can be achieved by a grid with the resolution of the image. If the contours are polygonal chains the grid should be defined so that their vertices lie on grid lines. Then it can be guaranteed that the marching cubes method interpolates the contours, that means they are part of the resulting surface.

A problem with regular grids is that their resolution cannot be controlled locally. An alternative is *adaptive subdivision.* For instance, each of the layers between two consecutive cutting planes may be partioned by a quadtree adapted to the contours of these two planes. A consequence, however, is that the cells may be of different sizes. This requires some sort of hierarchical marching cubes method which has to avoid cracks at common boundaries. For a discussion of this problem cf. [46].

Grids can be used to construct interpolations of a higher degree of smoothness using the implicit representation of surfaces (Eck [13]). In the implicit representation, a surface in space is the set of points $\mathbf{x} \in \mathbb{R}^3$ satisfying an equation $F(\mathbf{x}) = 0$, with $F : \mathbb{R}^3 \rightarrow \mathbb{R}$. Let each of the given contours be represented by a sequence of points $\mathbf{p}_i$, $i = 1, \ldots, n$. The $\mathbf{p}_i$ are three-dimensional points whose third coordinate denotes the location of the cutting plane of the contour. Thus the cutting planes are supposed to be perpendicular to the third axis of coordinate. We also assume that $\mathbf{p}_{i-1}$ and $\mathbf{p}_{i+1}$ are closer to $\mathbf{p}_i$ than to all other points $\mathbf{p}_j$.

Interpolation is performed so that the $\mathbf{p}_j$ lie on the surface. This means we are looking for a function $F : \mathbb{R}^3 \rightarrow \mathbb{R}$ with

$$F(\mathbf{p}) = 0, \ \mathbf{p} \in C, \ C \text{ the set of control vertices.}$$

To these points some more are added, as follows. A rectangular bounding box strictly containing the given contours is subdivided by a regular grid (Fig. 30). To each grid vertex $\mathbf{g}_{i,j,k}$, i, j the indices in the plane, k the index of a plane, a value $F_{i,j,k}$ is assigned which is defined as its shortest distance from all contour points on plane k. The distance is taken negative inside of the solid described by the contours, and positive outside. The set G contains all those vertices $\mathbf{g}_{i,j,k}$ satisfying the following three conditions:

$$F_{i,j,k} \cdot F_{i-1,j,k} < 0 \text{ or } F_{i,j,k} \cdot F_{i-1,j,k} < 0 \text{ or } F_{i,j,k} \cdot F_{i,j-1,k} < 0 \text{ or } F_{i,j,k} \cdot F_{i,j+1,k} < 0,$$

$$F_{i,j,k} \cdot F_{i,j,k-1} > 0 \text{ and } F_{i,j,k} \cdot F_{i,j,k+1} > 0,$$

$$|F_{i,j,k}| > g/5.$$

The first condition says that the selected points lie close to a contour line since there is a neighboring vertex lying on the opposite side of the contour. The second condition compares the vertices of different cutting planes. A vertex is accepted if the neighboring vertex immediately below respectively above lies on the same side of the contours. The first and the last plane in the stack are treated as if there were one additional plane before respectively behind with only positive function values. The third condition prevents that vertices lie too close to a contour. This reduces the influence of the error made by measuring the distances to the contours only in the planes and not also between the planes. g denotes the length of the shortest edge of a grid cell.

Thus the function of interpolation, F, has to satisfy the additional condition

$$F(\mathbf{g}_{i,j,k}) = F_{i,j,k}, \ \mathbf{g}_{i,j,k} \in G.$$

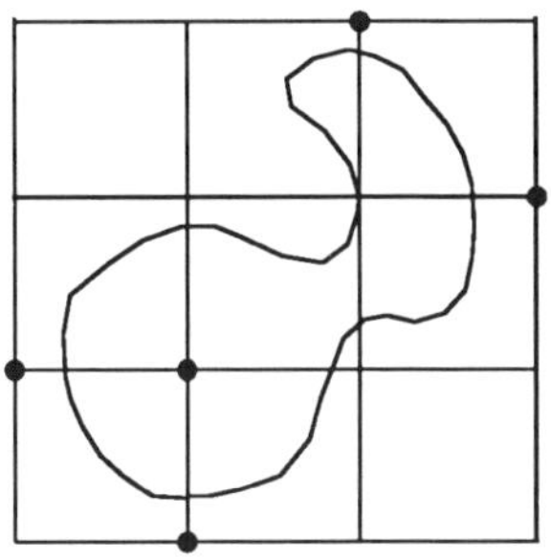

Fig. 30. Choosing additional points of interpolation.

The conditions for F define a scattered data interpolation problem for which the data points are given by points in $\mathbb{R}^3$ to which real values are assigned. Eck [13] suggests to use the approach of multiquadrics by Hardy [28],

$$F(\mathbf{x}) = \sum_{\mathbf{s}\in S} a(\mathbf{s})\sqrt{r(\mathbf{s}) + |\mathbf{x} - \mathbf{s}|},\ S \text{ the set of data points,}$$

setting

$$r(\mathbf{s}) = \min_{\mathbf{s}'\in S,\ \mathbf{s}'\neq\mathbf{s}} |\mathbf{s}' - \mathbf{s}|.$$

In our case, $S = C \cup G$. The equations form a linear system of equations for the $a(\mathbf{s})$ thus yielding the function F as solution.

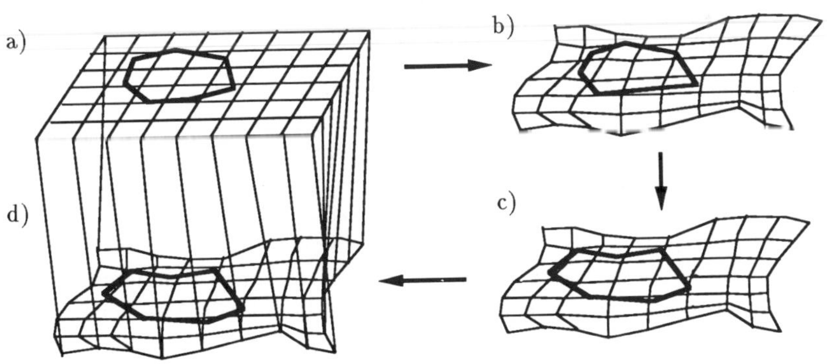

Fig. 31. Grids under deformation.

Like Delaunay interpolation, interpolation by grids implicitly follows a nearest neighbor heuristics. An improvement can be achieved by deformation, as follows (Fig. 31):

Algorithm Grids Under Deformation.

Input. Contours in two parallel planes.

Output. An interpolation of the given contours, if possible.

Procedure.

```
BEGIN
  fix a grid for the given contours of the first plane;
  determine a deformation for the first plane so that its contours
  match well with those of the second plane;
  apply the deformation to the grid of the first plane;
  insert the contours of the second plane into the deformed grid;
  construct the deformed spatial grids by connecting corresponding points
  of the original grid in the first plane and the deformed grid
  of the second plane;
  IF the deformed spatial grid has no twisted cells
  THEN
    calculate the surface of the solid by the marching square approach
  ELSE the approach does not yield a result
END.
```

The recognition of twisted spatial grid cells can be reduced to the test of tetrahedrons by subdividing the octahedral cells into tetrahedrons. A octahedral cell is accepted as nontwisted if all its tetrahedrons are not twisted. Again, in case of twisted cells the introduction of intermediate layers may help.

10 Acquisition of Data

In most applications, the contours are taken over into the computer as raster images. The contours are derived from the raster images by *segmentation*. The result of segmentation is a description of those regions of the image belonging to the investigated object. A possible description is a binary image representing the pixels that belong to the object by 1, and the others by 0. Another possibility is the description of the boundary of the object by a chain of pixels. Accordingly, *area-oriented* and *edge-oriented* algorithms of segmentation can be distinguished. In image processing numerous methods of segmentation were suggested which can be applied dependent on the type of image acquisition.

Most of the solutions for the SFCSC problem in this contribution assume a boundary representation as input. Algorithms are known for converting the area representation into the boundary representation (e.g. Pavlidis [53]). The pixel chain of a boundary representation may be seen as a polygonal chain whose vertices are the centers of the pixels. By point reduction (Sect. 4.1) chains with less vertices may be derived from them. For smooth interpolation, interpolating

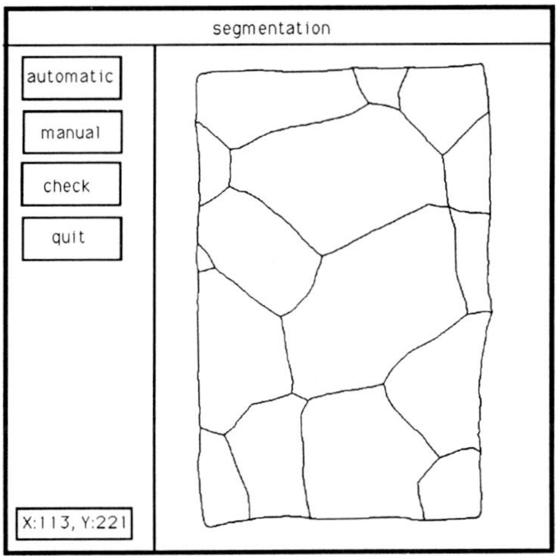

Fig. 32. User interface of a semi-automatic segmentation program.

and approximating curves of higher order may be used (e.g. Pavlidis [53], Farin [15]).

Several approaches of tomography yield a stack of images which are differently translated, rotated or even distorted. Giertsen, Halvorsen, Flood [22] describe a method of reconstruction of the correct location if the incorrect location resulted from rotation and translation. More serious distortions may be treated by the scattered data approach of deformation outlined in Sect. 4.3.2. As control points, structures visible in the images whose relative position from slice to slice is known may be used. Deformation may be applied on the original raster images or on the extracted contours.

If no points of reference are known, the reconstructed solid may be eventually smoothed globally if it is known from the application that the surface must be smooth. For example, the zig-zag artifacts of Fig. 4 may be eliminated in this way.

Raster images can also be used for the definition of shape criterions for interpolation. In image sequence analysis, several approaches for the assignment of corresponding points or regions in consecutive images were developed (Nagel [47, 48]). They can be used for assignment in the case of spatial contour interpolation, too. An advantage may be that the raster images contain more information than the segmented images and thus may give additional hints for the correct topological and geometrical assignment.

11 Interaction

As we have seen, the solution of the SFCSC problem usually is not unique. Thus the user has to express in some form which of the interpolations are useful. In

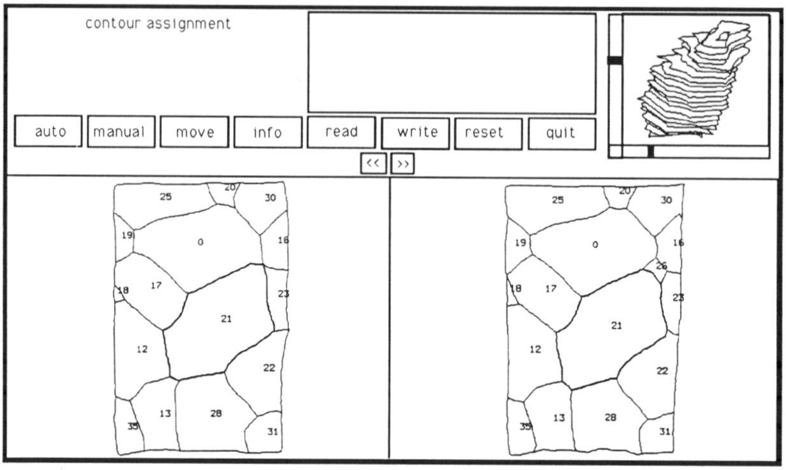

Fig. 33. User interface of a semi-interactive contour assignment program.

the preceding survey, numerous heuristics for this purpose were presented. A further possibility is interactive selection. One extreme case is that the user develops the interpolation completely "by hand". More desirable, however, is that the work is mainly done by the computer and that the user only interacts in critical situations or for post-processing. In the following interactive programs for segmentation, contour assignment, deformation, and triangulation are sketched.

11.1 Segmentation

Figure 32 shows the view on an interactive segmentation program, the user has when calling the program together with the file name of an image. The purpose of the program is to construct and manipulate a data structure describing the contours in the image by polygonal chains. It combines an algorithm of contour following with the possibility of explicit interactive input of vertices and edges.

In the working window the cross sectional image can be seen. On the left several function keys are arranged. The choice of `auto` starts the algorithm of contour following which gets a starting point by mouse click at the corresponding location of the working window. Edges can be drawn interactively using the mouse after choosing `manual`. The button `check` calls procedure testing whether the contours are indeed-simply closed and non-overlapping.

11.2 Topological Assignment

As a typical example of interactive topological assignment we use the program developed in the liver project. This program combines a heuristic assignment procedure with the possibility of interactive correction. Figure 33 shows the corresponding user interface. The two lower and larger working windows show the contours belonging to two consecutive cutting planes. The regions defined by the contours are labeled. Regions of different cutting planes belonging to the same three-dimensional cell have the same label. With `<<` and `>>` the stack of slices can be paged through up and down. In the upper right window the spatial representation of a single cell can be inspected supported by the possibility of arbitrary rotation. The single cell is selected by clicking one of its 2D-facets in one of the working windows in the manual mode. The information window describes the current mode. The buttons `auto` and `manual` are used for starting the automatic assignment procedure and for switching into the interactive mode. The heuristics of assignment used in the liver example is to assign contours whose centers are nearest neighbors and whose areas differ less than a given bound. Two facets of neighboring cutting planes are interactively assigned by first selecting a facet in one plane and then the corresponding facet in the other plane. A facet is selected by a mouse click in its inner. The mode `move` allows the free movement of the slices to correct possible distortions of the stack. The button `reset` deletes the current assignment of contours.

In general, the heuristics delivering exactly one assignment might be replaced by a procedure enumerating a whole family of possible solutions. Then the user should have a browsing facility for selecting the most acceptable alternative. There are numerous unsolved questions related to this. For example, the number of enumerated solutions should be not too large since selection might then take more time then direct modeling. It should be possible to interactively control the process of enumeration so that more likely cases are presented first.

11.3 Deformation

Figure 34 shows the interactive interface of a program implementing the deformation by the scattered data interpolation of Sect. 4.6.2. Its purpose is to specify the pairs of control points of a deformation. Besides some menus it consists of two large neighboring rectangular regions. The left region shows the image before deformation, in the example a digitized cross section of a blossom. The right region displays the goal image, that is the next slice which in our example is the same cross section at a later temporal state. By consecutively choosing a point in the left and a point in the right region through mouse clicks, pairs of an original point and its image under the deformation are defined. The pairs of control points are displayed using dashed lines. By paging cyclically through the lines with `<` and `>`, any of them can be selected and removed with the menu item `delete`. Note that the control points need not to lie on the given contours. The calculation of the deformation is initiated by selecting the menu item `transform`. The resulting deformed picture is then displayed in the right region, superimposed over the second image (cf. Fig. 34, bottom). The process of control point specification and display of the deformation can be iterated until a satisfiable shape is reached.

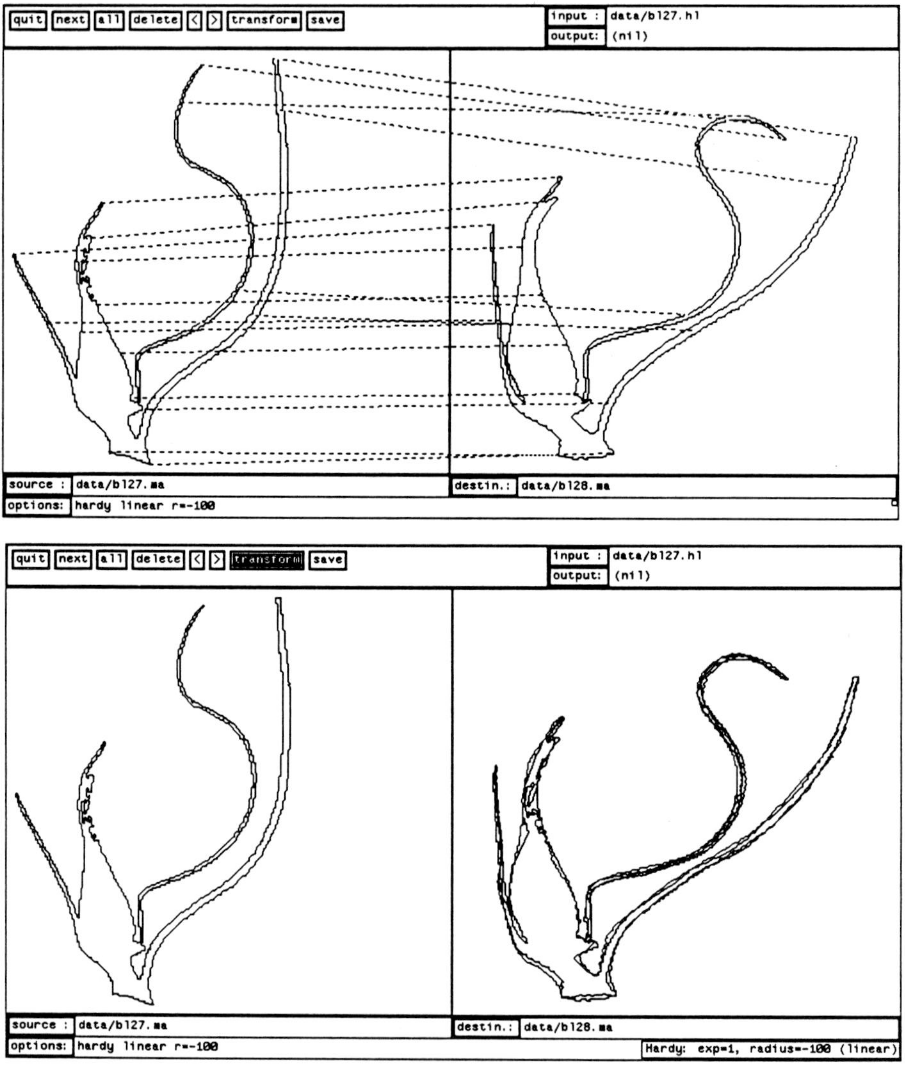

Fig. 34. Deformation of a cross section of a blossom. The upper picture shows two states of evolution of the blossom and an interactive assignment of the contours. In the right region of the picture below the original second state and the deformation of the first state calculated form the assignment are displayed simultaneously after selecting `transform`.

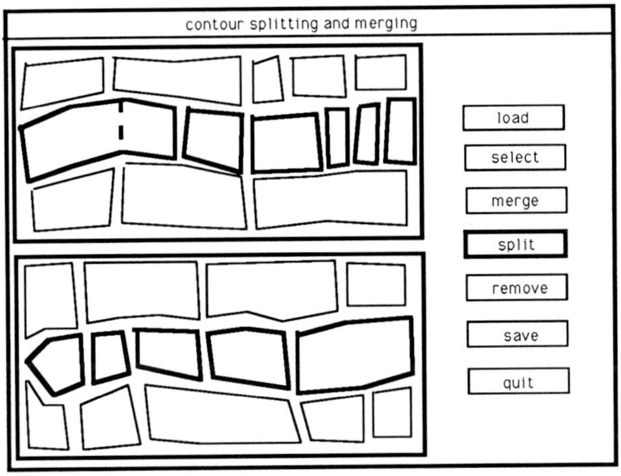

Fig. 35. Interactive contour splitting and merging.

The program may be augmented by an automatic assignment procedure based on shape criteria. This procedure may be started from the interactive environment and its assignment should be correctable interactively afterwards.

11.4 Triangulation

Surface triangulations without additional points consists of two steps, reduction on cylindrical situations by contour splitting and merging, and designing cylindrical triangulations.

Figure 35 shows the suggestion of a user interface for contour splitting and merging. It consists of two display areas showing the contours of two consecutive planes. It is assumed that the contours are already connected by an assignment graph. A connected component can be selected by a mouse click. The selected component is drawn fat. By choosing the menu `splitting` or `merging` the splitting or merging mode is reached. In this mode the at most two line segments required for these operations can be drawn directly in the windows. In the figure, an inserted splitting segment is drawn dashed. In the remove mode reached by `remove` the line segments belonging to a single split or merge operation can be removed by selecting them in the drawing area. Finally, the result can be saved.

The user interface of an interactive program for cylindrical triangulation may look like as shown in Fig. 36. For this program it is assumed that contour splitting and merging is already performed so that its application is restricted on cylindrical components of the assignment graph. Initially a component is loaded by selecting it interactively in a separate dialog not shown here which resembles that of contour selection in Sect. 11.2. Figure 36 shows two display areas separated by a menu area. The left area shows a 3D-representation of the current triangulation. By selecting suitable menus this can be rotated and scaled

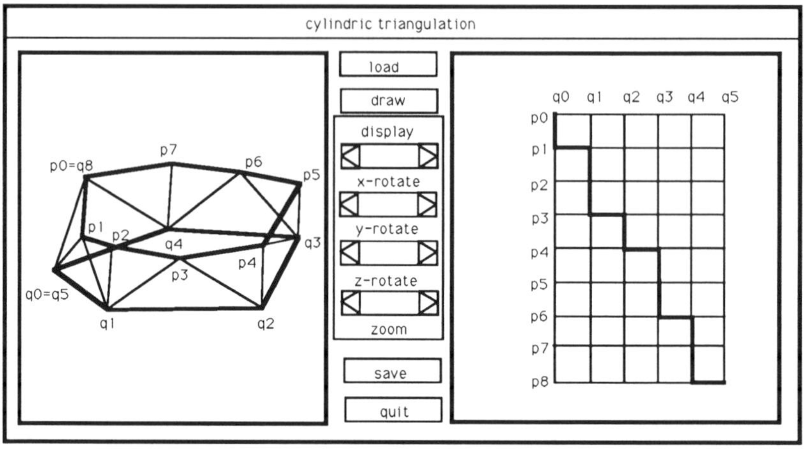

Fig. 36. Interactive cylindric triangulation.

and thus inspected from arbitrary views. The right area shows the toroidal graph belonging to the current triangulation. In the drawing mode obtained by `draw` the path drawn fat in the grid can be modified. The modification is immediately echoed in the left area thus offering an immediate impression of the actions performed by the user. The final result can be stored by `save`.

12 Concluding Remarks

Besides a survey of known approaches for solving the SFCSC problem several new ideas which were not investigated in detail yet were presented in this chapter. Following the organization of the material, this first concerns the enumeration of assignment graphs which should be extended to more general cases. A particular problem is the enumeration of just those assignment graphs which allow at least one penetration free triangulation. More sophisticated strategies than enumerating all topologically valid graphs and testing them for possible penetration-free triangulations are required in order to minimize the number of penetration tests to be performed. The goal is to test each pair of triangles at most once and to avoid unnecessary tests.

Most of the heuristics expressing quality of triangulation were applied only on the cylindrical case in the past. The question is for efficient algorithms solving e.g. the optimization problems also for the saddle point case.

The idea of pyramidal extrapolation is very preliminary. No efficient algorithms were developed yet.

Techniques of deformation were suggested in different contexts of interpolation, e.g. for improving surface triangulations by shape deformation, for introducing intermediate layers, for improving Delaunay interpolation, and for grid interpolation. Not all details are solved now and not all these suggestions are implemented yet.

References

[1] S. Abramowski, H. Müller (1991). *Geometric Modeling (in German).* B.I. Wissenschaftsverlag.

[2] H. Alt, B. Behrends, J. Blömer (1991). *Approximate Matching of Polygonal Shapes.* In *Proceedings of the 7th ACM Symposium on Computational Geometry*, pp. 186–193.

[3] H. Alt, M. Godau (1992). *Measuring the Resemblance of Polygonal Curves. Extended Abstract (Personal Communication).*

[4] E. Artzy, G. Frieder, G.T. Herman (1981). *The Theory, Design, Implementation and Evaluation of a Three-Dimensional Surface Detection Algorithm. Computer Graphics*, 15(3):1–24.

[5] R.E. Barnhill, G. Birkhoff, W. Gordon (1973). *Smooth Interpolation Over Triangles. J. Approx. Theory*, 8:114–128.

[6] J.D. Boissonnat (1988). *Shape Reconstruction from Planar Cross Sections. Computer Vision, Graphics and Image Processing*, 44:1–29.

[7] A. Bowyer (1981). *Computing Dirichlet Tessellations. The Computer Journal*, 24:162–166.

[8] L.-W. Chang, H.-W. Chen, J.R. Ho (1991). *Recontruction of 3D Medical Images: A Nonlinear Interpolation Technique for Reconstruction of 3D Medical Images. CVGIP: Graphical Models and Image Processing*, 53:382–391.

[9] H.N. Christiansen, T.W. Sederberg (1978). *Conversion of Complex Contour Line Definition into Polygonal Element Mosaics. Computer Graphics*, 12(3):187–192.

[10] P.N. Cook, S. Batnitsky (1981). *Three-Dimensional Recontruction From Serial Sections for Medical Applications. Proceedings of the 14th Hawaii International Conference on System Sciences*, 2:358–389.

[11] J. Duchon (1977). *Splines Minimizing Rotation Invariant Semi-Norms in Sobolev Spaces.* In W. Schempp, K. Zeller, editors, *Constructive Theory of Functions of Several Variables, Lecture Notes in Mathematics, Vol. 571*, pp. 85–100. Springer-Verlag, Berlin.

[12] N. Dyn, D. Levin, J.A. Gregory (1990). *A Butterfly Subdivision Scheme for Surface Interpolation with Tension Control. ACM Transactions on Graphics*, 9:160–169.

[13] M. Eck (1991). *Methods of Interpolation for the Reconstruction of 3D Surfaces from Planar Cross Sections (in German). CAD und Computergraphik*, 13(5):109–120.

[14] A.B. Ekoule, F.C. Peyrin, C.L. Odet (1991). *A Triangulation Algorithm from Arbitrary Shaped Multiple Planar Contours. ACM Transactions on Graphics*, 10:182–199.

[15] G.E. Farin (1988). *Curves and Surfaces for Computer Aided Geometric Design.* Academic Press, San Diego.

[16] R. Farwig (1986). *Rate of Convergence of Shepard's Global Interpolation Formula. Mathematics of Computation*, 46(174):577–590.

[17] I. Faux, M. Pratt (1979). *Computational Geometry for Design and Manufacture.* Ellis Horwood Limited (John Wiley & Sons), Chichester.

[18] R. Franke, G. Nielsen (1980). *Smooth Interpolation of Large Sets of Scattered Data. Int. Journal for Numerical Methods in Engineering*, 15:1691–1704.

[19] M. Fréchet (1906). *Sur quelques points du calcul fonctionnel. Rendiconti del Circolo Mathematico di Palermo*, 22:1–74.

[20] H. Fuchs, Z. Kedem, S.P. Uselton (1977). *Optimal Surface Reconstruction from Planar Contours. CACM*, 20:693–702.

[21] S. Ganapathy, T.G. Dennehy (1982). *A new general Triangulation Method for Planar Contours. Computer Graphics*, 16(3):69–75.

[22] C. Giertsen, A. Halvorsen, P.R. Flood (1990). *Graph-Directed Modelling from Serial Cross Sections. The Visual Computer*, 6:284–290.

[23] R.C. Gonzalez, P. Wintz (1987). *Digital Image Processing, 2nd ed.* Addison-Wesley, Reading, Mass.

[24] D. Gordon, J.K. Udupa (1989). *Fast Surface Tracing in Three-Dimensional Binary Images. Computer Vision, Graphics and Image Processing*, 45:196–212.

[25] W.J. Gordon, J.A. Wixom (1978). *Shepard's Method of "Metric Interpolation" to Bivariate and Multivariate Interpolation. Mathematics of Computation*, 32(141):253–264.

[26] L. Guibas, D. Salesin, J. Stolfi (1989). *Epsilon Geometry: Building Robust Algorithms for Imprecise Computations.* In *Proceedings 5th ACM Symp. on Computational Geometry*, pp. 208–217.

[27] L. Guibas, J. Stolfi (1985). *Primitives for the Manipulation of General Subdivisions and Computation of Voronoi Diagrams. ACM Transactions on Graphics*, 4:74–123.

[28] R.L. Hardy (1971). *Multiquadric Equations of Topography and other Irregular Surfaces. J. Geophys. Res.*, 76:1905–1915.

[29] G.T. Herman, D. Webster (1983). *A Topological Proof of a Surface Tracking Algorithm. Computer Vision, Graphics and Image Processing*, 23:162–177.

[30] C.M. Hoffmann, J.E. Hopcroft, M.S. Karasick (1988). *Towards Implementing Robust Geometric Compuations.* In *Proceedings 4th ACM Symp. on Computational Geometry*, pp. 106–117.

[31] C.M. Hoffmann, J.E. Hopcroft, M.S. Karasick (1989). *Robust Set Operations on Polyhedral Solids. IEEE Computer Graphics & Appl.*, 9(11):50–59.

[32] K. Kaneda, K. Harada, E. Nakamae, M. Yasuda, A.G. Sato (1987). *Reconstruction and Semi-Transparent Display Method for Observing Inner Structure of an Object consisting of Multiple Surfaces.* In T.L. Kunii, editor, *Computer Graphics 1987*, pp. 367–380. Springer-Verlag, New York.

[33] N. Kehtarnavaz (1988). *A Syntactic Semantic Technique for Surface Reconstruction from Cross Sectional Contours. Computer Vision, Graphics and Image Processing*, 42:399–409.

[34] N. Kehtarnavaz, R.J.P. De Figueiredo (1988). *A Framework for Surface Reconstruction from 3D Contours. Computer Vision, Graphics and Image processing*, 42:32–47.

[35] E. Keppel (1975). *Approximating Complex Surfaces by Triangulation of Contour Lines. IBM J. Res. Devel.*, 19:2–11.

[36] Y. L. Kergosien (1991). *Generic Sign Systems in Medical Imaging. IEEE Computer Graphics & Appl.*, 11(9):46–65.

[37] A. Klingert, W. Straub (1992). *A Robust and Efficient Triangulation Algorithm for Contours on Parallel Planes.* In *Report 11/92.* Institut für Betriebs- und Dialogsysteme, Universität Karlsruhe.

[38] F.P. Kuhl, C.R. Giardina (1982). *Elliptic Fourier Features of a Closed Contour. Computer Graphics and Image processing*, 18:236–258.

[39] P. Lancaster, K. Salkauskas (1981). *Surfaces Generated by Moving Least Squares Methods. Mathematics of Computation*, 37(155):141–158.

[40] W.C. Lin, S.Y. Chen (1989). *A New Surface Interpolation Technique for Reconstructing 3D Objects From Serial Cross Sections. Computer Vision, Graphics and Image Processing*, 48:124–143.

[41] W.C. Lin, C.C. Liang, C.T. Chen (1988). *Dynamic Elastic Interpolation for 3D Medical Image Recognition from Serial Cross Sections. IEEE Trans. Med. Imaging*, 7:225–232.

[42] W.E. Lorensen, H.E. Cline (1987). *Marching Cubes: A High Resolution 3D Surface Construction algorithm. Computer Graphics*, 21(4):163–169.

[43] K. Mehlhorn (1984). *Data Structures and Algorithms 3.* Springer-Verlag, Berlin.

[44] J. Milnor (1963). *Morse Theory.* Princeton University Press, New Jersey.

[45] M. Morse, S.S. Cairns (1969). *Critical Point Theory and Differential Topology.* Academic Press, San Diego.

[46] H. Müller, M. Stark . *Adaptive Generation of Surfaces in Volume Data. accepted for The Viusal Computer.*

[47] H.H. Nagel (1985). *Analysis and Interpretation of Image Sequences (in German). Informatik-Spektrum*, 8:201–214 and 312–327.

[48] H.H. Nagel (1986). *Image Sequences - Ten (Octal) Years - From Phenomenology towards a Theoretical Foundation.* In *Proc. 8th Int. Conf. Pattern Recognition*, pp. 1174–1185.

[49] B.K. Natarajan (1991). *On Piecewise Linear Approximations to Curves.* In *Report HPL-91-36.* Hewlett-Packard Laboratories, Palo Alto.

[50] G.M. Nielson (1979). *The Side-Vertex Method for Interpolation in Triangles. J. Approx. Theory*, 25:318–336.

[51] G.M. Nielson (1987). *A Transfinite, Visually Continuous, Triangular Interpolant.* In G.E. Farin, editor, *Geometric Modeling: Algorithms and New Trends*, pp. 235–246. SIAM.

[52] T. Ottmann, G. Thiemt, C. Ullrich (1987). *Numerical Stability of Geometric Algorithms.* In *Proceedings 3rd ACM Symp. on Computational Geometry*, pp. 119–125.

[53] T. Pavlidis (1982). *Algorithms for Graphics and Image Processing.* Springer-Verlag, Berlin.

[54] G. Reeb (1946). *On the Singular Points of a Completely Integrable Pfaff Form or a Numerical Function (in French). Comptes Rendus Acad. Sciences Paris*, 222:847–849.

[55] M.L. Rhodes (1979). *An Algorithmic Approach to Controlling Search in Three-Dimensional Image Data. Computer Graphics*, 13(2):134–142.

[56] D. Ruprecht, H. Müller (1991). *Free Form Deformation with Scattered Data Interpolation Methods.* In *Report 41/91, Institut für Informatik, Universität Freiburg, Germany.*

[57] P.T. Sander, S.W. Zucker (1987). *Tracing Surfaces for Surface Tracing.* In *IEEE International Conference on Computer Vision.*

[58] H.-J. Scherberger, H. Müller, O. Leder, H. Kurz (1992). *Visualization of Liver Tissue (in German). Informatik - Forschung und Entwicklung*, 7.

[59] M. Segal (1990). *Using Tolerances to Guarantee Valid Polyhedral Modeling Results. Computer Graphics*, 24(4):105–114.

[60] M. Shantz (1981). *Surface Definition for Branching Contour-Defined Objects. Computer Graphics*, 15(2):242–270.

[61] D. Shepard (1968). *A Two Dimensional Interpolation Function for Irregular Spaced Data.* In *ACM National Conference*, pp. 512–524.

[62] Y. Shinagawa, T.L. Kunii (1991). *Constructing a Reeb Graph Automatically from Cross Sections. IEEE Computer Graphics & Appl.*, 11:44–51.

[63] Y. Shinagawa, T.L. Kunii (1991). *The Homotopy Model: A Generalized Model for Smooth Surface Generation from Cross Sectional Data. The Visual Computer,* 7:72–86.

[64] Y. Shinagawa, T.L. Kunii, Y.L. Kergosien (1991). *Surface Coding Based on Morse Theory. IEEE Computer Graphics & Appl.*, 11(9):66–78.

[65] Y. Shinagawa, T.L. Kunii, Y. Nomura, T. Okuno, M. Hara (1989). *Reconstructing Smooth Surfaces From a Series of Contour Lines Using a Homotopy.* In R.A. Earnshaw, B. Wyvill, editors, *New Advances in Computer Graphics*, pp. 147–161. Springer-Verlag, Tokyo.

[66] A. Sunguroff, D. Greenberg (1978). *Computer Generated Images for Medical Applications. Computer Graphics*, 12(3):196–202.

[67] R. Tello, R.W. Mann, D. Rowell (1986). *Scheme for 3D Reconstruction of Surfaces from CT and MRI Images of the Human Body.* In *IEEE Conf. 1986*, pp. 216–219.

[68] K.D. Tönnies (1989). *Surface Triangulation by Linear Interploation in Intersecting Planes.* In *SPIE Vol. 1137, Science and Eng. of Medical Imaging*, pp. 98–105.

[69] J.K. Udupa, V.G. Ajjanagudde (1990). *Boundary and Object Labelling in Three-Dimensional Space. Computer Vision, Graphics and Image Processing*, 51:355–369.

[70] P.F. Watson (1981). *Computing the n-Dimensional Delaunay Tessellation with Application to Voronoi Polytopes. The Computer Journal*, 24:167–172.

[71] I. Weiss (1988). *3D Shape Representation by Contours. Computer Vision, Graphics and Image Processing*, 41:80–100.

[72] S.B. Xu, W.X. Lu (1988). *Surface Reconstruction of 3D Objects in CT. Computer Vision, Graphics and Image Processing*, 44:270–278.

Modeling and Visualizing Volumetric and Surface-on-Surface Data

Gregory M. Nielson

An introduction and overview of methods for modeling and visualizing trivariate data is presented. One variable is identified as being dependent on three other, independent variables. Data of this type arises often in practical problems of science and engineering The modeling portion is concerned with finding a mathematical relationship which represents the data. Visualization is concerned with using computer generated images to convey information so that the user can learn about the relationship. The discussion on modeling methods will concentrate on the two application areas of surface-on-surface and volumetric data. Volumetric data covers the case where the independent data values represent points in a three dimensional domain. The surface-on-surface case is where the independent data is restricted to lie on some surface such as the earth or the surface of an airplane wing. Mathematical models with volume or surface domains are developed, discussed and compared. Graphing and visualizing trivariate relationships is quite challenging. Extending methods which have proven to be successful in other situations is a starting point. Several resulting modifications and extensions for both the case of surface-on-surface and volumetric data are presented, discussed and compared. This includes some interactive techniques, isosurface algorithms, volume rendering and hypersurface projection graphs. Some new interrogation methods are also presented.

1 Introduction

In this chapter we present an introduction and overview of methods for modeling and visualizing trivariate data. We are concerned with the case where a single scalar variable is identified as being dependent on three other, independent variables. Data of this type arises often in practical problems of science and engineering. Often the three independent variables represent a position in three dimensional space or a position in a two dimensional domain at a particular time. In the next section, we describe several data sets which are the result of either measurements or a simulation based upon some mathematical model. It turns out that the data sets fall into two categories. One category is called volumetric data where the independent variables are free to be at any position in a contiguous, three dimensional domain. The second requires that the independent data sites lie on a two dimensional surface. This is called surface-on-surface data. After the section on data sets, the remaining portion of the chapter is divided into two main sections, one is concerned with visualization and the other with modeling. The modeling portion is concerned with finding a mathematical function which represents the relationship implied by the data. Visualization is concerned with using computer generated images to convey information so that scientists can understand relationships implied by the data.

Graphing and visualizing multivariate relationships is quite challenging. Extending methods which have proven to be successful in other situations is a starting point. Several resulting modifications and extensions for both the case of surface-on-surface and volumetric data are presented, discussed and compared. This includes some interactive techniques, isosurface algorithms, volume rendering and hypersurface projection graphs. We also discuss some interrogation methods for trivariate data which are now just beginning to be developed. Gaussian curvature has proven to be useful as a surface interrogation tool. This concept is extended to the case of volume domains and examples are shown.

Mathematical modeling is a very rich and broad topic which permeates most of science and engineering. Here we are mainly interested in modeling multivariate data so as to facilitate the visualization of this data. Consequently the form of the model is selected so that it can be effective and efficient for a wide variety of data sets rather than to take into account the underlying phenomenon of the physical system which produced the data.

2 Data

We describe eight different data sets. These are data sets that have been collected by us or provided by other scientists. We have tried to select the data sets included here so that they are typical and representative in as many ways as possible. All of the data sets have the property that there are three independent variables and a single, scalar dependent variable and therefore we can use $(x_i, y_i, z_i; F_i)$, $i = 1, \ldots, N$, to represent them. The three independent variables are x_i, y_i, z_i and F_i is the dependent variable. In some cases the three independent variables represent a position in space and in some other cases, they represent time and a two dimensional position. Each of the data sets falls into one of two categories. The first category is the case where the independent variables are unrestricted and allowed to lie at any position in a contiguous region of three dimensional space. We call this category, "volumetric data". The second is the case where the independent data is restricted to lie on some two dimensional subset of three dimensional space. This category is called "surface-on-surface data".

2.1 Pressure on a Wing Example

The first data set is related to the analysis of airflow over the surface of a wing. The data is the result of a simulation and is used in the design of a wing. The simulation involves the application of the key equations of fluid dynamics. These are the Navier-Stokes equations

$$r\left(\frac{\partial \mathbf{V}}{\partial t} + (\mathbf{V}\cdot\nabla)\mathbf{V}\right) = -\nabla p + m\nabla^2\mathbf{V} + \mathbf{F}, \tag{2.1}$$

where $\mathbf{V} = (u, v, w)$ represents the velocity vector and $p = p(x, y, z)$ is a scalar-valued function representing pressure. The scalar constant r is fluid density and m is the dynamic viscosity. The external forces are $\mathbf{F} = (X, Y, Z)$. In addition to these equations, there is the continuity equation,

$$\text{div } \mathbf{V} = \nabla\cdot\mathbf{V} = \frac{\partial u}{\partial x} + \frac{\partial v}{\partial y} + \frac{\partial w}{\partial z} = 0. \tag{2.2}$$

The overall problem is to solve for $\mathbf{V}$ and p as functions of time t and the three spatial variables (x, y, z). Boundary conditions for the wing design problem must be specified on the surface of the airplane. They are used to reflect the condition that the velocity of the flow is equal to the velocity of the airplane at this interface and at distances far from the airplane, the velocity must approximate the free stream conditions. In general, boundary conditions are extremely important. In order to include them properly, it is often necessary to have a mathematical representation of the boundary, which, in many physical models, is some geometric object. In this example, this is accomplished through the use of parametric surface patches. The surface of the wing is represented by a collection of patches of the form $P(u,v) = (X(u,v), Y(u,v), Z(u,v))$, $0 \le u \le 1$, $0 \le v \le 1$. Even though the numerical solution of the discretized versions of equations (2.1) and (2.2) will yield both the pressure, p, and the velocity vector, $\mathbf{V}$, at several locations, for this example we are only interested in the distribution of the pressure over the surface of the wing at one instance of time. In general terms the data (see Fig. 1) can be represented as $(x_i, y_i, z_i; F_i)$, $i = 1, \ldots, N$. But it is really more restricted than this type of representation indicates because there is some structure to the dependent data. We know that each data site lies on the surface of the wing and in this case we know that these values were obtained by evaluating the wing patch definition on a regular grid in the parameter domain. In other words, the data has the form $(x_{ij}, y_{ij}, z_{ij}; F_{ij})$, with $(x_{ij}, y_{ij}, z_{ij}) = W(u_i, v_j)$ where W is the parametric representation of the entire wing and u_i, $i = 1, \ldots, N_u$ and v_j, $j = 1, \ldots, N_v$ is a rectangular grid in the parameter domain.

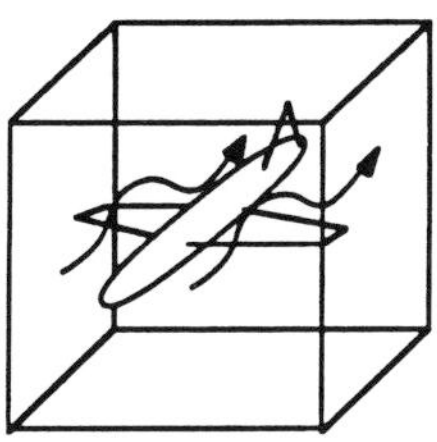

Location			Pressure
-132.1	38.5	6.1	0.164
-128.3	38.5	6.6	0.119
-116.8	38.6	7.5	0.067
⋮	⋮	⋮	⋮

Fig. 1. Pressure over a Wing Data courtesy of NASA, Ames.

2.2 CAT Scan Data Example

This next data set is typical of data produced by scanning devices in the medical field. The particular data that is indicated by the table of Fig. 2 was provided to us by the University of Pennsylvania. Each data sample $(x_i, y_j, z_k; D_{ijk})$ represents a density measurement D_{ijk} at the position (x_i, y_j, z_k) in three dimensional space. As is typical for medical data, the measurements are at locations on a uniform cuberille grid. This means there is a uniform partition in each of the three spatial variables. For this example we have

$$x_i = min_x + (i-1)D_x, \ i = 1, \ldots, 64$$

$$y_j = min_y + (j-1)D_y, \ j = 1, \ldots, 64$$

$$z_k = min_z + (k-1)D_z, \ k = 1, \ldots, 68$$

and $D_x = \dfrac{max_x - min_x}{63}$, $D_y = \dfrac{max_y - min_y}{63}$, $D_z = \dfrac{max_z - min_z}{67}$.

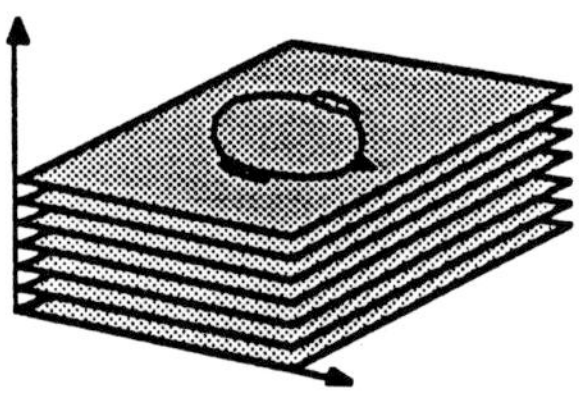

X_i	Y_j	Z_k	Density
0.000	0.000	0.000	243
0.000	0.000	0.015	175
⋮	⋮	⋮	⋮
0.000	0.000	1.000	186
0.000	0.016	0.000	187
⋮	⋮	⋮	⋮

Fig. 2. CAT Scan Data.

2.3 Precipitation on Earth Data Example

In this example, we are given yearly rainfall measurements at various measurement stations located throughout the world. These locations are mostly at or near major cities. For all intents and purposes, these locations can be considered to be random points on a sphere of radius one located at the origin. Thus we have the data, $(x_i, y_i, z_i; R_i)$, $i = 1, ..., N$, with the restriction that $x_i^2 + y_i^2 + z_i^2 = 1$. Another way to include the restriction that the data sites are on a unit sphere is to specify them with longitude and latitude angles so that $(x_i, y_i, z_i) = (\sin(\Theta_i)\cos(\Phi_i), \sin(\Theta_i)\sin(\Phi_i), \cos(\Theta_i))$ for some values of Θ_i and Φ_i.

Longitude	Latitude	Rainfall
43 19' 34"	23 36' 13"	14.6"
21 35' 59"	45 09' 36"	23.6"
43 19' 34"	23 36' 13"	14.6"
⋮	⋮	⋮

Fig. 3. Precipitation on Earth Data courtesy of US Weather Bureau.

2.4 Temperature Analysis Data Example

In the three previous examples, the independent variables represent a position in physical, three dimensional space. In this example, we have measurements that have been taken on a two dimensional domain at fixed periods of time. The daily high temperatures at 35 cities throughout the state of Arizona have been tabulated at weekly intervals for a period of one year. This implies some structure to the independent data which can be indicated in the representation $(t_i, x_j, y_j; F_{ij})$, $i = 1, \ldots, 52$, $j = 1, ..., 35$.

Time	Location		Temp.
Jan 2	2.16	-1.01	39.0
Jan 9	-1.74	2.50	62.0
Jan 16	-0.85	-1.46	65.0
⋮	⋮	⋮	⋮

Fig. 4. Arizona Temperature Analysis Data courtesy of D. Lane and J. Tvedt.

2.5 Flame Data Example

The next data set represents gas concentrations from an acoustically-driven forced flow. Because of the method of collection, the data has considerable structure. For each slice in the z-variable, the flame is scanned and samples are obtained at positions which emanate radially from the origin of a laser. This yields data that can be represented as

$$(r_i \cos(\Theta_j), r_i \sin(\Theta_j), z_k; C_{ijk})$$

where $r_i = \min_r + i\Delta r$, $\Theta_j = \min_\Theta + j\Delta\Theta$, $z_k = \min_z + k\Delta z$.

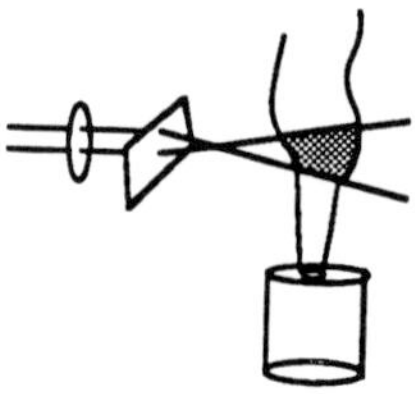

Location			Concentration
0.00	0.00	0.02	001
0.00	0.00	0.04	007
0.00	0.00	0.06	003
⋮	⋮	⋮	⋮

Fig. 5. Flame Data.

2.6 Well Log Data Example

In the geophysical sciences, it is quite common to collect measurements at various depths at several fixed locations. The locations (x_i, y_i) are usually positions on the earth and the depth may possibly vary from location to location. Also the total number of measurements per location may vary. This type of data can be represented as $(x_i, y_i, z_{ik}; M_{ik})$, $i = 1, \ldots, N$; $k = 1, \ldots, n_i$, where n_i is the number of measurements taken at location (x_i, y_i). If the number and value of the depth were the same for each location, then from a mathematical point of view this data would be similar to that described in Sect. 2.4.

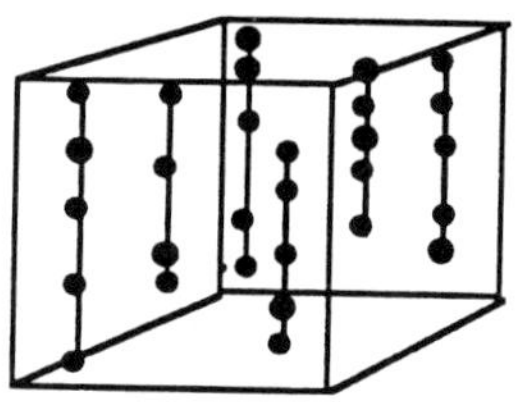

Location			Mineral
5.50	1.00	0.00	11.0
5.50	1.00	10.00	10.0
5.50	1.00	20.00	12.0
⋮	⋮	⋮	⋮

Fig. 6. Well Log Data.

2.7 Brain Data Example

The use of electroencephalograms (EEG) to visualize the activity of the brain has been used since the 1920's when Hans Berger first attached two electrodes to a scalp surface and visualized the first EEG on a cathode ray tube. Modern neuroscientists are interested in understanding the global activity of the brain by simultaneously analyzing the results of several electrodes placed over the scalp. The Brain Physics Group at Tulane University provided the data that is indicated in Fig 7. It represents voltage potential at 32 locations measured at one instance of time and so we have $(x_i, y_i, z_i; V_i)$, $i = 1, \ldots, 32$, with $(x_i, y_i, z_i) \in$ Scalp.

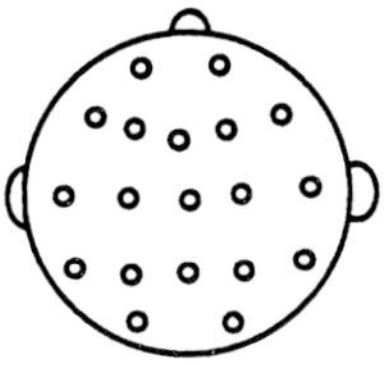

Location			Voltage
6.54	4.56	5.64	0.033
9.14	-3.14	1.38	0.086
9.45	2.12	1.19	0.310
⋮	⋮	⋮	⋮

Fig. 7. Brain data.

2.8 Spatial Sound Data Example

For this data set, sound level measurements have been taken at various locations in a room. The goal is to design and locate sound generating devices and sound absorbing objects so as to achieve an optimal distribution. The measurements are taken in a rather ad hoc manner. The technician attempts to get denser collections of measurements near the emitters and absorbers, but from a mathematical point of view it is probably best to consider the data to be completely random and so the data has the general representation $(x_i, y_i, z_i; F_i)$, $i = 1, \ldots, N$.

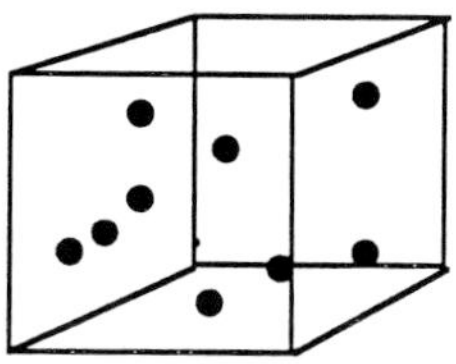

Location			Decibel
23.45	13.56	3.56	58.0
10.34	50.45	5.67	49.3
12.87	35.60	21.04	36.9
⋮	⋮	⋮	⋮

Fig. 8. Spatial sound data.

2.9 Summary of Data Examples

All of the data sets have independent data values that can be represented as points in 3D space and a single scalar value for the dependent variable. In some cases there is some structure or constraints on the independent data. A fundamental difference exists when the independent data is restricted to lie on a two dimensional surface. When the domain of the model is a 2D surface rather than a 3D region, the methods of defining and constructing the model are quite different. Also, methods for visualizing the models which result from the data are quite different. Data which has a model with a 2D surface domain is called surface-on-surface data. This includes the pressure over a wing data of Sect. 2.1. The surface domain in this case is the surface of the wing. The precipitation data of Sect. 2.3 is also of the category of surface-on-surface. The domain here is the surface of the earth. The brain data of Sect. 2.7 is surface-on-surface data with the scalp being the domain. The remaining examples fall into the category we call volumetric data. In these cases, the independent data sets lie in a contiguous region of 3D space.

Volumetric Data $(x_i, y_i, z_i; F_i)$, $i = 1, \ldots, N$

Well Log Data, Flame Data, Temperature Data,
Medical Data, Sound Data

Surface-on-Surface Data $(x_i, y_i, z_i; F_i)$, $(x_i, y_i, z_i) \in S$, $i = 1, \ldots, N$

Pressure on Wing, Rainfall Data, Brain Data

3 Visualization Methods

This section is concerned with methods for graphing and visualizing data of the type described in the previous section. It has two parts. The first is devoted to methods for visualizing volumetric data and the second is concerned with surface-on-surface data. Many of the methods that are generally available and which are described here assume that the data is available over a uniform grid, but as we have seen with the examples of the previous section, this is usually not the case. This is where the modeling of data comes in. A modeling function defined over the entire domain is determined so that it interpolates or approximates the data. This model can then be sampled on a uniform grid and this sampled data is used as input to a visualization tool. This important step of modeling is covered in the next section. In the present section, we are only concerned with visualizing the relationship and so, when necessary, we will assume that either the data has been provided over a uniform, cuberille grid or that some model is available for us to sample over a uniform grid.

3.1 Techniques for Visualizing Volumetric Models

We now discuss methods for visualizing volumetric data. The discussion is roughly in order of increasing complexity and decreasing possibility for interactive use.

3.1.1 Domain Decomposition Methods

The basic idea of this first technique, which is called "tiny cubes" is to place objects in the domain volume whose color (or some other attribute such as vibration or spinning speed) is determined by the value of the dependent variable F at the location of the object. The objects could be almost anything, but spheres and cubes are most typical. In the case of cubes, three resolution parameters are specified by the user: N_x, N_y and N_z. This leads to a total of $N_x N_y N_z$ color coded cubes to be displayed. In addition, the user specifies a value for the parameter, M, which controls the amount of open space and consequently the size of the cubes which are displayed. If we let the width, length and height of

index	R	G	B
$i = 0, 255$	i	$255 - i$	0
$i = 256, 511$	255	$i - 255$	0
$i = 512, 767$	255	255	$i - 512$

Fig. 9. Color table for example of Figure 11.

each "tiny" cube be denoted by $(\Delta x, \Delta y, \Delta z)$, then the lower-left-front corner of each cube is given by the coordinates

$$\mathbf{X}_i = \mathbf{X}_0 + (i-1)\Delta x(M+1), \ i = 1, ..., N_x$$

$$\mathbf{Y}_j = \mathbf{Y}_0 + (j-1)\Delta y(M+1), \ j = 1, ..., N_y$$

$$\mathbf{Z}_k = \mathbf{Z}_0 + (k-1)\Delta z(M+1), \ k = 1, ..., N_z$$

where $(\mathbf{X}_0, \mathbf{Y}_0, \mathbf{Z}_0)$ is the lower-left-front corner of the whole domain and

$$\Delta_x = \frac{x_{\max} - x_{\min}}{N_x(M+1) - M}, \ \Delta_y = \frac{y_{\max} - y_{\min}}{N_y(M+1) - M}, \ \Delta_z = \frac{z_{\max} - z_{\min}}{N_z(M+1) - M}.$$

The function value, F_{ijk} and the particular color table used will determine the color that is used at each vertex. The faces of the cubes are then Gouraud shaded. Example images are shown in Fig. A of the color illustrations of this chapter at the end of the book. An important feature of this method is the ability to interactively rotate the graph in order to obtain different views. This is necessary so that the kinetic depth effect can reveal the location of a particular "tiny cube". The ability of the workstation to interactively display the $6N_xN_yN_z$ polygons will limit the size of the resolution parameters; but this often will not be a problem since the resolution parameters are also limited by the complexity of the image. Resolution values beyond, say 15, yield images which are too complex to be understood.

The next method, which is called "vanishing cube", also associates a color with each data location, (x_i, y_j, z_k) and this color is based upon the value of the dependent variable, F_{ijk}, and the particular color table used. More precisely,

$$C_{ijk} = (R_{ijk}, G_{ijk}, B_{ijk}) = (R(F_{ijk}), G(F_{ijk}), B(F_{ijk}))$$

where $i = 1, \ldots, N_x$, $j = 1, \ldots, N_y$, and $k = 1, \ldots, N_z$ and the function notation $R(\ )$, $B(\ )$ and $G(\ )$ is used to represent the color code table. For example the table used for the top three images of Fig. A of the color illustrations of this chapter at the end of the book is shown in Fig. 9.

Once we have a color for each vertex, we can entirely color any of the planes parallel to the axes by using linear (or Gouraud) shading on each of the rectangles which comprise the plane. There are $N_xN_yN_z$ rectangles perpendicular to each axis for a total of $3(N_xN_yN_z)$ rectangles to be displayed. Of course, if we directly display these rectangles, all that will be viewed are those on the outer

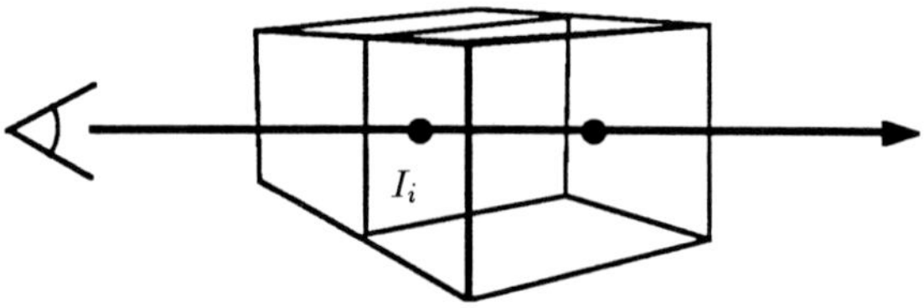

$$I := t_i I_i + (1 - t_i) I$$

Fig. 10. Transparency computations for vanishing cube method.

faces. In order to "see in" we compute an image based upon a simple model of transparency for the rectangles. The rectangles are all sorted by distance from the viewpoint and then displayed from back to front using an α-buffer. See Fig. A of the color illustrations of this chapter at the end of the book. The data for the bottom two images of Fig. A has been described in Sect. 2.5. While interaction is possible as far as varying the transparency factor, in order to rotate the graph in real time, the rectangles need to be sorted and this creates an extra computational burden which diminishes the possibility of interaction.

3.1.2 Slice Methods

The basic idea of slice-type methods is to simultaneously display three, two dimensional rectangular grid data sets, each one obtained by taking a slice through the domain by holding one of the independent variables constant. In function notation terms, we simultaneously display some type of graph of the three bivariate functions,

$$F_{\mathbf{x}}(y,z) = F(\mathbf{x},y,z), \;\; y_{min} \le y \le y_{max}; z_{min} \le z \le z_{max}$$

$$F_{\mathbf{y}}(x,z) = F(x,\mathbf{y},z), \;\; x_{min} \le x \le x_{max}; z_{min} \le z \le z_{max}$$

$$F_{\mathbf{z}}(x,y) = F(x,y,\mathbf{z}), \;\; x_{min} \le x \le x_{max}; y_{min} \le y \le y_{max}.$$

It is convenient to allow the user to interactively vary the fixed (but arbitrary) point $(\mathbf{x},\mathbf{y},\mathbf{z})$. One way to simulate a three dimensional locator, and move the mutual intersection point about using only an ordinary 2D-mouse, is a method called "triad mouse" described by Nielson and Olsen [23]. In the left portion of Fig. B of the color illustrations of this chapter at the end of the book, we show a color coded contour method for displaying the graphs of the rectangular grid. The data set is the one described in Sect. 2.5. The color table is shown in the color bar to the right.

The next slice type method, which is called "surface projection", uses a smooth shaded surface in order to display the three rectangular grid data sets. These three sets of surfaces could be located anywhere in the image, but it is convenient to have each of these graphs located on the face of a cube. In the example of Fig. B of the color illustrations, the values are scaled so that the minimum value is represented by a point on the cube and the maximum value is one unit in the direction normal to this plane.

3.1.3 Contour Methods

Contour plots for bivariate data are widely used. Terrain models from elevation data are often displayed with contour maps. Choropleths (color coded contour graphs) are used in connection with weather reports to show the distribution of temperature over a particular region. In this section we discuss some algorithms and techniques for computing the three dimensional analog of this type of visualization tool. For a trivariate model, $F(x, y, z)$, the isosurface is defined as

$$S_\alpha = \{(x, y, z) : F(x, y, z) = \alpha\}.$$

How this surface, S_α, is computed or approximated depends to a large extent upon the form of the model and the type of data which produced the model. In many algorithms, the data is contoured directly and so the model is only implicitly involved. It is typical to assume that the model, F, is defined in a piecewise manner over a decomposition of the domain. For example in the case of scattered trivariate data (x_i, y_i, z_i, F_i), $i = 1, \ldots, n$, we can first decompose the convex hull of $\{(x_i, y_i, z_i), i = 1, .., n\}$ into a collection of tetrahedra T_{ijkl} each having vertices p_i, p_j, p_k and p_l. We then assume that F is piecewise linear over each tetrahedron. If F is piecewise linear then the contour surface will be a polyhedron with facets consisting of triangles and planar quadrilaterals. For each tetrahedron, there are only three distinct cases to consider. See Fig. 11. If the dependent data values at all vertices are above (or below) the isovalue, then the contour surface does not intersect this cell. If there is one value above the contour level and three below, then the contour surface of this tetrahedron is a triangle. This is also the situation for the dual case of three vertices above and one below. If there are two values above and two below, then the contour surface is a planar quadrilateral. It is convenient to store this as two triangles so that the overall surface is represented as a collection of triangles. Computing the vertices is simple because of the linear variation along edges. For example a vertex V of S_α on the edge p_i to p_j is computed as

$$V = \left(\frac{\alpha - F_j}{F_i - F_j}\right) p_i + \left(\frac{F_i - \alpha}{F_i - F_j}\right) p_j.$$

If the data is located on a uniform grid, then there are some possible simplifications which can be used to speed up the whole process. These savings are mainly due to not having to compute the collection of indices which define the tetrahedra. The "marching cubes" method described in [11] process one voxel at a time. A voxel is just a "brick shaped" subset of the domain with vertices (x_a, y_b, z_c), $a = i, i+1$; $b = j, j+1$; $c = k, k+1$. See Fig. 12. The processing of each voxel includes the computation of triangles in the voxel which contribute to

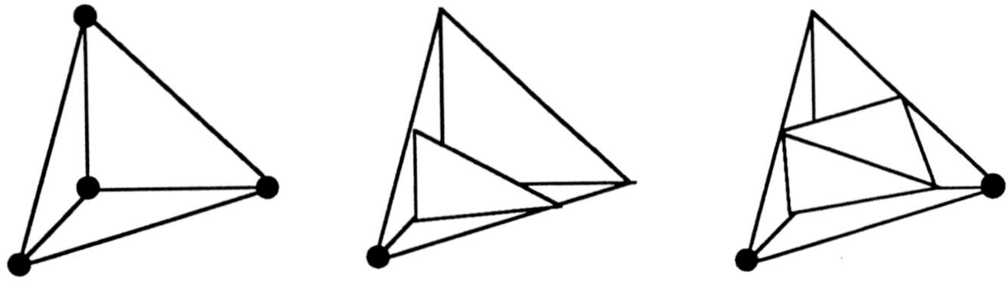

Fig. 11. The different cases for the contours for linear interpolation over a tetrahedron.

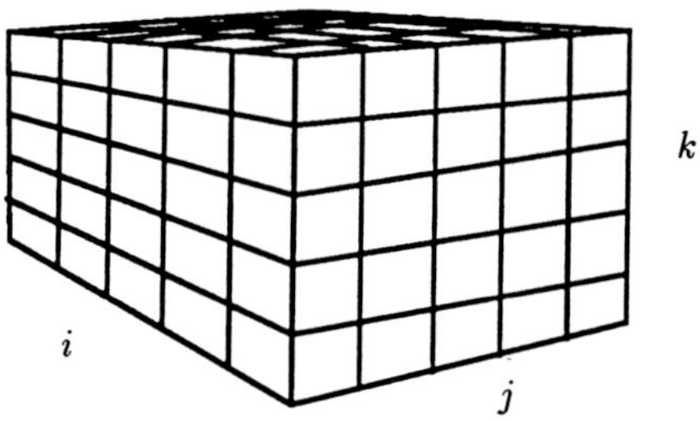

Fig. 12. Cuberille grid data and voxels.

the polyhedron approximation of S_α. The first basic assumption of the marching cubes method is that the model varies linearly along edges of the voxels. If the value of the dependent variable is greater than the isovalue at one endpoint of an edge and smaller at the other endpoint, then there will be a point on this edge which serves as a vertex for the triangulated approximation to S_α. Since there are eight vertices per voxel, there are potentially $2^8 = 256$ possible configurations to consider. This number can be reduced to 128 by taking into consideration the duality of greater than and less than. With this in mind, only configurations with at most 4 vertices with values above the isovalue have to be considered. This number can further be reduced by forming equivalence classes based upon rotations. See Fig. 13. The final 15 distinct configurations are shown in Fig. 14.

As it turns out, the mc method has a basic flaw which could lead to isosurfaces which have holes. This was first mentioned by Dürst [4]. Figure 15 further illustrates the problem. On the left is a configuration of type 6 and in the adjacent voxel we have a configuration which is a dual of type 3. The mc method would leave the "shaded hole" on the common face of these two voxels. Dürst recommended the simple fix of adding the triangles on this face. Nielson and Hamann [21] have discussed another way to alleviate the problem which is to consider other triangulations of the same vertices. For example, there would be two (three, including the Dürst proposal) possible ways of eliminating the hole of Fig. 15. These possibilities are shown in Fig. 16.

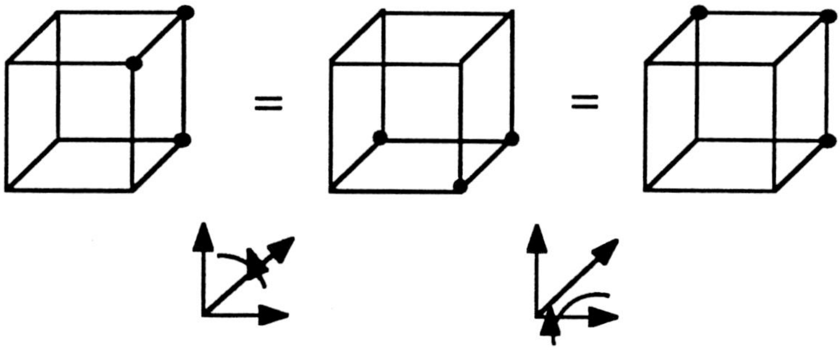

Fig. 13. Equivalence under rotations.

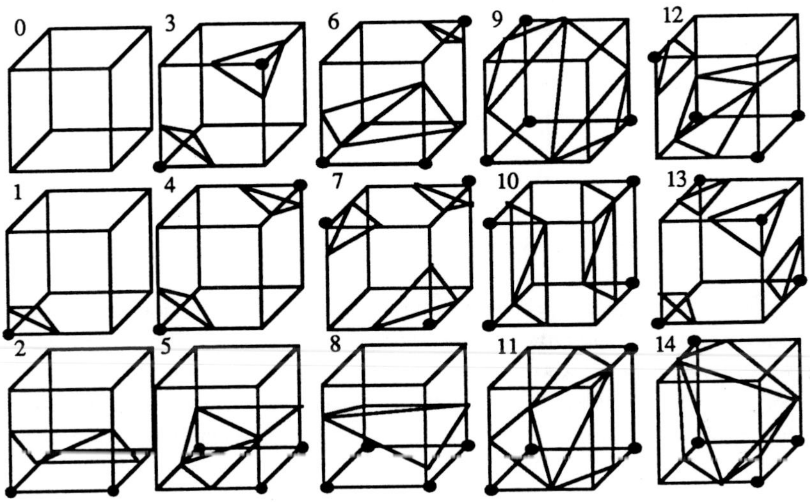

Fig. 14. The configurations of the mc method.

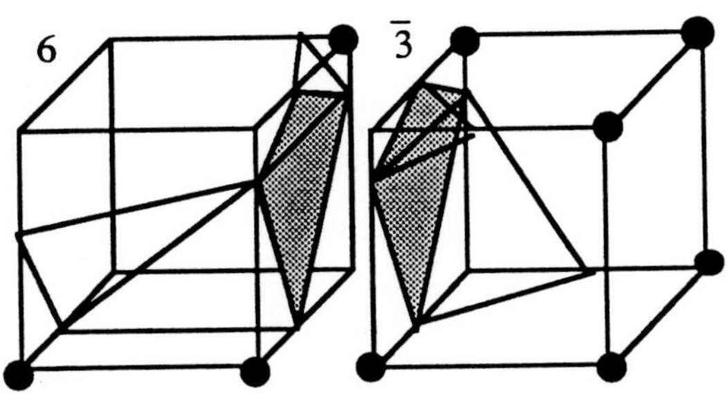

Fig. 15. The possible hole left by the mc method.

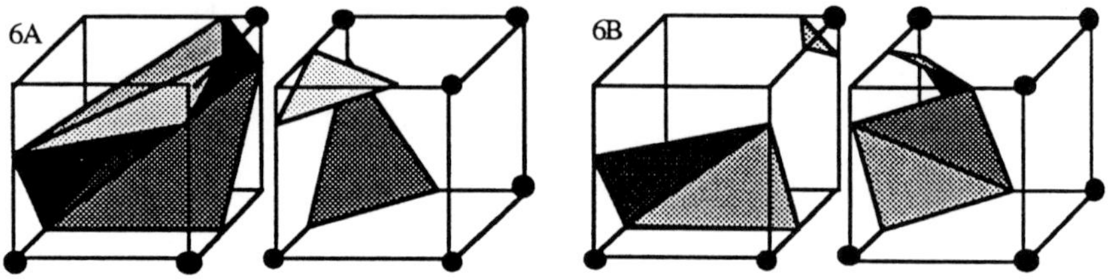

Fig. 16. Two possible ways to triangulate the edge vertices.

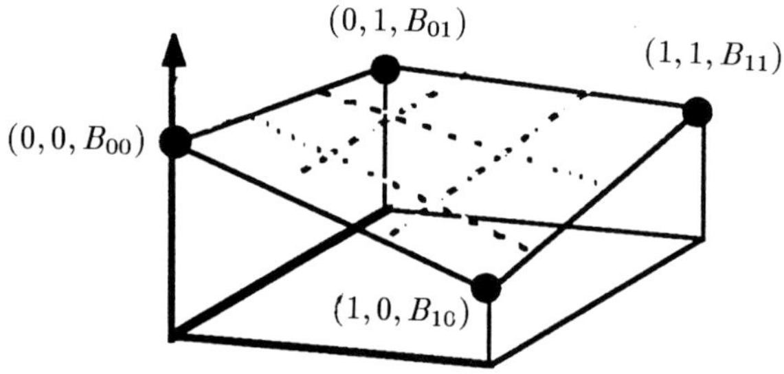

Fig. 17. Bilinear interpolation.

Nielson and Hamann [21] also discuss a criterion which can be used to make the choice from the different possible triangulations. The criterion is based upon the asymptotes of the contours of bilinear interpolation on the common face. With a change of variables, the domain of a face can be considered to be the unit square $\{(s,t) \ : \ 0 \leq s \leq 1, \ 0 \leq t \leq 1\}$ and so we have the following formula for bilinear interpolation:

$$B(s,t) = (1-s,s)\begin{pmatrix} B_{00} & B_{01} \\ B_{10} & B_{11} \end{pmatrix}\begin{pmatrix} 1-t \\ t \end{pmatrix}$$

where B_{00}, B_{01}, B_{10} and B_{11} represent the appropriate values of F_{ijk} at the four corner grid points.

It is easy to verify that the contour curves of B, $\{(s,t) \ : \ B(s,t) = \alpha\}$, are hyperbolas. Some possibilities as to how these contour hyperbolas and their asymptotes relate to the domain are shown in Fig. 18.

Ambiguous cases arise when both components of the hyperbola intersect the domain. The upper-left case of Fig 18 is such a situation. The "asymptotic decider" criteria for connecting vertices is based upon whether or not they are joined by a component of the hyperbolic arc. While it may not be immediately obvious, some simple algebraic calculations will show that this selection can be determined by comparing the contour value, α, with the value of the bilinear interpolant at the intersection point of the asymptotes. Based upon the notation of Figure 19, the test is:

IF $\alpha > B(S_\alpha, T_\alpha)$
THEN connect $(S_1, 1)$ to $(1, T_1)$ and $(S_0, 0)$ to $(0, T_0)$
ELSE connect $(S_1, 1)$ to $(0, T_0)$ and $(S_0, 0)$ to $(1, T_1)$

The asymptotes are easy to determine. They are $\{(s,t) \ : \ s = S_\alpha\}$ and $\{(s,t) \ : \ t = T_\alpha\}$, where

$$S_\alpha = \frac{B_{00} - B_{01}}{B_{00} + B_{11} - B_{01} - B_{10}}$$

and

$$T_\alpha = \frac{B_{00} - B_{10}}{B_{00} + B_{11} - B_{01} - B_{10}}.$$

Therefore, we have

$$B(S_\alpha, T_\alpha) = \frac{B_{00}B_{11} - B_{10}B_{01}}{B_{00} + B_{11} - B_{01} - B_{10}}.$$

The approach used in [2] to eliminate the problem of the original mc method is to subdivide each voxel until the problem goes away. Trilinear interpolation is used to compute the values of the dependent variable, F, at the new grid points. On a unit cube $\{(s,t,u) \ : \ 0 \le s \le 1, \ 0 \le t \le 1, \ 0 \le u \le 1\}$ the formula for trilinear interpolation is

$$T(s,t,u) = (1-u)\left[(1-s, s)\begin{pmatrix} B_{000} & B_{001} \\ B_{010} & B_{011} \end{pmatrix}\begin{pmatrix} 1-t \\ t \end{pmatrix}\right]$$

$$+u\left[(1-s, s)\begin{pmatrix} B_{100} & B_{101} \\ B_{110} & B_{111} \end{pmatrix}\begin{pmatrix} 1-t \\ t \end{pmatrix}\right].$$

We should point out some interesting issues that are not yet resolved with the mc method. As we have noted, several of the configurations of Fig. 14 can have alternate triangulations. Besides the alternate triangles associated with the asymptotic decider method, there are others which can alter the genus of the polyhedra approximation to S_α. For example, configuration 4 can be triangulated as shown in Fig. 20. The choice could be made on the basis that the genus of the polyhedron approximation is the same as that of the trilinear interpolant.

Another way to correct the problem of potential holes being produced by the mc method is to decompose each voxel into a collection of tetrahedra and then assume piecewise linear variation over each tetrahedra. Two types of possible decompositions are shown in Fig. 21. The top row shows decompositions into 5 tetrahedra and the bottom shows a decomposition into 6 tetrahedra. In order to maintain C^0 continuity in the case of 5 tetrahedra, the left and right decompositions have to be alternated from one voxel to the next. Even though the number

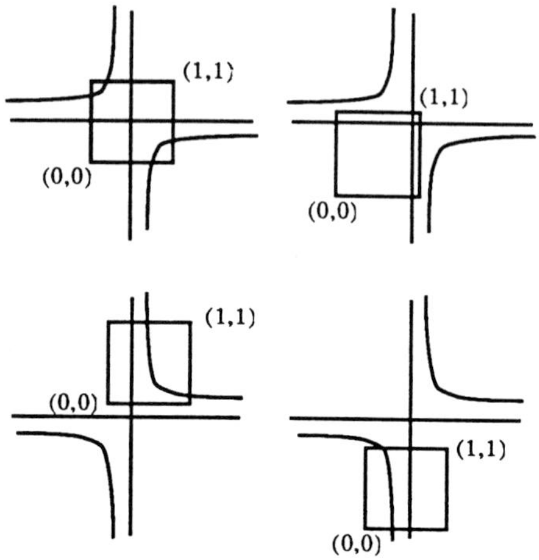

Fig. 18. Contours of bilinear interpolation.

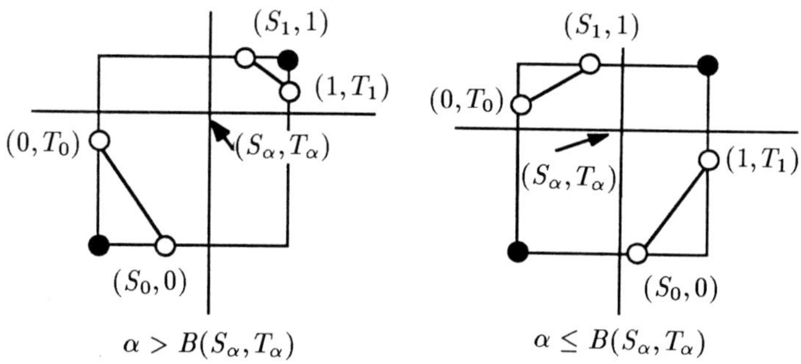

Fig. 19. Notation for asymptotic decider criterion.

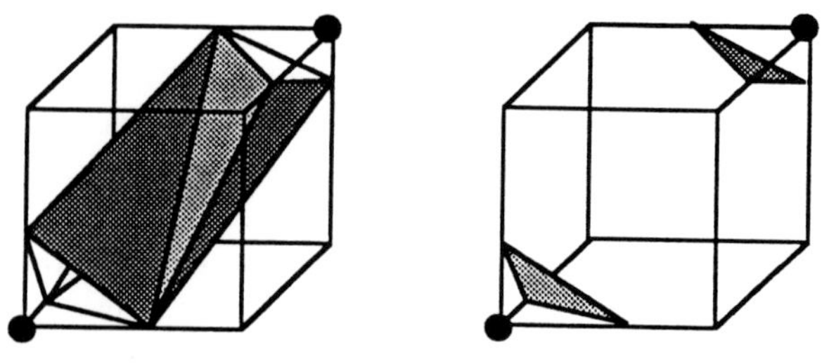

Fig. 20. Alternate triangulation of configuration 4.

of subdomains to consider is 5 or 6 times that of the original mc method, the number of triangles produced per subdomain is less in the tetrahedron case and so the total number of triangles is not 5 or 6 times that of the mc method. We have observed, on a variety of cases, that the methods based upon these tetrahedral decompositions produce approximately 150% to 250% more triangles than the original mc method.

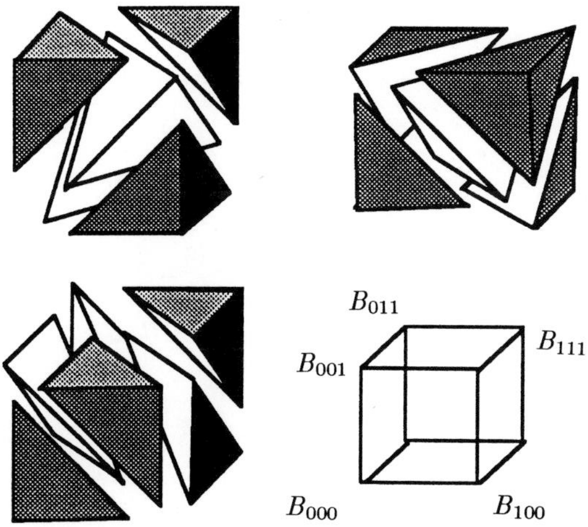

Fig. 21. Decomposition of the cube into tetrahedra.

The following formulas give the representation of the piecewise linear interpolant over the 5-tetrahedra decomposition of the upper left portion of Fig. 21. It is based on the use of barycentric coordinates.

3.1.4 Volume Rendering (Ray Casting) Methods

In this section, we discuss volume rendering techniques which are based upon ray casting. Potentially, each data point (voxel in the case of cuberille grid data) plays a role in determining the two dimensional image that is computed. In a nutshell, for each pixel of the image, a ray is cast through the volume and the values that this ray encounters are used to compute the intensity for this pixel. See Fig. 23. Most methods use the paradigm of "translucent jello". This is modeled by using the same simple transparency computations as used in the vanishing cube method mentioned above. See Fig. 10. We first assume that the dependent data values have been used to define an intensity function $I(x, y, z)$ and an opacity (transparency) function $t(x, y, z)$ which are both known at each point in the domain of the volumetric data. Along each ray, both intensity, I, and opacity, t, are sampled and then accumulated from back to front according to the simple transparency model. More precisely, if the ray for pixel ij has the initial point P_{ij} and direction D_{ij}, then it has the parametric form $P_{ij} + sD_{ij}$, $s > 0$. The

5-Tetrahedron Decomposition

IF $[(t-s-u)\geq 0]\wedge[u\geq 0]\wedge[s\geq 0]\wedge[(1-t)\geq 0]$

THEN $F(s,t,u):=(t-s-u)B_{010}+uB_{011}+sB_{110}+(1-t)B_{000}$

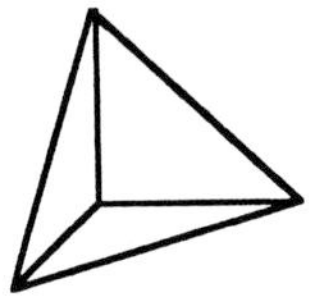

IF $[(s-t-u)\geq 0]\wedge[u\geq 0]\wedge[(1-s)\geq 0]\wedge[t\geq 0]$

THEN $F(s,t,u):=(s-t-u)B_{100}+uB_{101}+(1-s)B_{000}+tB_{110}$

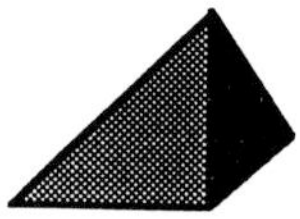

IF $[(u-s-t)\geq 0]\wedge[s\geq 0]\wedge[t\geq 0]\wedge[(1-u)\geq 0]$
THEN $F(s,t,u):=(u-s-t)B_{001}+sB_{101}+tB_{011}+(1-u)B_{000}$

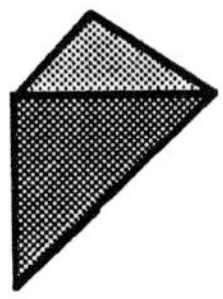

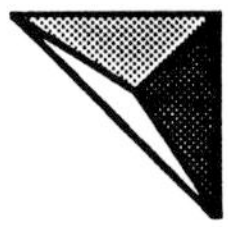

IF $[(s+t+u-2)\geq 0]\wedge[(1-s)\geq 0]\wedge[(1-t)\geq 0]\wedge[(1-u)\geq 0]$

THEN $F(s,t,u):=(1-s)B_{011}+(1-u)B_{110}+(1-t)B_{101}+(s+t+u-2)B_{111}$

IF $[(t-s-u)\leq 0]\wedge[(s-t-u)\leq 0]\wedge[(u-s-t)\leq 0]\wedge[(s+t+u-2)\leq 0]$

THEN $$F(s,t,u):=\frac{(s+u-t)}{2}B_{110}+\frac{(2-s-t-u)}{2}B_{000}+\frac{(t+u-s)}{2}B_{011}+\frac{(s+u-t)}{2}B_{101}$$

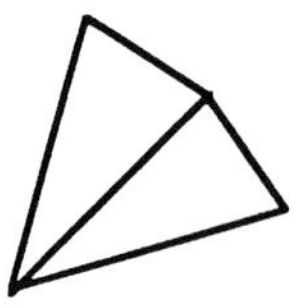

Fig. 22. Interpolation in a cube based upon 5-tetrahedra decomposition.

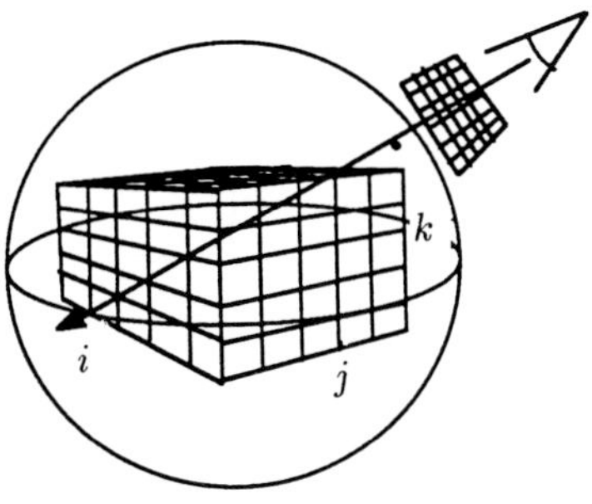

Fig. 23. Ray cast volume rendering.

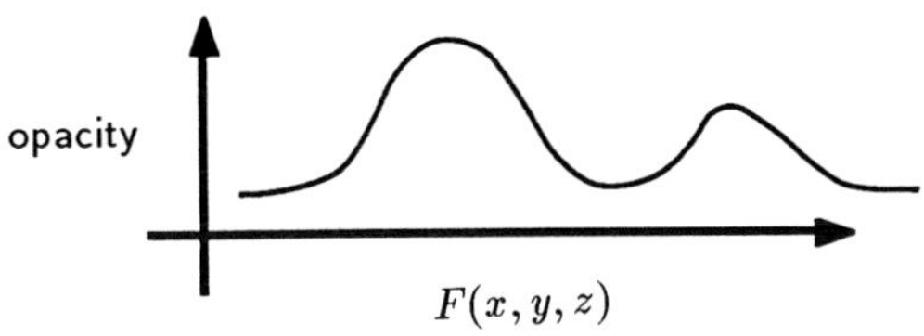

Fig. 24. Mapping from density to opacity.

sample points along the ray (with increasing distance from the display surface) are denoted by $s_1, s_2, \ldots, s_M$. The intensity for pixel ij is computed as

$$
\begin{aligned}
&I_{ij} := B_{ij} \\
&\textbf{FOR } k := M, M-1, \ldots, 1 \textbf{ DO} \\
&\quad I_{ij} := (1 - t(P_{ij} + s_k D_{ij}))I_{ij} + t(P_{ij} + s_k D_{ij})I(P_{ij} + s_k D_{ij})
\end{aligned}
$$

where B_{ij} represents the background intensity.

With this in mind, we will have a volume rendering algorithm as soon as we define the intensity function $I(x, y, z)$ and the opacity function $t(x, y, z)$. Typically these values will be computed at the independent data sites and then some type of interpolation method will be used to extend the definition to the entire domain. For medical data and other data resulting from scanning devices, the dependent data sites will be the vertices of the voxels of a cuberille grid and trilinear interpolation can be used. The interpolation scheme based on the 5-tetrahedron decomposition (discussed in Sect. 3.1.3) could also be used. The intensity function used by Levoy [10] is based upon the application of the well known and widely used illumination model of Phong [26]. At a point, say (x, y, z), where the intensity is to be estimated, the Phong model is applied to the contour surface passing through this point. The gradient $\nabla F = (F_x, F_y, F_z)$ provides the surface normal for the contour or isovalue surface.

Gradients can be estimated in a variety of ways, but a common method for cuberille grid data is to use second order differences; that is

$$\nabla F_{i,j,k} = \begin{pmatrix} F_x(x_i, y_j, z_k) \\ F_y(x_i, y_j, z_k) \\ F_z(x_i, y_j, z_k) \end{pmatrix} = \begin{pmatrix} \dfrac{(F_{i+1,j,k} - F_{i-1,j,k})}{2\Delta x} \\ \dfrac{(F_{i,j+1,k} - F_{i,j-1,k})}{2\Delta y} \\ \dfrac{(F_{i,j,k+1} - F_{i,j,k-1})}{2\Delta z} \end{pmatrix}.$$

Some adjustments have to be made on the boundaries of the domain. Either we can drop down to a first order divided difference or maintain the use of second order approximations with formulas such as

$$\nabla F_{1,N,k} = \begin{pmatrix} Fx(x_1, y_N, z_k) \\ Fy(x_1, y_N, z_k) \\ Fz(x_1, y_N, z_k) \end{pmatrix} = \begin{pmatrix} \dfrac{(-F_{3,N,k} + 4F_{2,N,k} - 3F_{1,N,k})}{2\Delta x} \\ \dfrac{(F_{1,N-2,k} - 4F_{1,N-1,k} + 3F_{1,N,k})}{2\Delta y} \\ \dfrac{(F_{1,N,k+1} - F_{1,N,k-1})}{2\Delta z} \end{pmatrix}.$$

Usually, we find that the better results provided by the second order approximations are worth the extra computational effort. Another approach to computing estimates for the gradient is based on trivariate, cubic interpolating splines. In this case, $F_x(x_i, y_j, z_k)$ is taken as the derivative of the cubic spline $S(x) = S_{jk}(x)$ which has the property that $S(x_i) = F_{i,j,k}$, $i = 1, \ldots, N$. The other partial derivatives are determined in an analogous manner. The computation of these derivatives requires the solution of a tridiagonal system of equations. For example, if we let $F_x(x_i, y_j, z_k)$ be denoted by $F'_{xi,j,k}$ then these values result from the solution of the tridiagonal system of equations

$$2F'_{x1,j,k} + F'_{x2,j,k} = 3(F_{2,j,k} - F_{1,j,k})$$

$$F'_{xi+1,j,k} + 4F'_{xi,j,k} + F'_{xi-1,j,k} = 3(F_{i+1,j,k} - F_{i-1,j,k}),\ i = 2, \ldots, N-1$$

$$F'_{xN-1,j,k} + 2F'_{xN,j,k} = 3(F_{N,j,k} - F_{N-1,j,k}).$$

Once the gradients have been estimated we are ready to apply the illumination model. In its simplest form, the Phong illumination model determines the intensity according to the formula

$$I = I_a k_a + I_\lambda [k_d (N, L)_+ + k_s (N, R)^n_{++}]$$

where I_a is a triple (red, green, blue) of intensity values for the ambient light source, I_λ is the intensity of the directed light source in the direction of L; k_a, k_d and k_s are object parameters (triples of r, g, b also) which specify the coefficient of ambient, diffuse and specular reflection (resp.). N is the surface normal and $R = 2(N, L)N - L$ is the reflected light vector. $(N, L)_+$ is taken to be (N, L)

if $(N,L)>0$ otherwise it is equal to zero. $(N,R)_{++}$ is (N,R) provided both (N,L) and (N,R) are non-negative and it is set to zero otherwise. The scalar n is called the Phong constant and it controls the shininess of the surface. One can also incorporate distances into this illumination model, but for this application, it usually adds very little.

We now consider the problem of defining the opacity function $t(x,y,z)$. How this is done is certainly dependent upon the application. In many applications, particularly those associated with medical imaging, the dependent data $F(x,y,z)$ represents the density of some physical material. This gives rise to the idea that there is a univariate mapping, H, called the *transfer function*, which is used to define the opacity according to $t(x,y,z) = H(F(x,y,z))$. If H is the characteristic function which is zero everywhere except at the value α, where it is one, then the volume rendered image will be the display of the contour surface $S_\alpha = \{(x,y,z) : F(x,y,z) = \alpha\}$. Levoy [10] describes two methods of defining $t(x,y,z)$ which both use $F(x,y,z)$ and $\|\nabla F(x,y,z)\|$. The first is aimed at displaying contour surfaces and takes into account possible aliasing artifacts and the fact that a contour surface most likely has some thickness to it. In this case,

$$t(x,y,z) = \begin{cases} 1 & \text{if } \|\nabla F\| = 0 \text{ and } F(x,y,z) = \alpha \\ 1 - \dfrac{1}{r}\dfrac{|\alpha - F(x,y,z)|}{\|\nabla F\|} & \text{if } \|\nabla F\| > 0 \text{ and } \dfrac{\alpha - F}{\|\nabla F\|} < r \\ 0, & \text{otherwise} \end{cases}$$

where r is the "thickness" of the contour. In the "region boundary surfaces" approach to the definition of the opacity $t(x,y,z)$ it is assumed that a certain number of opacities $t_1, t_2, ..., t_M$ have been assigned to the dependent variable values $F_1 < F_2 < \ldots F_M$. Any other value of $t(x,y,z)$ is computed by piecewise linear interpolation and then scaled by $\|\nabla F\|$. So we have

$$t(x,y,z) = \|\nabla F\| \left(t_{j+1}\frac{F(x,y,z) - F_j}{F_{j+1} - F_j} + t_j \frac{F_{j+1} - F(x,y,z)}{F_{j+1} - F_j} \right)$$

whenever $F_j \le F(x,y,z) \le F_{j+1}$.

3.1.5 Volume Interrogation Techniques

Sometimes the data indicates a relationship that is very smooth or slowly varying and this makes it difficult to detect certain qualitative changes using standard graphs. The sound data described in Sect. 2.8 is such an example. Graphing nearly constant data with contours will not be very effective. The contours will just wander about under the influence of the noise in the data. In this section, we describe a volume interrogation technique which could be helpful in analyzing and understanding relationships of this type. In order to motivate this technique, we describe a method that has been successful as a surface interrogation tool. The idea is to use Gaussian curvature as a texture image. The Gaussian curvature of a

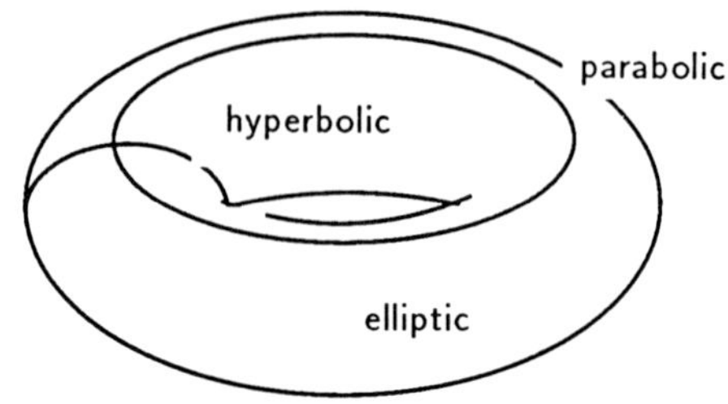

Fig. 25. Gaussian curvature on a torus.

parametric surface $S(u,v) = (X(u,v), Y(u,v), Z(u,v))$ at a point, $p = S(u,v)$, is $k = k_1 k_2$ where k_1 and k_2 are the principal curvatures at this point. The principal curvatures are the maximum and minimum curvatures of curves lying in planes passing through p and containing the normal at p. Not only does the magnitude of the Gaussian curvature reveal quantitative geometric information about the surface, but the sign of this value reveals some interesting qualitative information about the surface. Positive, negative and zero curvatures correspond to elliptic, hyperbolic and parabolic shapes. Locally, a tangent plane that is perturbed in the direction of the surface normal will have contours with the corresponding shape of these conics. Another implication of this is that in a local region of positive curvature, the surface can be reoriented so as to contain water, but in a region of negative curvature, this is impossible. See Fig. 25.

How does one extend the idea of Gaussian curvature as a surface texture map to volumetric data? In the case of a surface given as $(x, y, F(x,y))$, the principal curvatures, k_1 and k_2, can be computed as the eigenvalues of the 2×2 matrix

$$G = \frac{1}{N} \begin{pmatrix} F_{xx} & F_{xy} \\ F_{yx} & F_{yy} \end{pmatrix} \begin{pmatrix} 1 + F_x^2 & F_x F_y \\ F_y F_x & 1 + F_y^2 \end{pmatrix}^{-1}$$

where $N = \sqrt{1 + F_x^2 + F_y^2}$ and the notation F_x and F_{xy} is used to represent partial derivatives. This approach to Gaussian curvature allows an immediate extension to trivariate functions and volumetric data. The three "principal curvatures", k_1, k_2, and k_3 can be computed as the eigenvalues of the 3×3 matrix

$$G = \frac{1}{N} \begin{pmatrix} F_{xx} & F_{xy} & F_{xz} \\ F_{yx} & F_{yy} & F_{yz} \\ F_{zx} & F_{zy} & F_{zz} \end{pmatrix} \begin{pmatrix} 1 + F_x^2 & F_x F_y & F_x F_z \\ F_y F_x & 1 + F_y^2 & F_y F_z \\ F_z F_x & F_z F_y & 1 + F_z^2 \end{pmatrix}^{-1}$$

where $N = \sqrt{1 + F_x^2 + F_y^2 + F_z^2}$. Curvature for a trivariate relationship is now defined as $K = k_1 k_2 k_3$. In Fig. C of the color illustrations of this chapter at the end of the book, we show an example. The left image is a graph of the function $F(x,y,z) = \exp[-0.5(x2 + y2 + z2)]$. For this test function, we know there is a qualitative change taking place at the surface of the unit sphere. This is not easily discernible from the graph of the original function, but the graph of Gaussian curvature reveals this qualitative change quite nicely.

3.2 Techniques for Visualizing Surface-on-Surface Models

We now discuss the problem of visualizing the models which result from surface-on-surface data. We start with the case of a spherical domain. One approach to graphing a function whose domain is a sphere is to treat it as a bivariate function over the plane using latitude and longitude for the axes. An example is shown in Fig. 26. Unfortunately, this approach has some drawbacks. Relative distances between points on the spherical domain change drastically in different regions. For example, the front and back curves, which are constant, represent the value of the function at the north and south poles, respectively. The boundary curves on the left and right are equal because they represent the function evaluated along the same meridian. In order to alleviate these problems, we discuss some methods that deal more directly with the fact that the domain is a sphere.

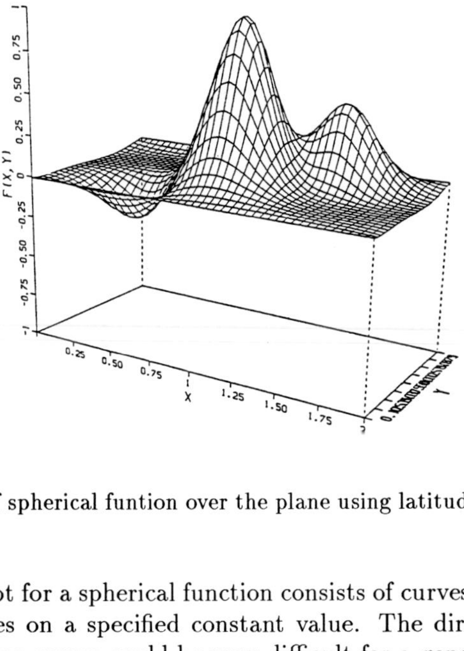

Fig. 26. Graph of spherical funtion over the plane using latitude and longitude as the axes.

A contour plot for a spherical function consists of curves on the sphere where the function takes on a specified constant value. The direct determination or calculation of these curves could be very difficult for a general function, F. The approach here is find an approximation to the contour curves by decomposing the sphere into a collection of subdomains and assuming that F has a simple representation over each of these subdomains. How well the piecewise approximation represents the true contour curve depends to a great extent on how the decomposition is accomplished. One possibility is based upon equal subdivision along longitude and latitude lines. This gives a collection of four sided regions which can be dealt with directly or split into triangular regions. We prefer to not use this type of decomposition because it gives such nonuniform treatment to different portions of the spherical domain. A course resolution along the equator

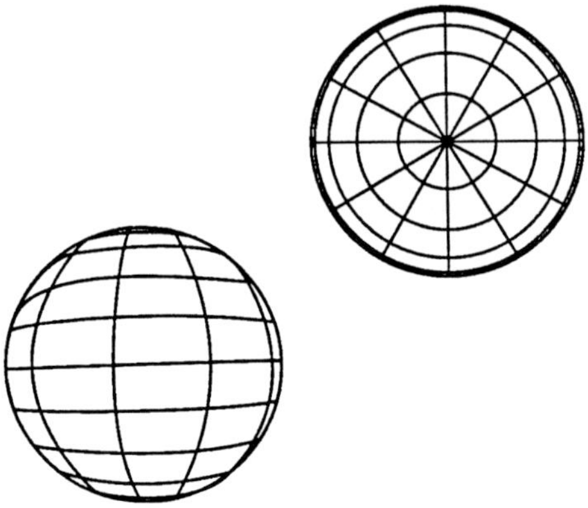

Fig. 27. Decomposition based upon uniform subdivision in latitude and longitude.

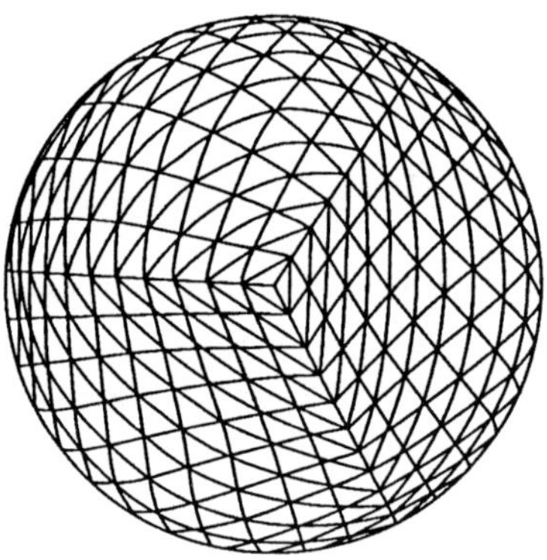

Fig. 28. Decomposition of sphere into a collection of spherical triangles.

corresponds to a fine resolution near the poles. This is illustrated in Fig. 27. It would be desirable to decompose the sphere into a collection of spherical triangles all of which are approximately the same size. With this goal in mind, we start with the four vertices of a tetrahedron and then uniformly subdivide the geodesic arcs joining these vertices. An example of this type of decomposition is shown in Fig. 28.

For a unit sphere, the four points of the tetrahedron are $(0,0,1)$, $(\sqrt{2}/3, -\sqrt{6}/3, -1/3)$, $(\sqrt{2}/3, \sqrt{6}/3, -1/3)$, $(-2\sqrt{2}/3, 0, -1/3)$. In order to compute equally spaced points on a geodesic arc joining two points P and Q we use the following parametric representation,

$$A(t) = \frac{P\sin(\Theta(1-t)) + Q\sin(\Theta t)}{\sin(\Theta)}, \quad 0 \le t \le 1 \tag{3.1}$$

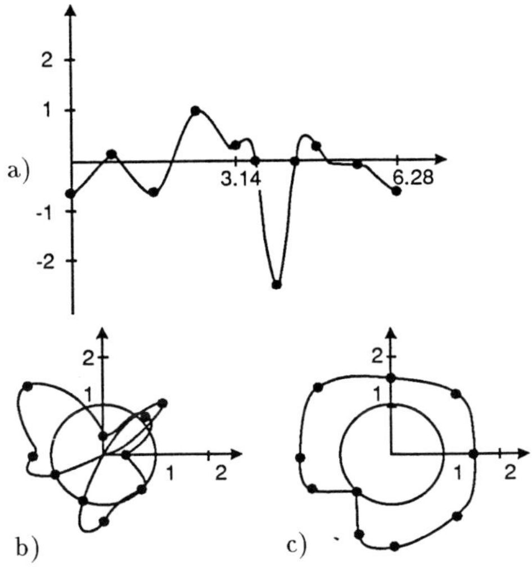

Fig. 29. Example of projecting a function defined over a circle.

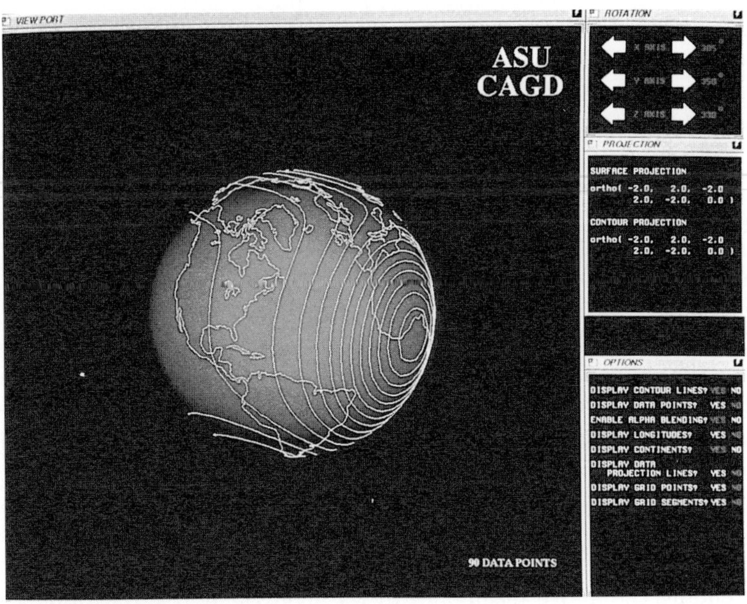

Fig. 30. Contour plot for function defined over a sphere.

where $\cos(\Theta) = P \cdot Q$. Equally spaced values of t will lead to points on the sphere which are equally spaced between P and Q. The function is now evaluated at the vertices of the triangular grid and assumed to vary linearly over each of the geodesic edges. If P and Q are two points of the spherical triangular grid and $F(P) > \alpha$, $F(Q) < \alpha$, then a point of the contour can be computed as

$$V = \frac{P \sin(\Theta(\frac{\alpha - F(Q)}{F(P) - F(Q)})) + Q \sin(\Theta(\frac{F(P) - \alpha}{F(P) - F(Q)}))}{\sin(\Theta)}. \tag{3.2}$$

Points on the contour are then joined by geodesic arcs. An example is shown in Fig. 30.

While contour plots are very useful, they are not always the best way to analyze a function because they don't easily show the geometric shape or the smoothness of the function. Continuity properties are not easily detectable, and different functions could have similar appearing contour plots. For example, a set of concentric circles could represent a sharp cone or a smooth hill. It is desirable to have a surface graph of the function in a form similar to that of a surface graph of a bivariate function defined over a plane. For a planar domain, the graph of a bivariate function $F(x, y)$ is displayed as a surface of (x, y, z) points satisfying $z = F(x, y)$, where (x, y) are points in the planar domain. This surface can be interpreted as projecting a distance $F(x, y)$ perpendicularly from the point (x, y) in the planar domain. Extending this to a spherical domain, we project a distance of $F(x, y, z)$ from the point (x, y, z) on the sphere in the direction of the normal vector to the sphere at (x, y, z). This definition of the graph of the function cannot be used without some modification, because if $F(x, y, z) < -2$, then the surface would project out from the sphere at the antipodal point $-(x, y, z)$. Figure 29 demonstrates this behavior by considering the simpler case of plotting a function defined over a circle. Figure 29a is a plot of a function on the interval $[0, 2\pi]$ and Fig. 29b is a plot of the corresponding function defined on the circle. Although the self-intersecting of the curves is not a major problem in this example, it is significantly more confusing in the surface case over the sphere. To avoid this problem, we scale the projection so that the minimum value gets mapped to the surface of the unit sphere and the maximum value gets mapped to a sphere of radius two units. This is illustrated in the case of a circle domain in Fig. 29c. To render the project surface graph of $F(x, y, z)$, we triangulate the sphere, evaluate F at all the vertices of the triangulation and then project out as we have described. Examples of this type of graph are shown in Fig. D of the color illustrations of this chapter at the end of the book.

Extending the above techniques to the more general surface domain leads to graphs as shown in the top portion of Fig. E of the color illustrations of this chapter at the end of the book. The transparent surface graph in the top left image of Fig. E of the color illustrations of this chapter at the end of the book is the result of projecting a distance proportional to $F(p)$ in a direction perpendicular or normal to p in the surface domain. Unfortunately, for nonconvex domains, this type of surface graph can have self-intersections which make it difficult to obtain very much geometrical information about this function. In the top right image of Fig. E of the color illustrations of this chapter at the end of the book, a radial projection was used from the center of the domain which does improve matters somewhat in this case. In general, for convex surface domains, the normal projection is preferred.

Another visualization tool for surface-on-surfaces is a method called the *hypersurface projection graph.* The motivation for this type of graph is based upon the following. The graph of a bivariate function $F(x, y)$ with some planar domain D, consists of the collection of 3D points $(x, y, F(x, y))$, where (x, y) is in D. Often this graph is rendered by displaying a network of 2D lines based upon the projection of the 3D points. In the case of surfaces-on-surfaces, we have a graph consisting of points in 4D space, $(x, y, z, F(x, y, z))$, where $(x, y, z) \in D$. When these points (or their rotated versions) are projected to 3D space using a parallel projection, we obtain points on a 3D surface. Three different projections can be used, $(F(x, y, z), y, z)$; $(x, F(x, y, z), z)$; and $(x, y, F(x, y, z))$. An example is shown in the bottom portion of Fig. E of the color illustrations of this chapter at the end of the book.

4 Modeling Methods

We now consider the topic of modeling trivariate data. For the discussion here, this will mean the construction of a function $F(x, y, z)$ which approximates the relationship implied by the data $(x_i, y_i, z_i; F_i)$, $i = 1, \ldots, N$. In some applications, the form of F and the parameters involved in the description of F may be of primary importance. Here, the main application of the model is to provide a means of graphing or visualizing the data. In this regard, we are not so interested in the form of F, but we are more interested in the results of the model and how easily it can be used to provide information for a method of visualizing trivariate relationships. This section is organized as follows. After some motivation which leads to some general structure for the modeling function, we review some techniques from univariate data modeling which have proven to be successful. From this, we identify two general ideas or approaches to solve the modeling problem. We call these two approaches "distance function" and "piecewise Hermite". We then discuss the four possibilities which result from the application of these two ideas to the two categories of data that we identified earlier; namely surface-on-surface data and volumetric data. The modeling process can be viewed as an operator that maps the data $(x_i, y_i, z_i; F_i)$, $i = 1, \ldots, N$ to the trivariate function $M(x, y, z)$. If we use the notation $\mathbf{X}$, $\mathbf{Y}$, $\mathbf{Z}$ and $\mathbf{F}$ to represent the data vectors $(x_1, x_2, \ldots, x_N)$, $(y_1, y_2, \ldots, y_N)$, $(z_1, z_2, \ldots, z_N)$ and $(F_1, F_2, \ldots, F_N)$ then this map can be written in the form

$$M(\mathbf{X}, \mathbf{Y}, \mathbf{Z}) = M[\mathbf{X}, \mathbf{Y}, \mathbf{Z}; \mathbf{F}](x, y, z).$$

There are two properties which are reasonable to assume for this mapping. First, if the dependent data is scaled, then it is reasonable to assume that the modeling function is scaled in a similar manner. For example, if the dependent data were to be doubled, then we might expect the approximation to also be doubled. Therefore, we assume that

$$M[\mathbf{X},\mathbf{Y},\mathbf{Z};a\mathbf{F}](x,y,z) = aM[\mathbf{X},\mathbf{Y},\mathbf{Z};\mathbf{F}](x,y,z). \tag{4.1}$$

Also, if the dependent data vector is decomposed into a sum, say $\mathbf{F} = \mathbf{G} + \mathbf{H}$, then it is reasonable to assume that

$$M[\mathbf{X},\mathbf{Y},\mathbf{Z};\mathbf{G}](x,y,z) + M[\mathbf{X},\mathbf{Y},\mathbf{Z};\mathbf{H}](x,y,z) = M[\mathbf{X},\mathbf{Y},\mathbf{Z};\mathbf{F}](x,y,z). \tag{4.2}$$

The two "linearity" assumptions of equations (4.1) and (4.2) imply that the modeling function is a linear combination of some basis functions. That is,

$$M[\mathbf{X},\mathbf{Y},\mathbf{Z};\mathbf{F}](x,y,z) = \sum_{i=1}^{N} a_i B_i(x,y,z)$$

for some suitably chosen basis functions B_i, $i = 1,\ldots,N$. The reason that this structure follows from the above two assumptions is that we can write an arbitrary dependent data vector

$$\mathbf{F} = f_1\mathbf{e}_1 + f_2\mathbf{e}_2 + \ldots + f_N\mathbf{e}_N$$

where $\mathbf{e}_i = (0,0,\ldots,1,\ldots,0)$ is the elementary basis vector with zeros everywhere except for a one in the ith position. So we have that

$$\begin{aligned}
M[\mathbf{X},\mathbf{Y},\mathbf{Z};\mathbf{F}](x,y,z) &= M[\mathbf{X},\mathbf{Y},\mathbf{Z};f_1\mathbf{e}_1 + f_2\mathbf{e}_2 + \ldots + f_N\mathbf{e}_N](x,y,z) \\
&= M[\mathbf{X},\mathbf{Y},\mathbf{Z};f_1\mathbf{e}_1](x,y,z) + M[\mathbf{X},\mathbf{Y},\mathbf{Z};f_2\mathbf{e}_2](x,y,z) \\
&\quad + \ldots + M[\mathbf{X},\mathbf{Y},\mathbf{Z};f_N\mathbf{e}_N](x,y,z) \\
&= f_1 M[\mathbf{X},\mathbf{Y},\mathbf{Z};\mathbf{e}_1](x,y,z) + f_2 M[\mathbf{X},\mathbf{Y},\mathbf{Z};\mathbf{e}_2](x,y,z) \\
&\quad + \ldots + f_N M[\mathbf{X},\mathbf{Y},\mathbf{Z};\mathbf{e}_N](x,y,z) \\
&= \textstyle\sum_{i=1}^{N}(f_i M[\mathbf{X},\mathbf{Y},\mathbf{Z};\mathbf{e}_i](x,y,z))
\end{aligned}$$

which is of the general form $\sum_{i=1}^{N} a_i B_i(x,y,z)$.

While it is nice to know this general structure, there still remains the problem of specifying the basis functions $B_i(x,y,z)$. In order to get some ideas of how to select these functions in the trivariate case, we take some motivation from a widely known and accepted set of basis functions used for univariate data; namely the cubic spline. So as to keep this discussion fairly self contained, we give a brief review of some of the basic properties of cubic, univariate splines. Given the data (x_i, F_i), $i = 1,\ldots,N$; $a < x_1 < x_2 < \ldots < x_N < b$; the cubic spline is the solution to the problem:

$$\text{Minimize} \int_a^b [F''(x)]^2 dx \tag{4.3}$$

subject to

$$F(x_i) = F_i, \ i = 1, \ldots, N.$$

If we accept the quantity $\int_a^b [F''(x)]^2 dx$ as some type of measure of the "smoothness" of a function, then the cubic spline is the "smoothest" function which passes through (interpolates) the data. It is no longer necessary to document the importance of the cubic spline. Spline functions have proven to be very useful and reliable tools in many areas of science and engineering. How can this concept be carried over to the problem at hand? In the remainder of this section, we discuss two ideas which are motivated by two different, but equivalent, representation of the cubic spline. They are both based upon the fact that the (natural) cubic spline, which solves the minimization problem stated above (see equation 4.3), is characterized by the following four conditions:

i) F, F' and F'' are continuous on the interval $[a, b]$.

ii) $F(x_i) = F_i$, $i = 1, \ldots, N$.

iii) F is piecewise cubic on $[x_1, x_N]$. That is, on each interval, $[x_i, x_{i+1}]$, $i = 1, \ldots, N-1$, F is a cubic polynomial.

iv) F is linear on $[a, x_1]$ and $[x_N, b]$ and so $F''(x_1) = 0 = F''(x_N)$.

The first representation of the natural cubic spline is

$$F(x) = \sum_{i=1}^{N} c_i |x - x_i|^3 + a + bx.$$

The basis functions 1, x, and $|x - x_i|^3$, $i = 1, \ldots, N$, all have continuous second derivatives. So the only property $F(x)$ lacks in this form is that it does not necessarily interpolate the data and that it does not necessarily satisfy the end conditions $F''(x_1) = F''(x_N) = 0$. If these conditions are imposed, we obtain the system of equations,

$$\begin{pmatrix} & & & 1 & x_1 \\ & |x_i - x_j|^3 & & 1 & x_2 \\ \vdots & \vdots & \vdots & \vdots & \vdots \\ & & & 1 & x_N \\ 1 & 1 & 1 & 0 & 0 \\ x_1 & x_2 & x_N & 0 & 0 \end{pmatrix} \begin{pmatrix} c_1 \\ c_2 \\ \vdots \\ c_N \\ a \\ b \end{pmatrix} = \begin{pmatrix} F_1 \\ F_2 \\ \vdots \\ F_N \\ 0 \\ 0 \end{pmatrix}. \tag{4.4}$$

This system can be solved to yield the coefficients $c_1, c_2, \ldots, c_N$, a and b. We will refer to the ideas covered in this approach to the cubic spline as the "distance function" approach.

A different way to compute the natural cubic spline is based on the piecewise, Hermite representation,

$$F(x) = (1-x)^2(2x+1)F(x_i) + x^2(3-2x)F(x_{i+1})$$
$$+x(1-x)^2F'(x_i) + x^2(x-1)F'(x_{i+1}) \tag{4.5}$$
$$x \in [x_i, x_{i+1}],\ i = 1, \ldots, N-1.$$

In this form, F satisfies conditions ii), iii) and the first two continuity conditions of i). If we force F to have a continuous second derivative and to satisfy the end conditions of iv) then we get a tridiagonal system of equations in the unknowns, $F'(x_i)$, $i = 1, \ldots, N$;

$$2F'(x_1) + F'(x_2) = 3\left(\frac{F_2 - F_1}{x_2 - x_1}\right)$$

$$\lambda_i F'(x_{i-1}) + 2F'(x_i) + \mu_i F'(x_{i+1}) =$$
$$3\mu_i\left(\frac{F_{i+1} - F_i}{x_{i+1} - x_i}\right) + 3\lambda_i\left(\frac{F_i - F_{i-1}}{x_i - x_{i-1}}\right),\ i = 2, \ldots, N-1$$

$$2F'(x_{N-1}) + F'(x_N) = 3\left(\frac{F_N - F_{N-1}}{x_N - x_{N-1}}\right)$$

where $\lambda_i = (x_{i+1} - x_i)/(x_{i+1} - x_{i-1})$ and $\mu_i = 1 - \lambda_i$.

The form of the representation of equation (4.5) is called the "piecewise Hermite" form of the cubic spline. Henceforth we will use this phrase to refer to the idea of representing the model in a piecewise manner with functions which match position and first order derivatives on the boundaries of the subdomains. We have identified two ideas which we now set out to exploit for the two problem domains: volumetric data modeling and surface-on-surface data modeling. These four possibilities are covered in the remainder of this section.

4.1 Distance Function Approach

4.1.1 Distance Function Approach to Volumetric Data

There is a straightforward generalization to volumetric data of the distance function approach to natural cubic splines. We simply use a modeling function of the form

$$F(p) = \sum_{i=1}^{N} c_i \|p - p_i\|^3 + a + bx + cy + dz$$

where $p = (x, y, z)$ and $p_i = (x_i, y_i, z_i)$. The system of equations analagous to (4.4) is

$$\begin{pmatrix} & & & 1 & p_1 \\ & |p_i - p_j|^3 & & 1 & p_2 \\ \vdots & & & \vdots & \vdots \\ & & & 1 & p_N \\ 1 & 1 & 1 & 0 & 0 \\ p_1^t & p_2^t & p_N^t & 0 & 0 \end{pmatrix} \begin{pmatrix} c_1 \\ c_2 \\ \vdots \\ c_N \\ a \\ b \\ c \\ d \end{pmatrix} = \begin{pmatrix} F_1 \\ F_2 \\ \vdots \\ F_N \\ 0 \\ 0 \\ 0 \\ 0 \end{pmatrix}. \tag{4.4}$$

An example of this method is shown in Fig. F of the color illustrations of this chapter at the end of the book. The independent data sites $p_i = (x_i, y_i, z_i)$ are chosen at random in a unit cube and the dependent values, $F_i = F(p_i)$, are taken from a given function $F(p)$. By using an underlying function to produce the dependent data, we can test the effectiveness of the method to reproduce certain shapes and behaviors.

When the number of data points is very large or the dependent data is noisy, then global methods based upon distance functions may not be a practical way of modeling the data. Distance function methods of interpolation require that we solve a linear system of equations of the size $(N + 4) \times (N + 4)$ where N is the number of data points. These systems can be very badly conditioned. In fact, as one independent data point approaches another, the condition number tends to infinity. Even if the data sites are quite uniformly distributed, condition numbers which will swamp single precision are not uncommon for data sets of size $N > 500$. For trivariate data, many applications lead to much larger data sets and so some alternatives are needed. We discuss two here. One is to find an approximation by the method of least squares and the other is to make use of "localizing" or "mollifying" functions. We discuss the method of least squares first. The modeling function is of the form

$$F(p) = \sum_{j=1}^{M} c_j \|p - q_j\|^3 + a + bx + cy + dz$$

and $c_1, \ldots, c_M$; a, b, c and d are chosen so as to minimize

$$\sum_{i=1}^{N} \left(\sum_{j=1}^{M} c_j \|p_i - q_j\|^3 + a + bx_i + cy_i + dz_i - F_i \right)^2 .$$

The singular valued decomposition is used to compute these values. See Nielson and Dierks [19] for more discussion on this topic.

In order to completely define the basis functions, the knots, q_i, must be specified. A simple approach is to choose these values to be uniformly distributed on a cuberille grid. Another approach is to try to select the knots so that they are distributed somewhat in the same fashion as the data sites. The "Venezia

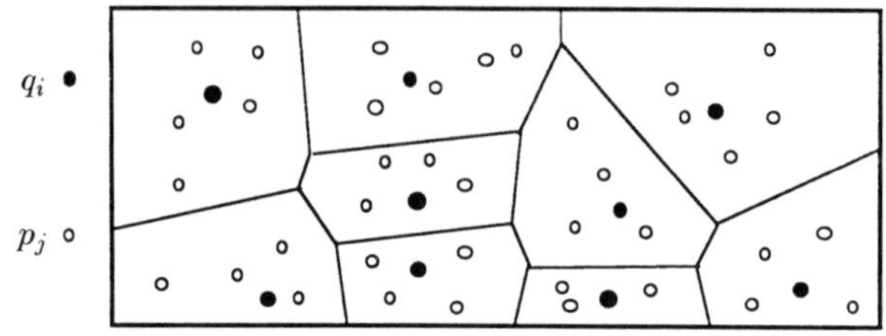

Fig. 31. The solid circles are the centroids of the points lying in their respective Theissen regions.

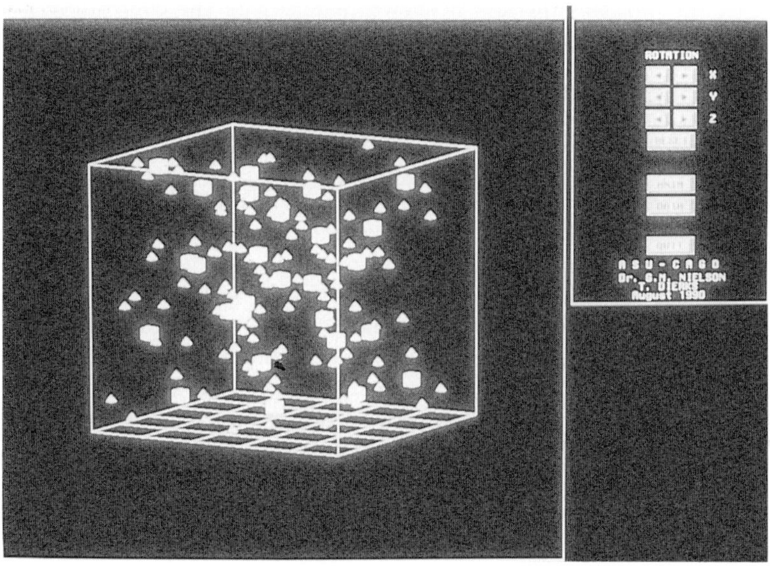

Fig. 32. The 27 knots (boxes) are selected to be close to the 125 data sites (pyramids).

criteria" characterizes the q_i by requiring them to be the centroids of the points lying in its Theissen region. The Theissen region for a point $q_k \in \{q_i : i = 1, \ldots, M\}$ is the set of points closer to q_k than any other q_j, $j \neq k$. In Fig. 31, the q_i are indicated by filled circles and the p_i are shown as empty circles. In order to compute this type of distribution in the case of volumetric data, a three dimensional analogue of a two dimensional method proposed by Franke and McMahon (see Nielson and Dierks [19]) can be used. This is an iterative algorithm which starts with an initial configuration of knots and then moves these values so as to reduce the distance between the knots and the data sites. An example from Nielson and Dierks [19] is shown in Fig. 32.

A natural question is whether or not the knot distribution by the Venezia criteria improves matters. Nielson, Dierks [19] have performed some empirical studies intended to help answer this question. Random sets of points of size 64,

Knots	Data Points	RMS for F1 Uniform Knots	RMS for F1 Optimal Knots	RMS for F3 Uniform Knots	RMS for F3 Optimal Knots
27	64	0.112753	0.072651	0.052302	0.054745
	125	0.082485	0.062220	0.030959	0.061313
	250	0.077981	0.062639	0.031308	0.041224
	500	0.074303	0.058589	0.027663	0.039659
	1000	0.074683	0.058582	0.027586	0.040051
64	125	0.080449	0.041400	0.031496	0.025203
	250	0.055728	0.031252	0.025691	0.019582
	500	0.047361	0.028616	0.020503	0.015929
	1000	0.045759	0.033946	0.020309	0.014903
125	250	0.036270	0.017625	0.012581	0.007528
	500	0.017666	0.011974	0.006065	0.008275
	1000	0.014678	0.012729	0.005791	0.005016
216	500	0.020848	0.007201	0.010843	0.003717

Fig. 33. RMS errors for multiquadric method with uniform and optimal knots.

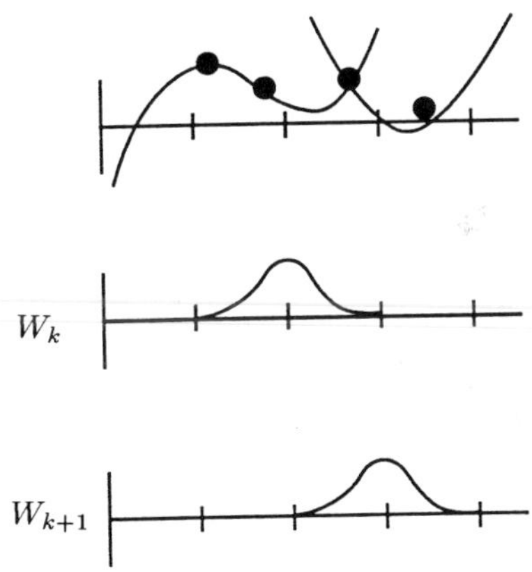

Fig. 34. Localizing functions.

	r200	l200	s200
F1	.0163/.0162	.0321/.0311	.0389/.0391
F2	.0121/.0120	.0141/.0137	.0175/.0176
F3	.0101/.0091	.0082/.0082	.0282/.0275
F4	.0015/.0010	.0034/.0028	.0031/.0022
F5	.0036/.0032	.0171/.0167	.0139/.0131
F6	.0031/.0031	.0040/.0032	.0015/.0020

Fig. 35. RMS errors for local version/global version.

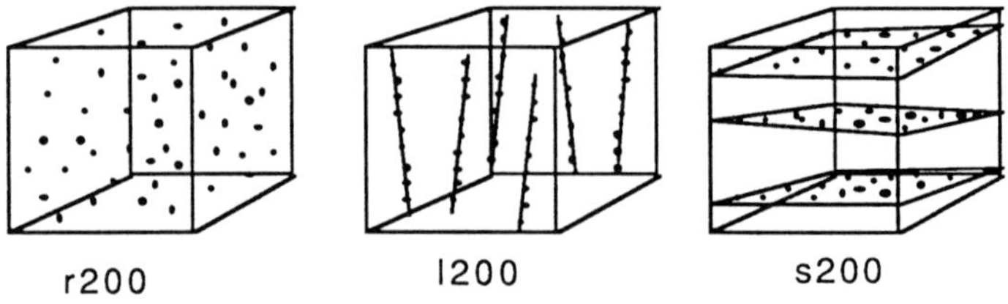

Fig. 36. Data configurations.

125, 250, 500 and 1000 were generated in a unit cube. The associated dependent data was obtained by evaluating a known trivariate function at these points. Next, the least squares approximations using 27, 64, 125 and 216 knots were computed. These approximations were computed both for uniform knots and for the knots that resulted from the Venzia criteria. Typical results are those of Fig. 33 where the RMS errors for two functions are given. In most cases, the optimal knots did improve matters, but not a great deal. The two functions used are given in equation (4.6). Actually, the results of Fig. 33 are for the multiquadric method (see Franke and Nielson [8]), but they are typical and similar to those for the distance method described here.

We now move to our second approach to modeling large data sets when using distance function models. It is based upon the use of "localizing" functions which are smooth and have a small region of support. The basic ideas can easily be understood by considering the univariate case illustrated in Fig. 34. The functions w_k are nonzero only on two intervals and have the property that $\sum w_k(x) = 1$ for any x in the domain. Local interpolants F_k are computed so that $F_k(x_i) = F_i$ for all data points x_i in the support of w_k. From this, it is easy to see that

$$F(x) = \sum w_k(x) F_k(x)$$

has the property that $F(x_i) = F_i$ for all x_i in the union of the support of the w_k which has been arranged to be the entire original domain. Any method can be used to obtain the local interpolants F_k; including the distance function methods discussed above or the piecewise Hermite methods to be discussed in Sect. 4.2. The localizing function w_k can be easily constructed with piecewise cubics (bicubics or tricubics). Usually the number and support of the w_k's is selected so that the number of data points per support of w_k is roughly constant. Fig. 35. Examples of this method in the trivariate case can be found in [29]. A natural question that arises is how the localized version compares to the original distance function method. In Fig. 35, we give the RMS errors of these two methods. The dependent data values are computed from six functions, all with differing characteristics, and for three different data sets. The data sets are all of the same size ($N = 200$), but their distributions are quite different. Figure 36 indicates the different distributions. The RMS errors for the global method are compared with those of the localized version in

$$\begin{aligned}
F_1(x,y,z) &= 0.75\cdot\exp\left[-\frac{(9x-2)^2+(9y-2)^2+(9z-2)^2}{4}\right] \\
&+ 0.75\cdot\exp\left[-\frac{(9x+1)^2}{49}-\frac{(9y+1)}{10}-\frac{(9z+1)}{10}\right] \\
&+ 0.5\cdot\exp\left[-\frac{(9x-7)^2+(9y-3)^2+(9z-5)^2}{4}\right] \\
&-0.2\ \cdot\exp\left[-(9x-4)^2-(9y-7)^2-(9z-5)^2\right] \\
F_2(x,y,z) &= [\tanh(9z-9x-9y)+1]/9 \qquad (4.6)\\
F_3(x,y,z) &= (1.25+\cos(5.4y))\cos(6z)/(6+6(3x-1)^2) \\
F_4(x,y,z) &= \exp\left[-\tfrac{81}{16}[(x-0.5)^2+(y-0.5)^2]+(z-0.5)^2\right]/3 \\
F_5(x,y,z) &= \exp\left[-\tfrac{81}{4}[(x-0.5)^2+(y-0.5)^2+(z-0.5)^2]\right]/3 \\
F_6(x,y,z) &= \sqrt{64-81[(x-0.5)^2+(y-0.5)^2+(z-0.5)^2]/9}-0.5
\end{aligned}$$

4.1.2 Distance Function Approach to Surface-on-Surface Data

The first surface-on-surface domain we consider is a sphere. The sphere is not only one of the simplest surface-on-surface domains, but it is also very important in some applications because the earth is approximated by a sphere. The data described in Sect. 2.3 is such an example. Without loss of generality, we can assume that the domain is a unit sphere located at the origin,

$$S=\{(x,y,z)\ :\ x^2+y^2+z^2=1\}.$$

The independent data sites $p_i=(x_i,y_i,z_i)$ are points on the sphere. There is a very natural generalization of the distance function $|x-x_i|$ to the domain of a sphere and that is to use $r(p,p_i)$ which represents the distance from p to p_i as measured on the sphere. This distance is the length of the shortest path on the sphere between p and p_i. In general, the shortest path on a surface between two points is called the geodesic curve and the length of it would be taken to be the geodesic distance between two points on a surface. For the sphere, there is a simple formula for computing the geodesic distance. It is $r(p,p_i)=\cos^{-1}(p,p_i)$, where (p,p_i) is the dot product. With this in mind, it seems natural to use the basis function $[r(p,p_i)]^3$ in place of $|x-x_i|^3$, but there is a problem that arises which requires some modification. This problem has to do with the continuity of the first derivative of these functions. This is more fully understood if we look at a cross section as shown in Fig. 37. The point p_i is taken to be $(1,0)$ and we have graphed $r(p,p_i)^3$ as a radial function. We can see there are problems at the

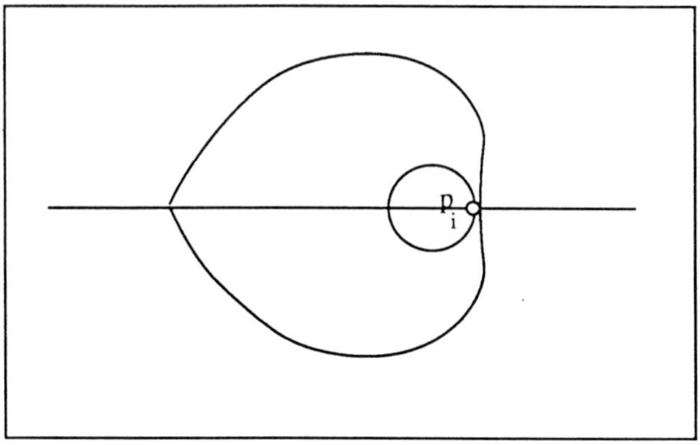

Fig. 37. Discontinuity of geodesic distance function at antipodal point.

antipodal point to p_i. From both the top and bottom the function is increasing in distance from the origin. Thus, it is clear that only the function (and not its higher order derivatives) is continuous. If continuity beyond the function itself is not a problem, then these basis functions can be used without modification. But if C^1 is needed, then some adjustments must be made. One possibility is to "round off" the backside with a piecewise definition

$$r(p, p_i) = \begin{cases} \cos^{-1}(p, p_i) & (p, p_i) \geq 0 \\ H(\cos^{-1}(p, p_i)) & (p, p_i) < 0 \end{cases}$$

where $H(\pi/2) = \pi/2$, $H(\pi) = \pi$, $H'(\pi/2) = 1$, $H'(\pi) = 0$. One possible choice is $H(s) = \pi - 4s + (8/\pi)s^2 - (4/\pi^2)s^3$.

For the more general case where the domain D is an arbitrary closed surface, very few techniques have been developed. One rather general yet simple approach is the *domain mapping method* of [5]. In a nutshell, this method involves mapping the surface domain D to a sphere, solving a corresponding interpolation problem on the sphere and then mapping back to D for a solution. The surface domain D does not need to be convex, but it is assumed to be topologically equivalent to a sphere. Usually D is defined by a map, B, from a domain A. The surface-on-surface interpolant $F(p)$ which satisfies $F(p_i) = F_i$ is constructed using the following steps.

Step 1. For $i = 1, \ldots, N$, find a_i in A such that $B(a_i) = p_i$ in D.

Step 2. Find a mapping E from A onto the surface of the unit sphere so that for s in S and p in D, the mapping $p = BE^{-1}(s)$ is a one-to-one map from S onto D.

Step 3. For $i = 1, \ldots, N$, compute points $s_i = E(a_i)$ in S.

Step 4. Construct the scattered data interpolant $G(s)$ on the sphere S which satisfies $G(s_i) = F_i$ for $i = 1, \ldots, N$.

Step 5. For p in D, find a point a in A such that $B(a) = p$ and define $F(p) = G(E(a))$.

Steps 1 and 5 are potentially difficult problems for a general domain D because they may involve the inversion of a nonlinear map B. However, for the evaluation of the interpolant F in Step 5 over all of D, as opposed to the evaluation at a single fixed point, we can avoid the inversion by evaluating $F(B(a)) = G(E(a))$ over all a in A. In fact, if the domain D is only known by the mapping $B(a)$, for a in A, then to actually know that a point p is in D, we must know some point a in A such that $B(a) = p$. In this case, no inversion of the mapping B is necessary and the term "find" in Steps 1 and 5 can be replaced by "let". How this general approach is implemented depends to a great extent upon how D is defined. We discuss three cases.

D is a mapping of a rectangle

A common situation is when D is a parametric surface defined by the periodic map $B(u, v)$, where (u, v) is in the rectangle A given by $[u_{\min}, u_{\max}) \times [v_{\min}, v_{\max}]$. We assume that B is one-to-one except at the preimage of two points d_1 and d_2 in D. The points d_1 and d_2 are called the polar points. Without loss of generality, we assume that $B(u, v_{\max}) = d_1$ and $B(u, v_{min}) = d_2$ for all $u_{\min} \leq u < u_{\max}$. A mapping E from A to the unit sphere S in Step 2 which establishes a one-to-one correspondence between S and D is given by

$$E(u, v) = (-\cos(v') \sin(u'), \cos(v') \cos(u'), \sin(v'))$$

where

$$u' = \frac{u - u_{\min}}{u_{max} - u_{min}} 2\pi \text{ and } v' = \frac{v - v_{\min}}{v_{\max} - v_{\min}} \pi - \frac{\pi}{2}.$$

For all $u_{\min} \leq u < u_{\max}$, we have $E(u, v_{\max}) = (0, 0, 1)$ and $E(u, v_{\min}) = (0, 0, -1)$. If we let $E^{-1}(s)$ be the set of points in A such that $D(a) = p$, then the composition $BE^{-1}(s)$ is a one-to-one map from S onto D. The upper two images of Fig. 26 show a domain D defined as a surface of revolution of a planar curve; thus D is defined by a parametric map of a rectangle.

D is a mapping of a sphere

A useful method for defining a closed surface domain D is to project out radially from a sphere. Let A be the surface of the sphere with center c and radius r. Suppose that the domain surface D is defined by

$$D = \{p = a + h(a)(a - c)/r \ : \ \text{for } a \text{ in } A\}, \tag{4.7}$$

where $h(a)$ is some non-negative function on A. For our general five step technique, simply define the mappings $B(a) = a + h(a)(a - c)/r$ and $E(a) = (a - c)/r$. Furthermore, given a point p in D, we can easily compute the point a in A that satisfies $B(a) = p$ as the intersection of the ray joining c to p with the sphere A.

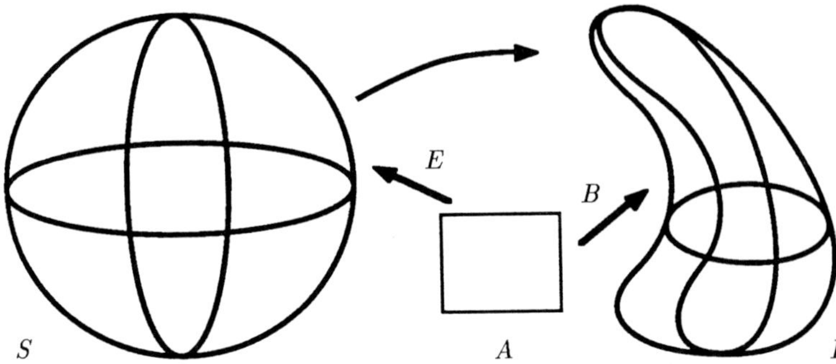

Fig. 38. Domain mapping approach to modeling of surface-on-surface data.

D is only known at discrete points

There are instances where the domain D is not known precisely; it is known only that the sampled data points p_i are on some surface. If the set of points p_i has a "star-like" property with respect to some central point c, then we can develop a surface which passes through the p_i and use this surface as an approximation to the unknown surface domain D. If the points p_i are triangulated into a polyhedron, then we assume there exists a central point c such that each ray joining c to p_i intersects the polyhedron only at p_i. We do not assume that the topology of the polyhedron is known, and we make no direct use of any triangulation of the points p_i. Our approach is to find a surface that passes through the p_i which is a radial projection of a sphere. Supposing that we have found a central point c, let $r = \min[\|p_i - c\|]$ for all i and let A be the sphere with center c and radius r. Compute $h_i = \|p_i - c\| - r$ and let a_i be the intersection of the ray joining c and p_i with the sphere A. Define the mapping $E(a) = (a - c)/r$ from A to the unit sphere S and let $s_i = E(a_i)$. Construct the interpolant $H(s)$ over the unit sphere S that satisfies $H(s_i) = h_i$, for $i = 1, \ldots, N$. Define D using equation (4.7) and $h(a) = H(E(a))$. With D now represented as a radially projected function defined over a sphere, we can now apply the method above where D is a mapping of a sphere.

4.2 Piecewise Hermite Approach

In many ways the Minimum Norm Network (MNN) method [14] involves the ideas of the piecewise Hermite approach to cubic splines. We will describe this method and then indicate how it is extended to the case of surface-on-surface data and volumetric data. The MNN method was developed for bivariate scattered data where we have the data (x_i, y_i, F_i), $i = 1, \ldots, N$, with the independent data sites arbitrarily located in some planar domain. There are three steps to this method:

i) The convex hull of the points $p_i = (x_i, y_i)$ is decomposed into a collection of triangles.

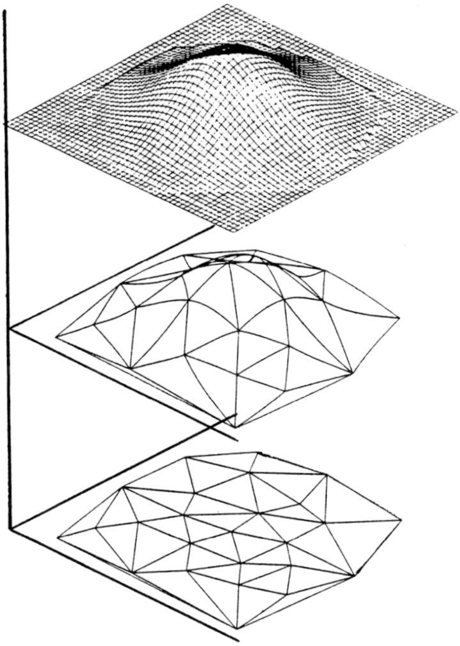

Fig. 39. The three steps of the MNN method.

ii) An interpolating curve network defined over the edges, which has certain minimization properties, is computed.

iii) The curve network is filled-in to complete the definition of the modeling function by the use of a C^1 triangular interpolant.

We now briefly describe each of these steps.

Triangulation Step of MNN Method

The triangle with vertices p_i, p_j and p_k is denoted by T_{ijk} and the list of triples which represents the triangulation is denoted by I_t. The edge joining p_i and p_j is denoted by e_{ij} and $N_e = \{ij \ : \ ijk \text{ in } I_t \text{ for some } k\}$ is used to refer to the collection of all edges. Formally, the definition of a triangulation requires:

i) No triangle T_{ijk} is degenerate. That is, the three vertices are not collinear.

ii) The interior of any two triangles do not intersect.

iii) No edge of a triangle T_{ijk} contains any vertices other than V_i, V_j, and V_k.

iv) The union of the all triangles is the convex hull.

There are many possible triangulations of the convex hull. It is usually the case that long skinny triangles are avoided. These are triangles with very large angles or ones with very small angles. There are two types of optimal triangulations which have been discussed quite widely: the max-min and min-max. Both of these optimal triangulations have a similar method of characterization. Associated with each triangulation there is a vector with n_t entries representing either the largest or smallest angle of each triangle. The entries of each vector are ordered and then a lexicographic ordering of the vectors is used to impose an ordering on the set of all triangulations. In the case of the min-max criteria, A_i is the largest angle of a triangle and the entries of each vector, A_t, are ordered so that

$$A_t = (A_1, A_2, \ldots, A_{n_t}),\ A_i \geq A_j,\ i < j.$$

The smallest of these vectors based on their lexicographic ordering associates with the optimal triangulation. In the case of the max-min criteria, a_i, is the smallest angle and the entries of each vector are ordered the other way so that

$$a_t = (a_1, a_2, \ldots, a_{n_t}),\ a_i \leq a_j,\ i \leq j.$$

The largest of these vectors represents the optimal triangulation in the max-min sense. In Fig. 40, six data points are shown which have a total of ten possible triangulations which are shown in Fig. 41. Based upon the max-min criteria we have the ordering

$$\tau_1 < \tau_2 < \tau_5 < \tau_3 < \tau_0 < \tau_6 < \tau_7 < \tau_4 < \tau_8 < \tau_9$$

and so τ_9 is the optimal triangulation in this sense. It is interesting to note that each triangulation is obtainable from one with a smaller associated vector by swapping the diagonal of a convex quadrilateral. In fact, this basic operation forms the basis of the algorithm of Lawson for computing the optimal triangulation. Lawson [9] showed that any triangulation can be obtained from any other triangulation by a sequence of these operations. Furthermore, he proved that if the choice of the diagonal is made on the basis of the max-min criteria for the quadrilateral only, eventually the global optimal triangulation will be obtained. In other words, for this criterion, a local optimum is a global optimum. It is interesting to note that this is not the case for the min-max criterion. This same example points this out. Based upon the min-max criteria, τ_4 is optimal and τ_8 is a local minimum. Locally optimal swaps of diagonals from τ_8 would never lead to τ_4. More details on this example and related results can be found in [16].

Another important property of max-min triangulation, especially for our discussion here, is its relationship to the Dirichlet tessellation. The Dirichlet tessellation is a partition of the plane into regions R_i, $i = 1, \ldots, N$, called Thiessen regions. The Thiessen region R_k consists of all points in the plane whose closest point among p_i, $i = 1, \ldots, n$, is p_k. A Dirichlet tessellation is usually illustrated by drawing the boundaries of the Thiessen regions. An example is shown in the right image of Fig. 42. In the left image of Fig. 42 is shown the max-min

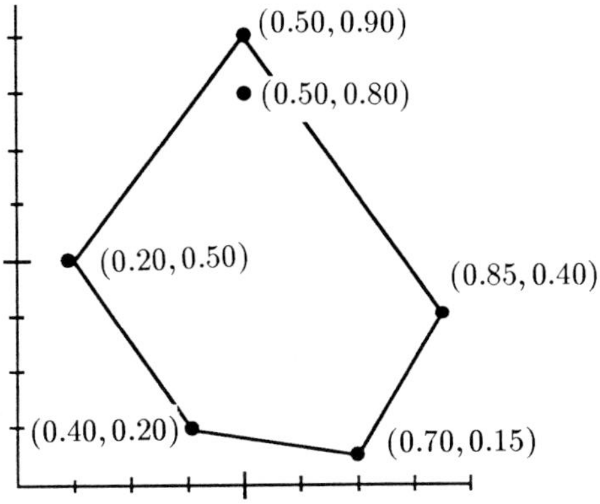

Fig. 40. Six points.

triangulation. It is dual to the Dirichlet tessellation in that the edges of this optimal triangulation join vertices which share a common Thiessen region boundary. Because of this relationship between the Dirichlet tessellation and the optimal max-min triangulation, we can extend the idea of max-min triangulation to any domain where we can compute the distance between two points. The sphere provides a good example. Here the distance between two points p and q is easily computed as $\cos^{-1}(p,q)$ so the Dirichlet tessellation is also easy to compute. An example is shown in the left images of Fig. 43. The right image depicts the triangulation which is dual to this tessellation.

Network Step of MNN Method

Once the convex hull of the p_i, $i = 1, \ldots, N$, has been triangulated, the next task is to estimate the derivatives $F_x(p_i)$ and $F_y(p_i)$. A variety of methods can be used. One of the simplest is to just average the derivatives of the linear interpolant for each triangle which includes p_i as a vertex. Unfortunately, this simple approach does not work very well. Improved methods involve local least squares methods, but probably the most effective method is that of the MNN method which is motivated by the minimal variation characterization of univariate cubic splines mentioned above. Analogous to the quantity of equation (4.3) there is

$$\sigma(F) = \sum_{ij \in N_e} \int_{e_{ij}} \left(\frac{d^2 F}{d^2 e_{ij}} \right)^2 ds_{ij}$$

where ds_{ij} represents the element of arc length on the curve consisting of the line segment e_{ij}. In the univariate case when we minimize the integral of the second derivative squared subject to the interpolation requirements, we find that the solution is a piecewise cubic curve and, if the curve is represented in Hermite form, the derivatives are computed as the solution of a linear system of equations. In the bivariate MNN method, the solution of the minimum of $\sigma(F)$ is also

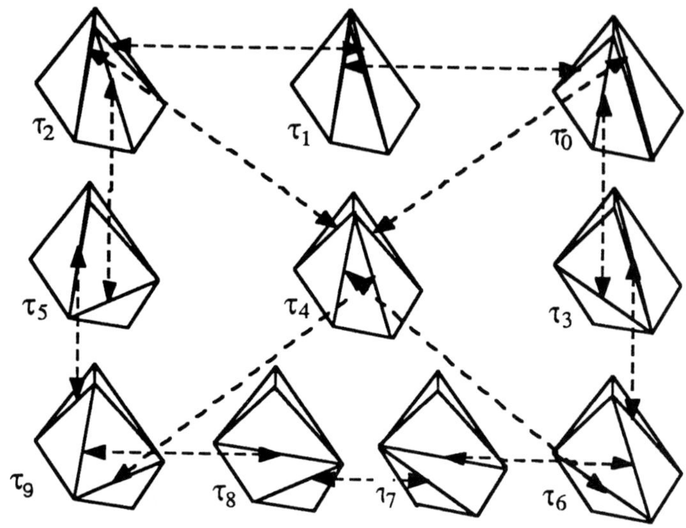

Fig. 41. Ten triangulations of six points.

piecewise cubic and the minimization process leads to the following systems of equations:

$$\sum_{ij\in N_i} \frac{(x_j - x_i)}{\|e_{ij}\|^3}[(x_j - x_i)S_x(V_i) + (y_j - y_i)S_y(V_i)$$

$$+\frac{(x_j - x_i)}{2}S_x(V_j) + \frac{(y_j - y_i)}{2}S_y(V_j) + \frac{3}{2}(F_i - F_j)] = 0$$

$$\sum_{ij\in N_i} \frac{(y_j - y_i)}{\|e_{ij}\|^3}[(x_j - x_i)S_x(V_i) + (y_j - y_i)S_y(V_i)$$

$$+\frac{(x_j - x_i)}{2}S_x(V_j) + \frac{(y_j - y_i)}{2}S_y(V_j) + \frac{3}{2}(F_i - F_j)] = 0 \qquad (4.8)$$

where $N_i = \{ij \ : \ ij, ji \in N_e\}$. This system is sparse and can easily be solved using iterative methods. More discussion on this can be found in [14].

Triangle Patch Step of MNN Method

The final step of the MNN method consists of extending the curve network by using a C^1 triangular patch. A 9-parameter, C^1 triangular patch is a bivariate function, F, defined over the domain of a triangle, T, with vertices V_i, V_j, and V_k which has the property that F interpolates to given function values and first order derivative values at the vertices. Furthermore, it is often assumed that the derivative across a boundary edge varies linearly along this edge. This will guarantee that when two of these triangular patches join and share common data at the vertices they will join with a continuous cross boundary derivative all along the common edge. This would lead to a surface that has continuous

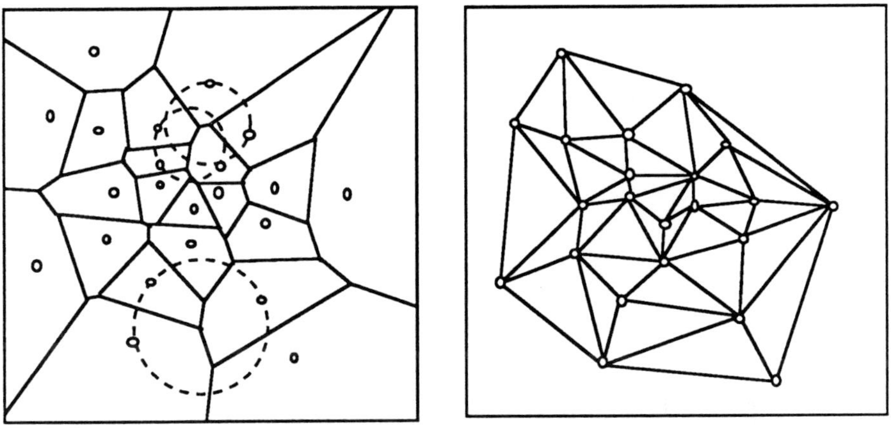

Fig. 42. The Dirichlet tessellation and its dual triangulation.

first order derivatives everywhere. A commonly used 9-parameter, C^1 patch, and the one that is used in the original MNN method, is

$$\begin{aligned} C_\Delta[F](x,y) = \textstyle\sum_{(i,j,k)\in I} \; & \{F(V_i)[b_i^2(3-2b_i) + 6wb_i(b_k\alpha_{ij} + b_j\alpha_{ik})] \\ & + F'_k(V_i)[b_i^2 b_k + wb_i(3b_k\alpha_{ij} + b_j - b_k)] \\ & + F'_j(V_i)[b_i^2 b_j + wb_i(3b_j\alpha_{ik} + b_k - b_j)]\}, \end{aligned}$$

$$F'_k(V_i) = (x_k - x_i)F_x(V_i) + (y_k - y_i)F_y(V_i),$$

$$F'_j(V_i) = (x_j - x_i)F_x(V_i) + (y_j - y_i)F_y(V_i),$$

$$w = \frac{b_i b_j b_k}{b_i b_j + b_i b_k + b_j b_k},$$

$$I = \{(i,j,k),(j,k,i),(k,i,j)\},$$

$$a_{ij} = \frac{\|e_{jk}\|^2 + \|e_{ik}\|^2 - \|e_{ij}\|^2}{2\|e_{ik}\|^2}.$$

This 9-parameter, C^1 interpolant is a discretized version of a transfinite, C^1, triangular interpolant which is described in [13].

4.2.1 Piecewise Hermite Approach to Surface-on-Surface Data

Now that we have covered the basics of the MNN method, we set out to show how it can be extended to the case of surface-on-surface data. As before, we first discuss the case of the domain being a sphere. Here we give only a brief description. More details are given by Nielson and Ramaraj [25]. The first step requires the decomposition of the sphere into a collection of spherical triangles consisting of edges which are geodesic arcs. As we noted above this can be accomplished by using the triangulation which is dual to the Dirichlet tessellation

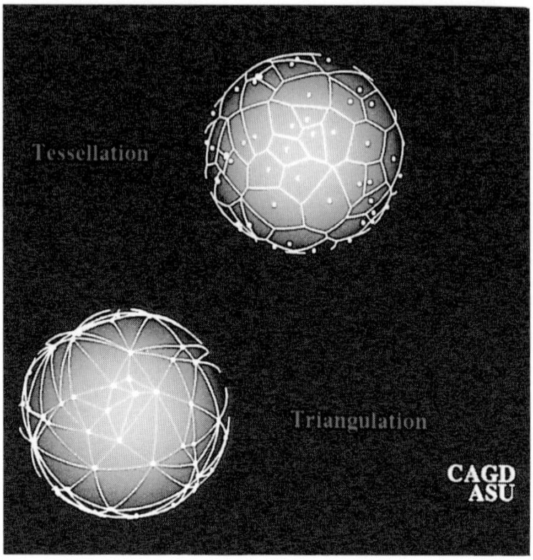

Fig. 43. Spherical tessellation and triangulation.

of the sphere. The ideas of the interpolating curve network extend quite naturally to the spherical case. Over each geodesic arc of the triangulation, the model is a cubic polynomial of the argument consisting of geodesic distance. A similar characterization to that of the original MNN method yields a similar system of linear equations which can be solved in order to obtain the first order derivatives of the model at each data site. Rather than the partial derivatives $\partial F/\partial x = F_x$ and $\partial F/\partial y = F_y$ which are used in the original planar MNN method, for the spherical version, derivatives in the direction of latitude and longitude are used. These are denoted by $F_\theta = \partial F/\partial\theta$ for latitude and $F_\phi = \partial F/\partial\phi$ for longitude. F_θ and F_ϕ serve as "basis" directional derivatives in that any other directional derivative can be written as a linear combination of these two derivatives. In the planar case, a directional derivative in the direction from (x_i, y_i) to (x_j, y_j) can be written as $\partial F/\partial[p_j - p_i] = (x_j - x_i)F_x + (y_j - y_i)F_y$. For the spherical case, the directional derivative in the direction from p_i to p_j is

$$\frac{\partial F}{\partial[p_j - p_i]} = \frac{\partial\theta}{\partial[p_j - p_i]}F_\theta + \frac{\partial\phi}{\partial[p_j - p_i]}F_\phi$$

The quantities $\partial\theta/\partial[p_j - p_i]$ and $\partial\phi/\partial[p_j - p_i]$ are more complicated than the simple $(x_j - x_i)$ or $(y_j - y_i)$ for the planar case. Formulas can be found in Appendix A of Ramaraj [28]. Analogous to the systems of equations (4.8) of the planar MNN we have for the sphere,

$$\sum_{ij\in N_i} \frac{1}{\|p_j - p_i\|}\frac{\partial\theta}{\partial[p_j - p_i]}(p_i)\left[\frac{\partial\theta}{\partial[p_j - p_i]}(p_i)S_\theta(p_i) + \frac{\partial\phi}{\partial[p_j - p_i]}(p_i)S_\phi(p_i)\right.$$
$$\left. + \frac{1}{2}\frac{\partial\theta}{\partial[p_j - p_i]}(p_i)S_\theta(p_j) + \frac{1}{2}\frac{\partial\phi}{\partial[p_j - p_i]}(p_i)S_\phi(p_j) + \frac{3}{2}\frac{F_i - F_j}{\|p_j - p_i\|}\right] = 0,$$

$$\sum_{ij\in N_i} \frac{1}{\|p_j - p_i\|} \frac{\partial \phi}{[p_j - p_i]}(p_i) \left[\frac{\partial \phi}{\partial [p_j - p_i]}(p_i) S_\phi(p_i) + \frac{\partial \theta}{\partial [p_j - p_i]}(p_i) S_\theta(p_i) \right.$$

$$\left. + \frac{1}{2} \frac{\partial \phi}{\partial [p_j - p_i]}(p_i) S_\phi(p_j) + \frac{1}{2} \frac{\partial \theta}{\partial [p_j - p_i]}(p_i) S_\theta(p_j) + \frac{3}{2} \frac{F_i - F_j}{\|p_j - p_i\|} \right] = 0,$$

which characterize a curve network which minimizes

$$\sigma(F) = \sum_{ij\in N_e} \int_{e_{ij}} \left(\frac{\partial^2 F}{\partial^2 [p_j - p_i]} \right)^2 ds_{ij}.$$

The spherical triangular patch used by Nielson and Ramaraj [25] is based upon the use of the so called side-vertex (see Nielson [12]) interpolants which uses univariate Hermite interpolation along rays emanating from a vertex and joining to the opposing edge. See Fig. 44. We assume that position and derivative

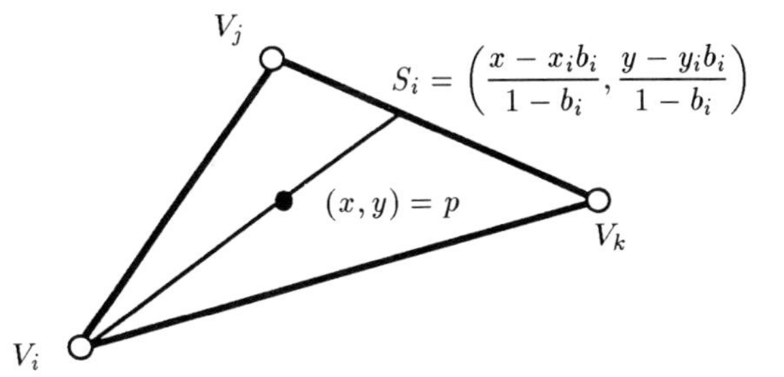

Fig. 44. Side-vertex method notation.

information is available on the entire boundary of a triangle with vertices V_i, V_j, and V_k. Let p be a point in this triangle with the barycentric coordinates b_i, b_j and b_k. The Hermite interpolant emanating from V_i is given by

$$S_i[F](p) = b_i^2(3 - 2b_i)F(Vi) + b_i^2(b_i - 1)F'(V_i)$$
$$+ (1 - b_i)^2(2b_i + 1)F(S_i) + b_i(1 - b_i)^2 F'(S_i)$$

where

$$F'(V_i) = \frac{(x - x_i)F_x(V_i) + (y - y_i)F_y(V_i)}{1 - b_i}$$

and

$$F'(S_i) = \frac{(x - x_i)F_x(S_i) + (y - y_i)F_y(S_i)}{1 - b_i}.$$

$S_i[F]$ has the property that it interpolates to the boundary data provided by F at V_i and on the entire opposing edge e_{kj}. It also matches first order derivatives on this edge and at V_i. It does not necessarily interpolate F or its derivatives on the other two edges. In order to have an interpolant for the entire boundary of the triangular domain, we form the convex combination

$$S[F] = \frac{b_j^2 b_k^2 S_i[F] + b_i^2 b_k^2 S_j[F] + b_j^2 b_i^2 S_k[F]}{b_i^2 b_j^2 + b_j^2 b_k^2 + b_i^2 b_k^2}$$

which has the property that it matches F and its first order derivatives on the entire boundary of the triangular domain.

The interpolant $S[F]$ is actually a transfinite triangular interpolant. This means that if any function values and derivatives values are given on the entire boundary of the triangular domain, then S will interpolate to all of this data. In our case, we only know the function and its first order derivatives at the three vertices of the triangular domain. The transfinite information required by S is itself obtained by interpolation. Function values along an edge are obtained by applying univariate Hermite interpolation along this edge. Cross boundary derivatives (derivatives in a direction perpendicular to an edge) are determined by a simple linear interpolation between these values at the vertices. In this case, where the boundary information has been discretized as just mentioned, it is possible to obtain a final interpolant with simpler weights in the convex combination. Namely,

$$S[F] = \frac{b_j b_k S_i[F] + b_i b_k S_j[F] + b_j b_i S_k[F]}{b_i b_j + b_j b_k + b_i b_k}. \tag{4.9}$$

The ideas behind the construction of S extend very easily to a spherical triangle domain. The basic side-vertex operator for the sphere is

$$S_i(p) = t^2(3-2t)F(V_i) + (1-t)^2(2t+1)S(M_i)$$
$$+t^2(t-1)\|V_i - M_i\|\frac{\partial F}{\partial[V_i - M_i]}(V_i) + t(1-t)^2\|V_i - M_i\|\frac{\partial F}{\partial[V_i - M_i]}(M_i)$$

where M_i is the point of intersection of the geodesic edge e_{ij} and the plane containing V_i , p and the origin, and $t = \|p - M_i\|/\|V_i - M_i\|$ and $\| \ \|$ means geodesic distance. A convex combination will lead to an interpolant matching both position and derivative values on the entire boundary of the spherical triangle domain. The convex combination used by Nielson and Ramaraj is the same as that of equation (4.9) and b_i, b_j and b_k are the barycentric coordinates of the planar triangle with vertices V_i, V_j and V_k. (It has been conjectured that barycentric coordinates for spherical triangles do not exist.)

Pottman [27] has discussed the generalization of the MNN method to the more general situation where we have a two dimensional surface domain, D which contains the independent data sites, $p_i = (x_i, y_i, z_i)$, $i = 1, \ldots, N$. The problem is to construct a function F defined on all of D so that $F(p_i) = F_i$, $i = 1, \ldots, N$. Rather than work directly with the surface domain, D, it is replaced with a piecewise triangular approximation, $D^* = \bigcup_{ijk \in N_t} T_{ijk}$, where T_{ijk} is the three dimensional triangle with vertices V_i, V_j and V_k. Most likely this approximation is obtained in an adaptive manner with more vertices near points of high curvature of D. It is required that the vertices of D^* include the independent data sites, p_i, $i = 1, \ldots, N$.

The minimum norm network in this context is piecewise cubic and defined over the edges of D^*. It is characterized as minimizing the quantity

$$\sigma(F) = \sum_{ij \in N_e} \int_{e_{ij}} \left(\frac{d^2 F}{d^2 e_{ij}}\right)^2 ds_{ij}$$

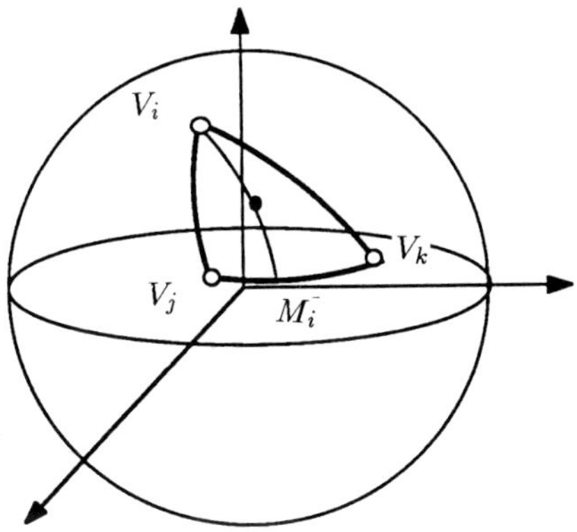

Fig. 45. Spherical side-vertex triangular interpolant.

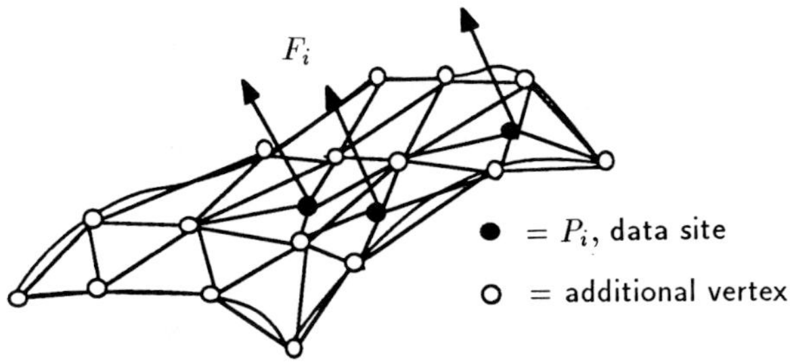

Fig. 46. Generalized MNN method.

subject to $F(p_i) = F_i$. As with the original MNN, the derivatives which define this network are computed as the solution of a linear system of equations and the surface is extended to the entire domain by the use of a C^1 triangular interpolant and, in this case, a variation of the side-vertex method is used again.

4.2.2 Piecewise Hermite Approach to Volumetric Data

We now consider the topic of applying the ideas of the piecewise Hermite approach to volumetric data. As with the univariate cubic spline and the planar MNN method, there are three steps: 1) Decomposition of domain into subdomains, 2) Computation of derivatives and 3) Definition of final interpolant by means of a Hermite type of interpolant defined over subdomains. For our discussion here, we make no assumptions about the volumetric data $(x_i, y_i, z_i; F_i)$, $i = 1, \ldots, N$ other than the dependent data sites are all distinct.

Analogous to the intervals for the cubic spline and the triangulation of the convex hull in the case of the MNN method, there is the decomposition of the convex hull of the points $p_i = (x_i, y_i, z_i)$, $i = 1, \ldots, N$, into a collection of tetrahedra, T_{ijkl}. This "tetrahedronization" is represented as a collection of 4-tuples $= n_t$ and is characterized as the dual of the Dirichlet tessellation of the points p_i, $i = 1, \ldots, N$. This means that $ijkl$ will be in n_t provided the circumscribing sphere for the points p_i, p_j, p_k and p_l contains no other data points. The next step involves the computation of all first order derivatives at the data sites. These values are denoted by $F_x(p_i)$, $F_y(p_i)$ and $F_z(p_i)$. Analogous to the MNN, the following quantity is minimized:

$$\sigma(F) = \sum_{ij \in N_e} \int_{e_{ij}} \left(\frac{d^2 F}{d^2 e_{ij}} \right)^2 ds_{ij}$$

where $N_e = \{ij \ : \ e_{ij}$ is an edge of one of the tetrahedra$\}$. The solution is a piecewise cubic network such that

$$\sum_{ij \in N_i} \frac{(x_j - x_i)}{\|e_{ij}\|^3} [(x_j - x_i)S_x(V_i) + (y_j - y_i)S_y(V_i) + (z_j - z_i)S_z(V_i)$$

$$+ \frac{(x_j - x_i)}{2} S_x(V_j) + \frac{(y_j - y_i)}{2} S_y(V_j) + \frac{(z_j - z_i)}{2} S_z(V_j) + \frac{3}{2}(F_i - F_j)] = 0$$

$$\sum_{ij \in N_i} \frac{(y_j - y_i)}{\|e_{ij}\|^3} (x_j - x_i)S_x(V_i) + (y_j - y_i)S_y(V_i) + (z_j - z_i)S_z(V_i)$$

$$+ \frac{(x_j - x_i)}{2} S_x(V_j) + \frac{(y_j - y_i)}{2} S_y(V_j) + \frac{(z_j - z_i)}{2} S_z(V_j) + \frac{3}{2}(F_i - F_j)] = 0$$

$$\sum_{ij \in N_i} \frac{(z_j - z_i)}{\|e_{ij}\|^3} [(x_j - x_i)S_x(V_i) + (y_j - y_i)S_y(V_i) + (z_j - z_i)S_z(V_i)$$

$$+ \frac{(x_j - x_i)}{2} S_x(V_j) + \frac{(y_j - y_i)}{2} S_y(V_j) + \frac{(z_j - z_i)}{2} S_z(V_j) + \frac{3}{2}(F_i - F_j)] = 0.$$

This linear system of equations is completely analogous to those of the original, planar MNN method and it can be solved in a very similar fashion by using an iterative method. Once the first order derivatives at each data site have been computed, we are ready to fill in the model with a C^1 tetrahedral interpolant. This requires the definition of an interpolant, F, defined on T_{ijkl} that will match given position and derivative information at all four vertices. The method we describe here is a three dimensional generalization of the side-vertex interpolant described above [12]. It is called the face-vertex method [24] and in its transfinite version, we assume that position and derivative information is available at all locations on the four faces which make up the boundary of the tetrahedron T_{ijkl}. The basic face-vertex operator is defined as

$$S_i[F](p) = b_i^2(3 - 2b_i)F(V_i) + b_i^2(b_i - 1)F'(V_i)$$
$$+(1 - b_i)^2(2b_i + 1)F(S_i) + b_i(1 - b_i)^2 F'(S_i) \quad (4.10)$$

where

$$F'(V_i) = \frac{(x - x_i)F_x(V_i) + (y - y_i)F_y(V_i) + (z - z_i)F_z(V_i)}{1 - b_i}$$

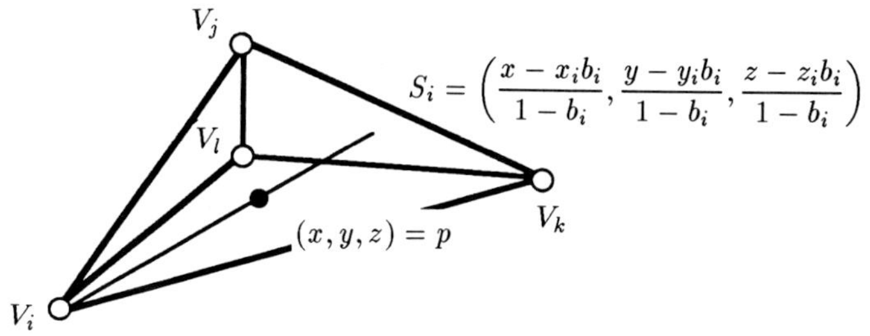

Fig. 47. The face-vertex method notation.

and

$$F'(S_i) = \frac{(x - x_i)F_x(S_i) + (y - y_i)F_y(S_i) + (z - z_i)F_z(V_i)}{1 - b_i}.$$

The point S_i is the intersection point of the ray from V_i through p and the face opposite V_i and the derivatives are taken in the direction of this same ray. If we form the convex combination

$$S[F] = \frac{b_j^2 b_k^2 b_l^2 S_i[F] + b_i^2 b_k^2 b_l^2 S_j[F] + b_j^2 b_l^2 b_i^2 S_k[F] + b_j^2 b_k^2 b_i^2 S_l[F]}{b_j^2 b_k^2 b_l^2 + b_i^2 b_k^2 b_l^2 + b_j^2 b_k^2 b_i^2 + b_i^2 b_j^2 b_k^2}$$

then $S[F]$ will match position and derivative values on the entire boundary of T_{ijkl}. When we actually use the face-vertex interpolant in this application, we only know position and derivative values at the four vertices. In order to apply the operator of equation (4.10), we need to define position and derivatives on the entire boundary of T_{ijkl}. The process of defining the transfinite boundary information from a finite (discrete) amount of data (usually given at the vertices) is called "discretization". In the case of the present face-vertex method, we describe how to do this in two steps. First we assume that information is known on all the edges of the tetrahedra and we describe how to extend it to the entire boundary. Secondly, we describe how to discretize the edge information itself. If we know both position and derivative information on the edges, then we can use any C^1 transfinite planar triangular interpolant to define position values on the interior points of the face triangles. For example, the side-vertex method itself could be used. Specifying position information on a face also implies some information about the derivatives on the interior of a triangle. Namely, all directional derivatives in a direction parallel to the face triangle are determined and so, in order to completely specify all derivatives, we need only provide a definition for the derivative perpendicular to the face. For this we use the C^0 version of the side-vertex interpolant which interpolates position data only and not derivatives, but we apply it to the edge data consisting of derivatives normal to a face. The C^0 side-vertex interpolant has the simple form

$$A[F] = (1-b_i)F(S_i) + (1-b_j)F(S_j) + (1-b_k)F(S_k) - b_iF(V_i) - b_jF(V_j) - b_kF(V_k).$$

We now describe the second step of the discretization which is how to compute edge information when only the point and derivative values are known at the four vertices. For position only on an edge, we simply use univariate cubic Hermite interpolation. This will also specify one directional derivative on the edge; namely $\partial F/\partial e_{ij}$ which will vary as a quadratic polynomial. In order to get a C^1 join from one tetrahedron to the next, the other two directional derivatives must vary linearly along this edge. This is accomplished by specifying the gradient, ∇F, by the relationship

$$\nabla F_{ij}(p) = (1-t)\nabla F_i + t\nabla F_j + [\frac{\partial F}{\partial e_{ij}}(p) - ((1-t)\nabla F_i + t\nabla F_i, e_{ij})]e_{ij}$$

where $\nabla F_i = (F_x(p_i), F_y(p_i), F_z(p_i))$ and $t = \|p-p_i\|/\|p_j-p_i\|$. This interpolation of the gradient is consistent with the value $\partial F/\partial e_{ij}$ already specified because $(\nabla F_{ij}(p), e_{ij}) = \partial F/de_{ij}$ and it also has the property that for $(n, e_{ij}) = 0$,

$$(\nabla F_{ij}(p), n) = (1-t)(\nabla F_i, n) + t(\nabla F_j, n),$$

and so we have linear interpolation for any derivative in a direction perpendicular to e_{ij}. This completes the definition of the 16-parameter, C^1, tetrahedron interpolant which is based upon the face-vertex interpolant.

Acknowledgements

This work was supported by the North Atlantic Treaty Organization under grant RG 0097/88. The time and effort of the students in CSC 573 during the spring semester of 1992 at ASU is gratefully acknowledged. The contributions of many colleagues, at ASU and elsewhere, is greatly appreciated.

References

[1] R. Carlson, T. Foley (1991). *The parameter R2 and multiquadric interpolation. Computers and Mathematics with Applications*, 21(9):29–42.

[2] H.E. Cline, W.E. Lorensen, S. Ludke, C.R. Crawford, B. C. Teeter (1988). *Two algorithms for the reconstruction of surfaces from tomograms. Medical Physics.*

[3] T. Dierks (1990). *Analysis and visualization of scattered volumetric data,*. In *M. S. Thesis.* Arizona State University, Tempe, AZ.

[4] M. Dürst (1988). *Additional Reference to "Marching Cubes". Computer Graphics,* 22(2).

[5] T. Foley, R. Franke, H. Hagen, G.M. Nielson, D. Lane (1990). *Interpolation of scattered data on closed surfaces. Computer Aided Geometric Design,* 7:303–312.

[6] T. Foley, G.M. Nielson, D. Lane, R. Ramaraj (1990). *Visualizing Functions over a Sphere. IEEE Computer Graphics & Appl.,* 10(1):32–40.

[7] T.A. Foley, D. Lane, G. Nielson (1990). *Towards Animating Ray-Traced Volume Visualization. Visualization and Computer Animation Journal,* 1:2–8.

[8] R. Franke, G. Nielson (1990). *Scattered Data Interpolation and Applications: A Tutorial and Survey.* In H. Hagen, D. Roller, editors, *Geometric Modelling: Methods and Their Application.* Springer-Verlag, Berlin.

[9] C. Lawson (1977). *Software for C1 surface interpolation.* In J.R. Rice, editor, *Mathematical Software III*, pp. 161–194. Academic Press, New York.

[10] M. Levoy (1988). *Display of Surfaces from Volume Data. IEEE Computer Graphics & Appl.*, 8(3):29–37.

[11] W.E. Lorensen, H.E. Cline (1987). *Marching Cubes: A High-Resolution 3D Surface Construction Algorithm. Computer Graphics*, 21(4):163–169.

[12] G.M. Nielson (1979). *The side-vertex method for interpolation in triangles. Journal of Approx. Theory*, 25:318–336.

[13] G.M. Nielson (1980). *Minimum norm interpolation in triangles. SIAM Journal Numer. Analysis*, 17:46–62.

[14] G.M. Nielson (1983). *A method for interpolating scattered data based upon a minimum norm network. Mathematics of Computation*, 40:253–271.

[15] G.M. Nielson (1987). *Coordinate free scattered data interpolation,.* In C. Chui, F. Utreras, L. Schumaker, editors, *Topics in Multivariate Approximation*, pp. 175–184. Academic Press, NY.

[16] G.M. Nielson (1987). *An example with a local minimum for the MinMax ordering of triangulations.* In *Technical Report TR-87-014.* Arizona State University Computer Science.

[17] G.M. Nielson (1991). *On the Topic of Interactive Scientific Visualization.* In D. Thalman, N. M. Thalman, editors, *New Trends in Animation and Visualization*, pp. 135–150. Wiley.

[18] G.M. Nielson (1991). *Visualization in Scientific and Engineering Computing. IEEE Computer*, 24(9):58–66.

[19] G.M. Nielson, T. Dierks (1991). *Modelling and Visualization of Scattered Volumetric Data.* In *SPIE Conference Proceedings 1459, San Jose.*

[20] G.M. Nielson, T. Foley, B. Hamann, D. Lane (1991). *Visualization and modelling of scattered multivariate data. IEEE Computer Graphics & Appl.*, 11(?):47–55.

[21] G.M. Nielson, B. Hamann (1990). *The asymptotic decider: Resolving the ambiguity in marching cubes.* In *Proceedings of Visualization '91*, pp. 83–91. IEEE Computer Society Press, Los Alamitos, California.

[22] G.M. Nielson, B. Hamann (1990). *Interactive techniques for visualizing volumetric data.* In *Proceedings of Visualization '90*, pp. 45–50. IEEE Computer Society Press, Los Alamitos, California.

[23] G.M. Nielson, D. Olsen (1987). *Direct manipulation techniques for 3D objects using 2D locator devices.* In *Proceedings of Workshop on Interactive 3D Graphics, Chapel Hill, NC, Oct. 23-24, '86*, pp. 175–182. ACM, NY.

[24] G.M. Nielson, K. Opitz (1992). *The face-vertex method for smooth interpolation in tetrahedra.* In *Technical Report, TR-92-013.* Arizona State University Computer Science.

[25] G.M. Nielson, R. Ramaraj (1987). *Interpolation over a sphere. Computer Aided Geometric Design*, 4:41–57.

[26] B.T. Phong (1975). *Illumination of computer generated pictures. Communications of ACM*, 18(6):311–317.

[27] H. Pottmann . *Interpolation on surfaces using minimum norm networks.* manuscript.

[28] R. Ramaraj (1986). *Interpolation and display of scattered data over a sphere.* In *M. S. Thesis.* Arizona State University.

[29] J. Tvedt (1991). *A software system for comparison of trivariate scattered data interpolation methods.* In *M. S. Thesis.* Arizona State University,.

Curve and Surface Interrogation

*Hans Hagen, Stefanie Hahmann, Thomas Schreiber,
Ernst Gschwind, B. Wördenweber, Y. Nakajima*

Free-form curves and surfaces are very important for sophisticated CAD/CAM systems. Apart from the geometric modelling aspect of these curves and surfaces, the analysis of their quality is a necessary tool in the design and construction process. The purpose of this paper is to give a critical survey on curve and surface interrogation methods and to present generalized focal surfaces as a new surface interrogation tool.

1 Introduction

In the scope of this article, we will discuss various visualization techniques, which have the goal of identifying unwanted curvature regions (inflection points, dents, etc.). There is no one optimal method for all kinds of applications. We present some methods, which we think are most appropriate for certain applications.

First we will discuss the reflection line method. This method simulates the so-called light cage used in automobile industry by computer, and is a very effective tool to test the aesthetic quality of a surface. The isophote method analyzes surfaces by determining lines of equal light intensity; silhouettes are special isophotes. This technique is used to test geometric continuity between surface patches. The convexity of curves and surfaces can be visualized by orthotomics or polarity methods. To detect undesired curvature situations on a surface we recommend the use of generalized focal surfaces. We introduce here this new method and show its effectiveness by displaying various examples.

More details and aspects on curve and surface interrogation algorithms can be found in [2, 3, 5, 6, 7, 8, 9].

2 Reflection Line Method

The reflection line method determines unwanted dents by emphasizing irregularities in the reflection line pattern of parallel light lines.

Let $X(u,w)$ be a representation of the surface to investigate, and let $N(u,w)$ be the normal vector of the surface. Furthermore a *light line* L is given in parameter form:

$$L(t) = L_0 + t \cdot \vec{s} \tag{2.1}$$

where ($t \in \mathbb{R}$), and point A is a fixed eye point.

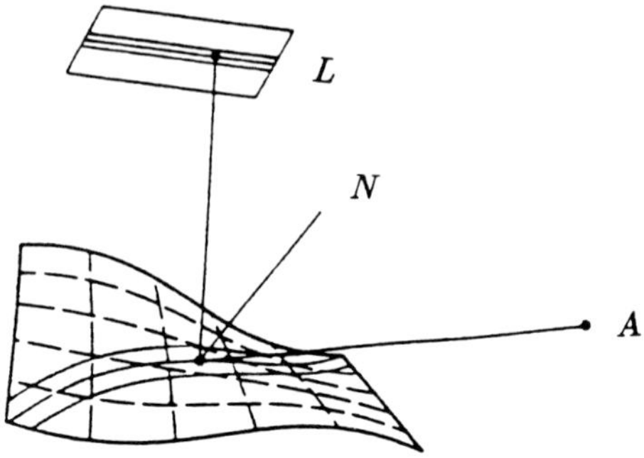

Fig. 1. Reflection line method.

The reflection line is the projection of the line L on the surface X, which can be seen from the eye point A, if the light line L is reflected on the surface (Fig. 1).

From geometric dependencies the following reflection condition is easily seen:

$$\frac{\vec{a}}{\|\vec{a}\|} + \frac{\vec{b}}{\|\vec{b}\|} = 2N(u,w) < N(u,w), \frac{\vec{b}}{\|\vec{b}\|} > = 2N(u,w) < N(u,w), \frac{\vec{a}}{\|\vec{a}\|} > \qquad (2.2)$$

where $\vec{a} = P - A$, $\vec{b} = L - P$ and $<,>$ is the scalar vector product. In order to evaluate the surface one uses a set of reflection lines with the direction $\vec{s}_0$. One now steps along each curve of the set and obtains, for a fixed eye point A, the following non-linear system of equations for the unknown parameters u and w of the reflection point to be determined:

$$\vec{b} + \lambda\vec{a} = 2N(u,w) < N(u,w), \vec{b} > \quad \text{with} \quad \lambda := \frac{\|\vec{b}\|}{\|\vec{a}\|}. \qquad (2.3)$$

These three non-linear equations can be reduced to two by eliminating λ; this system can then be solved by numerical methods, but the existence and unambiguity of solutions has to be ensured by an appropriate choice of the eye point A (see [7]).

Figure 2 shows the reflection line analysis of a hair dryer. The surface has irregularities which can be clearly recognized in the bundle of reflection lines.

3 Isophotes

This method analyzes surfaces by lines of equal light intensity, the isophotes. If $X(u,w)$ is a parametrization of the surface and $\vec{l}$ the direction of a parallel lighting, then the isophote condition is given by:

$$< N(u,w), \vec{l} >= c = \text{const.} \qquad (3.1)$$

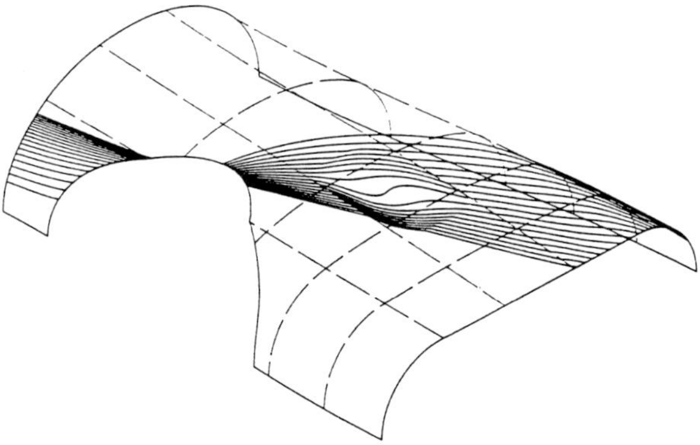

Fig. 2. Reflection line analysis of a hair dryer.

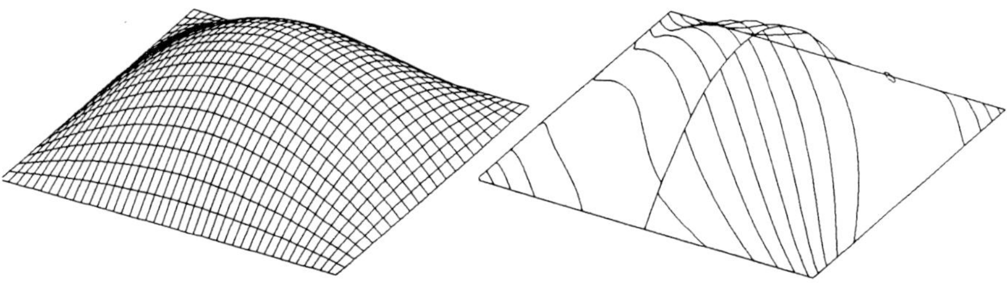

Fig. 3. C^0-continuous surface analyzed with isophotes.

where $N(u,w)$ is the normal vector of the surface $X(u,w)$.

Note that silhouettes are special isophotes ($c = 0$) with respect to the light source. If the surface is C^r-continuous, then the isophotes are C^{r-1}-continuous curves. Figure 3, 4 and 5 show surfaces, which don't seem to contain visual irregularities. The gaps in the isophotes of Fig. 3 prove that the surfaces are only C^0-continuous; whereas in Fig. 4 the discontinuities in the isophotes show the G^1-continuity of the surface; finally the continuous isophotes of Fig. 5 mark G^2-continuous surfaces.

Isophotes are constructed with the isophote conditions, whereby in general various different values for c are tested. The equations of type (3.1) are solved numerically.

This method has some severe drawbacks:

1. Sometimes properties of the isophotes may not be clearly recognized, if the observer has an unfortunate line of sight. Either the surface must be rotated, or the observation point must be changed.

2. One can test 1000 isophote directions, there can still be a gap in the 1001 direction.

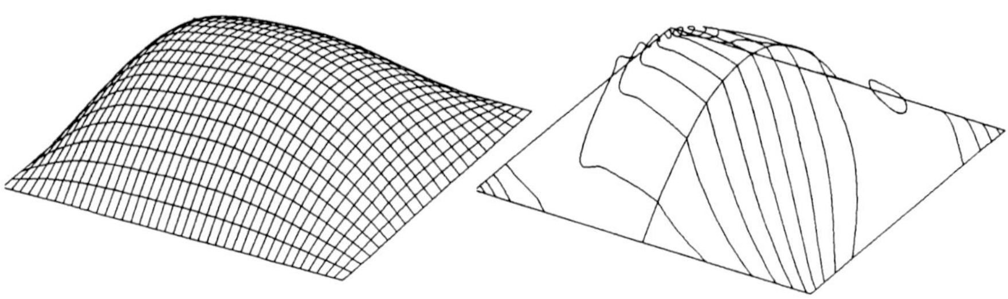

Fig. 4. G^1-continuous surface analyzed with isophotes.

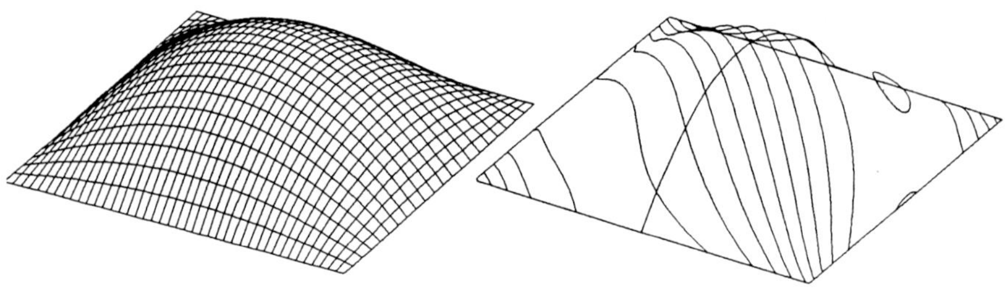

Fig. 5. G^2-continuous surface analyzed with isophotes.

3. No human being likes to sit in front of a graphics terminal for a longer period of time and to watch isophotes coming up.

What we need is an automatic process. There is an automatic way to test the continuity across the boundaries of a patchwork [9]. We give a short description of this algorithm; for the necessary fundamentals of differential geometry see the Appendix.

The envelope of the tangent planes along a curve y on the surface is a developable, ruled surface Φ. At each point p of y, the tangent $\dot{y}$ is a conjugate direction to the corresponding ruling of Φ. Conjugate directions are conjugate diameters of the Dupin indicatrix. Such conjugate directions satisfy the symmetric bilinear equation:

$$h_{11}\Delta u\Delta\tilde{u} + h_{12}(\Delta u\Delta\tilde{v} + \Delta\tilde{u}\Delta v) + h_{22}\Delta v\Delta\tilde{v} = 0 \tag{3.2}$$

This relation degenerates at parabolic points, because the asymptotic direction (the direction in which the normal section curvature vanishes) is the conjugate to itself, but also conjugate to all other directions. At planar points, we have this degeneration for each (tangent) direction.

Let X^1 and X^2 denote the two G^2 surface patches of the patchwork X, G^1-continuously linked together along the common boundary curve y (Fig. 6). Since both X^1 and X^2 have y as a surface curve, and the tangent planes along y are unique, the Dupin indicatrices have a common diameter, but in general there are no further common elements.

We now consider an isophote passing through P. The tangents t_i of c at P with respect to X^i are conjugate to the orthogonal projection f of the light ray onto the tangent plane. In general the isophote c shows a tangent discontinuity at P if the Dupin indicatrices of X^1 and X^2 are not equal, but we have to avoid the situation $f = \dot{y}$ or $f = y'$. These considerations lead to the following algorithm for displaying the curvature situation across the boundary curves of a number of surface patches:

1. give f
2. calculate the conjugate directions t_1 and t_2 and the angle α between t_1 and t_2
3. vary $f(\Delta u = \cos\varphi; \Delta w = \sin\varphi)$ and calculate $\alpha_{\max}$
4. **if** $\alpha_{\max} = \pi$
   then the surface is G^2-continuous
   else display curvature discontinuity.

More details about this algorithm can be found in [9].

Figures 7 and 8 show the analysis of two surfaces, composed of two patches, with the generalized isophote method. The plot on the right hand side of Fig. 7 indicates curvature discontinuity along the common boundary curve whereas the surface in Fig. 8 shows curvature continuity.

4 Orthotomics

Starting with a planar curve $X(t)$, we choose a point P, which is not on $X(t)$, and which is not on any tangent of $X(t)$, cf. Fig 9. If P is reflected by a tangent of $X(t)$, one obtains the point Y_2.

Varying t within the parameter range of $X(t)$, a curve $Y_2(t)$ describes the so-called 2-*orthotomic* of $X(t)$ with respect to P. If the factor 2 is replaced by a factor k, then we speak of a k-*orthotomic*. The parameter form is

$$Y_k(t) = P + k < X(t) - P\ ,\ N(t) > N(t) \tag{4.1}$$

where $N(t)$ denotes the normal vector of the curve.

The following fact is important for applications:

(4.2) Let $X(t)$ be a regular planar curve and let P be a point not on the curve or on any tangent of the curve. Then the k-orthotomic curve $Y_k(t)$ of $X(t)$ with respect to P has a singularity in $t = t_0$, if and only if $X(t)$ has an inflection point in $t = t_0$.

Figure 10 shows a set of Bezier curves. The corresponding 10-orthotomics displayed in Fig. 11 show that the two bottom curves are not convex.

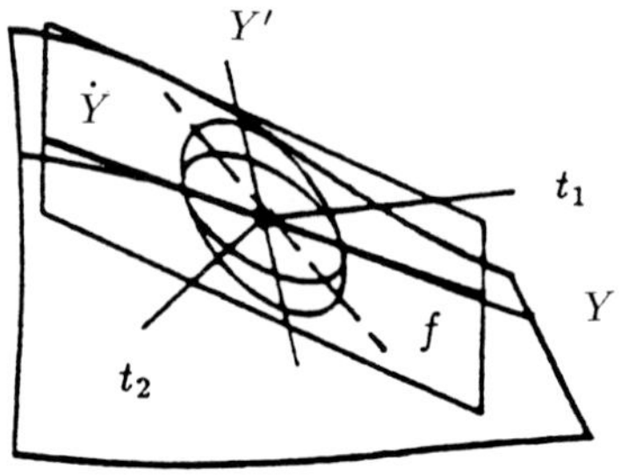

Fig. 6. Generalized isophote method.

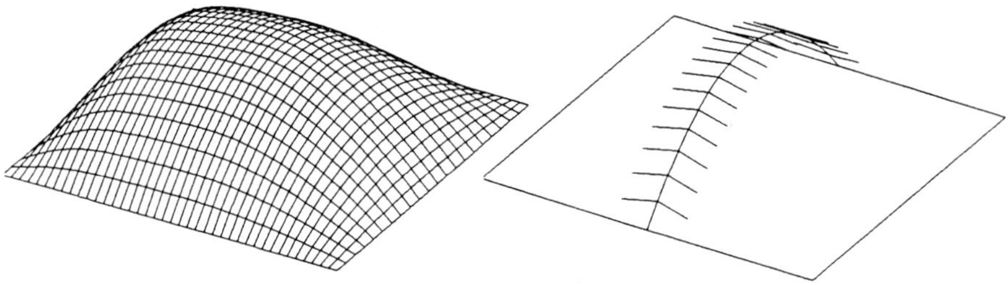

Fig. 7. G^1-continuous surface and its analysis by the generalized isophote method.

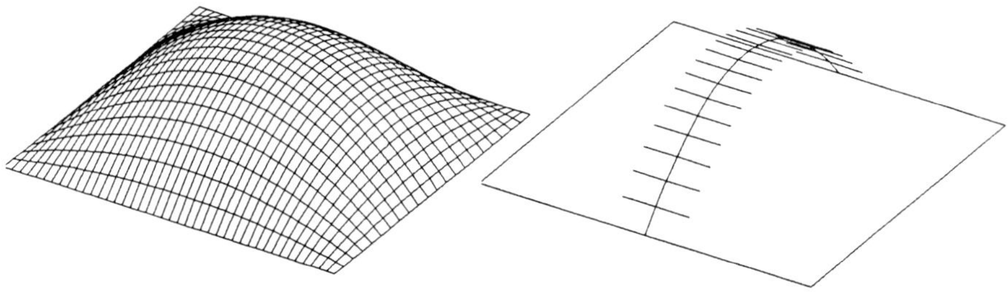

Fig. 8. G^2-continuous surface and its analysis by the generalized isophote method.

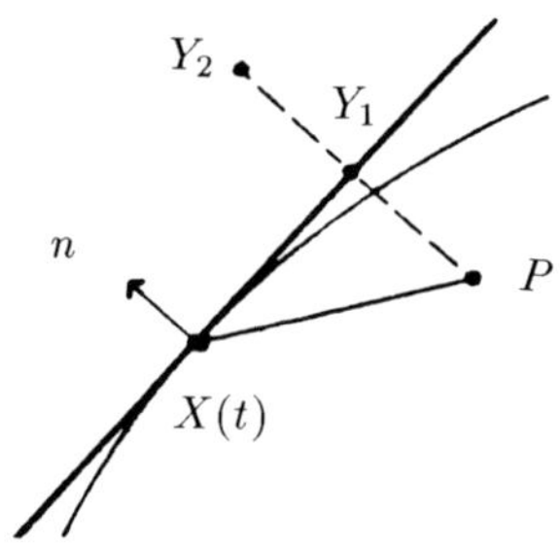

Fig. 9. K-orthotomics.

Fig. 10. A set of Bezier-curves.

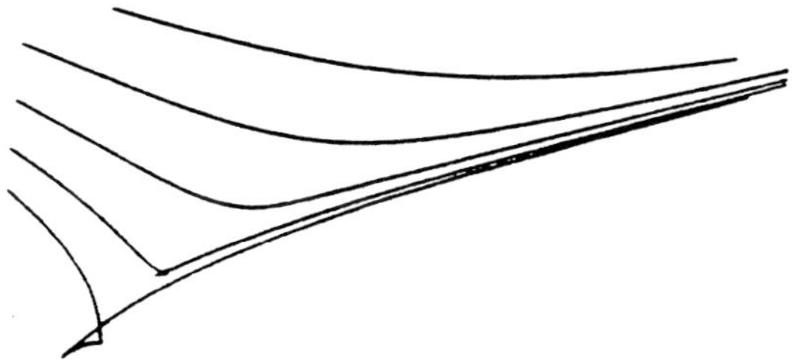

Fig. 11. 10-orthotomics of the Bezier-curves.

This principle can also be applied to surfaces. Starting with a surface $X(u,w)$ and a point P, which is not on X, and which does not lie on any tangent plane of X, P is reflected by a tangent plane of X and this length is multiplied by a factor k. The parameter form of the k-orthotomic surface of X with respect to P is:

$$Y_k(u,w) = P + k < X(u,w) - P\ ,\ N(u,w) > N(u,w) \qquad (4.3)$$

where $N(u,w)$ denotes the normal vector of the surface.

An analogous fact is important for applications:

(4.4) Let $X(u,w)$ be a regular surface and let P be a point not on the surface or on any tangential plane of the surface. The k-orthotomic surface $Y_k(u,w)$ of $X(u,w)$ with respect to P has a singularity in (u_0,w_0), if and only if the Gaussian curvature of X vanishes, or changes its sign at this point.

To demonstrate this method we consider the bicubic tensor product Bezier surface $X^{3,3}(u,w)$ of Fig. 12. The surface has a change of sign for the Gaussian curvature in one corner region, cf. 12, right. The corresponding orthonomic surface in Figure 13 shows the corresponding singularity.

For more details see [6].

5 Polarity Method

We will describe the principle of this method using a planar curve $X(t) = (x(t),y(t))$, cf. Fig 14. An arbitrary point (parameter $t = t_0$) on this curve is mapped by the polarity at the unit circle onto the straight line $x(t_0)+\eta y(t_0)+1 = 0$.

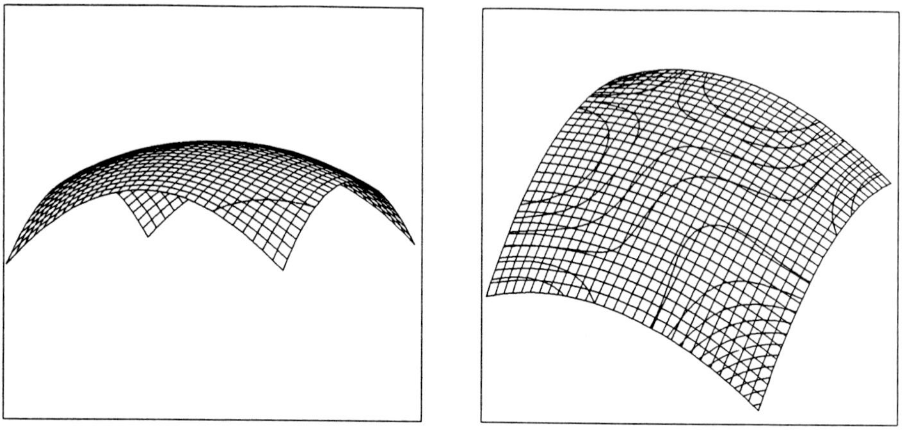

Fig. 12. $X^{3,3}(u,w)$ with contour of Gaussian curvature = 0. The right hand side shows equally spaced Gaussian curvature contours on the surface.

If t is varied within its definition range, the created set of straight lines envelope a polar curve $P(t)$ of $X(t)$. Differentiation and elimination lead to the parameter form of $P(t)$:

$$P(t) = (\xi(t), \eta(t)) = \left(\frac{-\dot{y}}{x\dot{y} - \dot{x}y}, \frac{\dot{x}}{x\dot{y} - \dot{x}y}\right) \tag{5.1}$$

The polar curve of a curve in $3D$ is determined in an analogous way.

In the case of a surface in parametric representation $X(u,w) = (x(u,w), y(u,w), z(u,w))$ the result of a polarity at the unit sphere is a polar surface $P(u,w)$ with the parameter form:

$$P(u,w) = N \cdot \frac{\det(N, X_u, X_w)}{\det(X, X_u, X_w)} = \frac{[X_w, X_u]}{\det(X, X_u, X_w)} \tag{5.2}$$

[,] denotes the cross product.

The following facts are important for applications:

(5.3) If the planar curve $X(t)$ has an inflection point for $t = t_0$, then the polar curve $P(t)$ has a singularity for $t = t_0$.

(5.4) If the surface $X(u,w)$ has a root or change of sign in the Gaussian curvature at (u_0, w_0), then the polar surface has a singularity here.

In Fig. 15, the inflection points W_1 and W_2 of the original curve are mapped into the cusps of the polar curve and both *touching* points of the double tangent T correspond to the self-intersection point of the polar curve.

Figure 16 shows the polar surface of the bicubic Bezier surface of Fig. 12. The polar surface looks similar to the orthotomic surface, because the center of

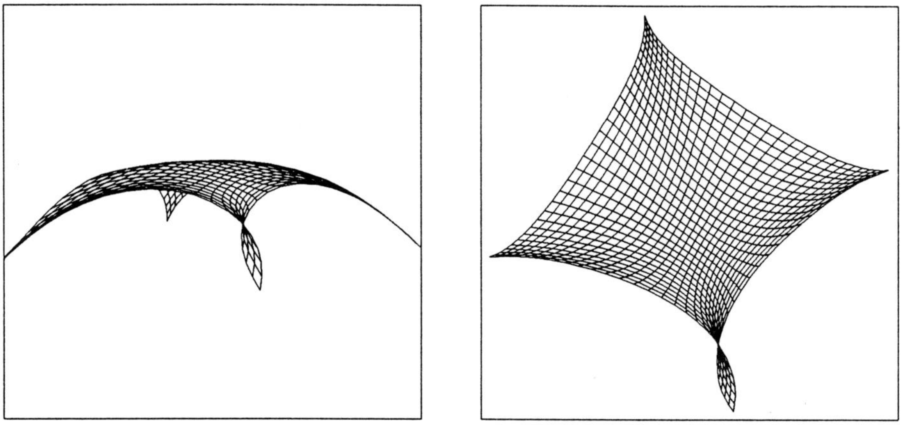

Fig. 13. The orthotomic surface of $X^{3,3}(u,w)$.

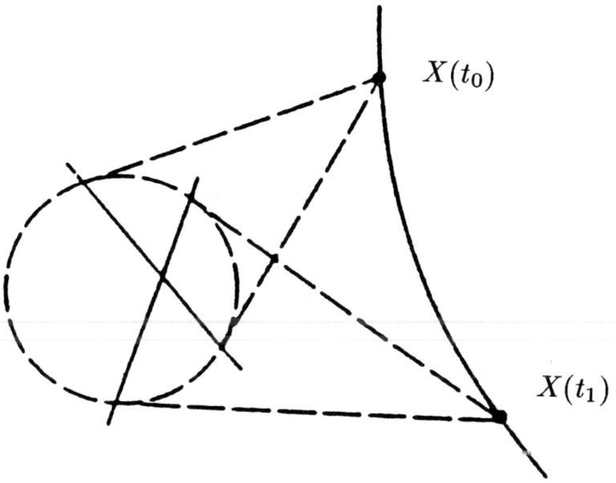

Fig. 14. Polarity method.

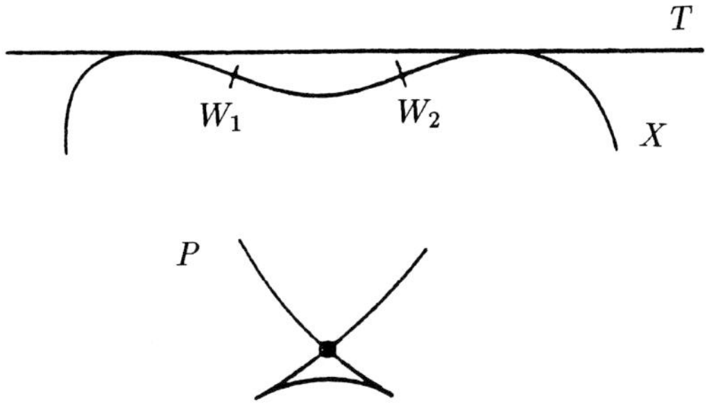

Fig. 15. Polarity analysis of a planar curve.

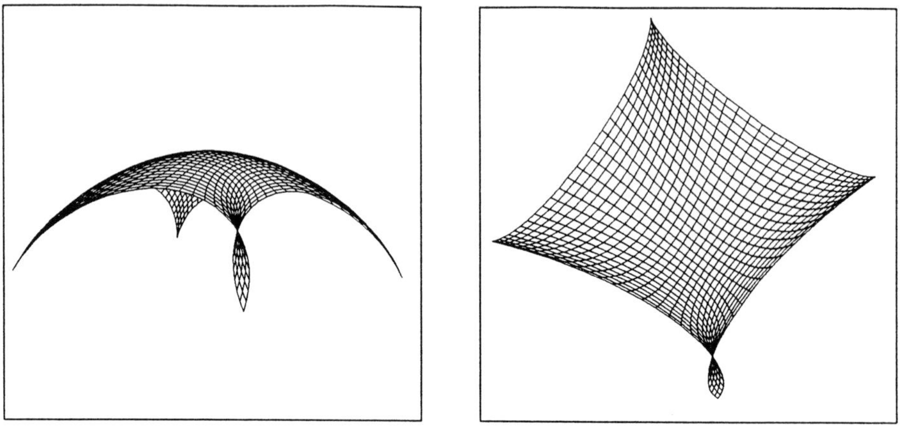

Fig. 16. Polar surface of $X^{3,3}(u,w)$ (see Chapter 4).

polarity is chosen to be equal with the projection point of the orthotomic analysis carried out in section 4.

For more informations about the polarity method see [5].

6 Focal Surfaces

Focal surfaces are special line congruences. Line congruences have been introduced in the field of visualization by Hagen and Pottmann (see [10]). They can be used to visualize the pressure and heat distribution on an aeroplane, temperature, rainfall, ozone over the earth's surface, etc. However, the setting in this paper is different. Here, focal surfaces are used as a surface interrogation tool to analyze the "quality" of the surface before further processing of the surface, for example in a NC-milling operation.

We represent surfaces parametrically as vector valued functions $X(u,w)$, cf. Fig. 17.

Given a set of unit vectors $E(u,w)$ one defines a *line congruence*:

$$C(u,w) = X(u,w) + D(u,w)E(u,w) \tag{6.1}$$

where $D(u,w)$ is called the signed distance between $X(u,w)$ and $E(u,w)$.

If $E(u,w) = N(u,w)$, then $C = C_N$ is a normal congruence. A *focal surface* $F(u,w)$ is a special normal congruence with $D(u,w) = \kappa_1^{-1}(u,w)$ or $D(u,w) = \kappa_2^{-1}(u,w)$:

$$F(u,w) = X(u,w) + \kappa_i^{-1}(u,w)N(u,w)\ ,\ i = 1,2 \tag{6.2}$$

Considering fundamental facts from differential geometry, it is obvious that the centers of curvature of the normal section curves at a particular point on a surface fill out a certain segment of the normal vector at this point. The extremities of these segments are the centers of curvature of two principal directions.

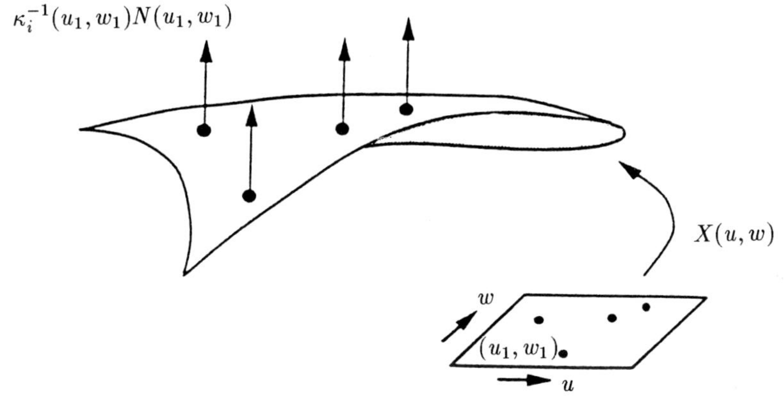

Fig. 17. Normal congruence [11].

These two points are called the *focal points* of this particular normal. This terminology is justified by the fact that a line congruence can be considered as the set of lines touching two surfaces, the focal surfaces of the line congruence (Fig. 18). The points of contact between a line of the congruence and the two focal surfaces are the focal points of this line. It turns out that the focal points of a normal congruence are the centers of curvature of the two principal directions (see [11] and [4]).

The sphere is the only surface for which the two sheets of the focal surface degenerate into a point and the Dupin cyclides are the only surfaces whose focal surfaces degenerate into curves.

We introduce a generalization of the "classical" focal surface concept to achieve a new curve and surface interrogation tool.

For a planar parametric curve $X = X(t)$ one chooses

$$S(t) = X(t) + a\, f(\kappa(t))\, N(t) \quad \text{with} \quad a \in \mathbb{R} \tag{6.3}$$

as the "*variable curvature offset*" (VCO), where f is a scalar factor depending on the curvature $\kappa = \kappa(t)$ (some examples are $f = \kappa$, $f = \kappa^2$ or $f = 1/\kappa$).

Using $S(t) = X(t) + a\,\kappa(t)\,N(t)$ as VCO (with the same parametrization as $X(t)$), we show, that these curves visualize certain properties of the original curve X. Such properties are

(1) inflection points

(2) G^1, G^2 discontinouity

(3) curvature behaviour.

Figure 19 shows 4-th degree Bezier segments and the corresponding VCO with $a = 25000$. The left curve has no inflection points whereas an inflection point of the curve $X(t)$ is marked by an intersection of S and X at this parameter.

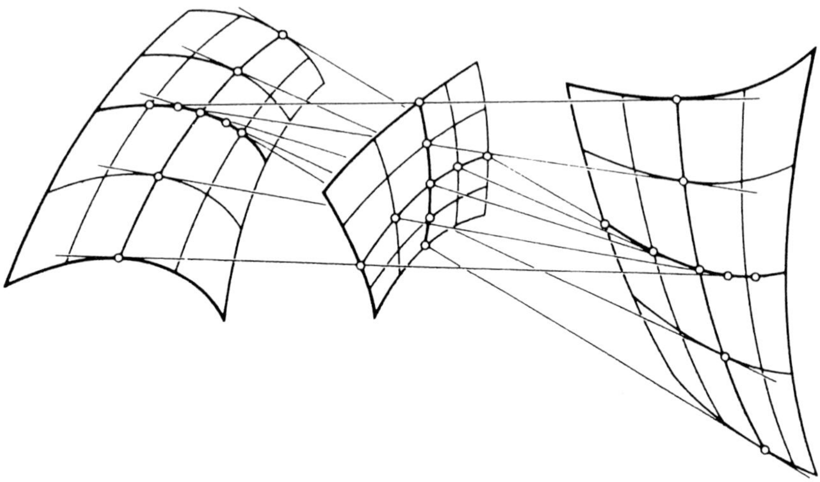

Fig. 18. Surface with focal surfaces.

In Fig. 20, G^2 respectively G^1 continuity of two curve segments are visible in the curvature offsets by G^1 respectively C^0 discontinuity at the corresponding point.

In Fig. 21, one can see the advantage of this method versus curvature plots: the VCO pinpoints the unwanted situations. In the case of inflection points, the intersection is the pinpoint; the connecting normal marks the G^2-discontinuity.

The generalized focal surfaces have the following form:

$$S(u,w) = X(u,w) + a\,F(\kappa_1,\kappa_2)\,N(u,w) \quad \text{with} \quad a \in I\!R \tag{6.4}$$

where the scalar factor f now depends on the the principle curvatures $\kappa_1 = \kappa_1(u,w)$, $\kappa_2 = \kappa_2(u,w)$. Analogously to curves, this VCO makes G^2-discontinuities visible, cf. Fig. A of the color illustrations of this chapter at the end of the book.

It is important to know whether a certain region of the surface is convex (which is equivalent to a non-negative Gaussian curvature), non-convex or whether it contains flat points ($\kappa_1 = \kappa_2 = 0$).

In the example of Fig. B of the color illustrations of this chapter at the end of the book we test the convexity using the Gaussian curvature as a VCO factor $f(u,w) = \kappa_1(u,w) \cdot \kappa_2(u,w)$. Both surfaces intersect at points of vanishing Gaussian curvature; the original surface is red.

To test a surface for flat regions, the VCO factor $f(u,w) = \kappa_1^2(u,w) + \kappa_2^2(u,w)$ is useful, as the focal analysis of Fig. C of the color illustrations of this chapter at the end of the book shows: a flat point exists wherever both surfaces touch.

This generalized focal surface

$$S(u,w) = X(u,w) + a\,(\kappa_1^2(u,w) + \kappa_2^2(u,w))\,N(u,w) \tag{6.5}$$

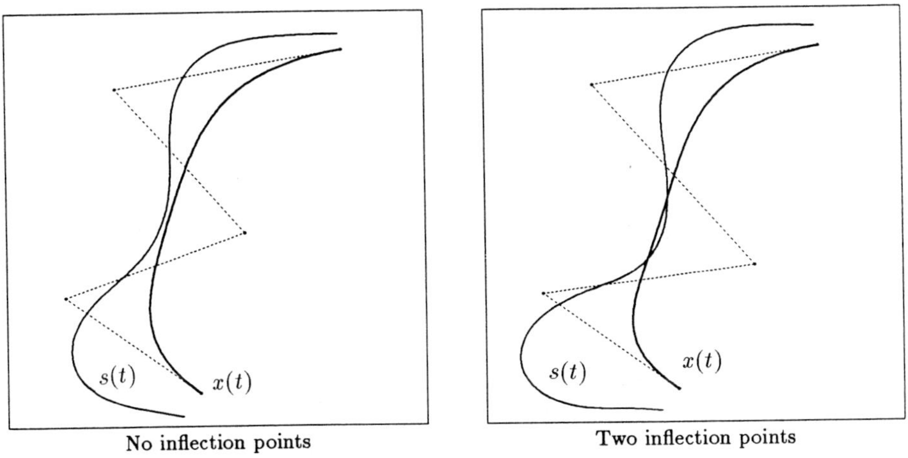

Fig. 19. 4-th degree Bezier segments, $a = 25000$.

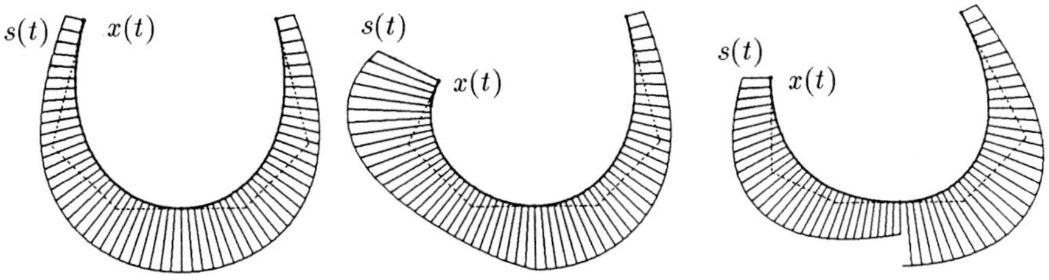

G^{∞}-continuous curve $\quad$ G^2-continuous curve $\quad$ G^1-continuous curve

Fig. 20. Two cubic Bezier segments with normals.

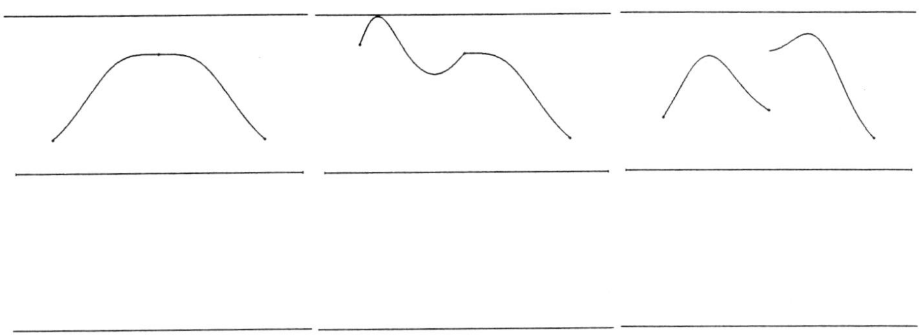

Fig. 21. Curvature plots of Fig. 20.

may also be applied for the analysis of smoothed surfaces. Figure D of the color illustrations of this chapter at the end of the book shows a bicubic rational Bézier surface together with its VCO. These two surfaces have the same control polygons. The first has all weights set to 1; in the second, optimal weights are calculated by minimizing the energy integral $\int_S(\kappa_1^2 + \kappa_2^2)ds$ [1].

References

[1] H. Hagen, G.-P. Bonneau . *Variational design of smooth bezier surfaces.* In *to be published in Computing.*

[2] H. Hagen, Th. Schreiber, E. Gschwind (1990). *Methods for surface interrogation.* In *Proceedings Visualization '90*, pp. 187–193.

[3] R. Hartwig, H. Nowacki (1982). *Isolinien und Schnitte in Coonschen Flächen.* In H. Nowacki, editor, *Geometrisches Modellieren, Informatik-Fachberichte 65*, pp. 329–394. Springer Verlag.

[4] J. Hoschek (1971). *Liniengeometrie.* BI Zürich.

[5] J. Hoschek (1984). *Detecting regions with undesirable curvature. CAGD*, 1:183–192.

[6] J. Hoschek (1985). *Smoothing of curves and surfaces. CAGD*, 2:97–105.

[7] R. Klass (1980). *Correction of local surface irregularities using reflection lines. CAD*, 12:73–77.

[8] Th. Poeschl (1984). *Detecting surface irregularities using isophotes. CAGD*, 1:163–168.

[9] H. Pottmann (1988). *Eine Verfeinerung der Isophotenmethode zur Qualitätsanalyse von Freiformflächen. CAD und Computergraphik*, 4:99–109.

[10] H. Pottmann, H. Hagen, A. Divivier (1991). *Visualization functions on a surface. Journal of Visualization and Computer Animation*, 2:52–58.

[11] K. Strubecker (1959). *Differentialgeometrie III.* De Gruyter Berlin.

Appendix: Fundamentals of Differential Geometry

A parametrized C^r-surface is a C^r-differential map $X : U \rightarrow E^3$ of an open domain $U \in R^2$ into the Euclidean space E^3 , where $X_1 := \frac{\partial x}{\partial u}$ and $X_2 := \frac{\partial x}{\partial w}$ are linearly independent.

The two-dimensional linear subspace $T_P X$ of E^3 generated by the span $\{X_1, X_2\}$ is called the tangent space of X at P. The unit normal field N is given by

$$N := \frac{[X_1, X_2]}{||[X_1, X_2]||}. \qquad (A.1)$$

The moving frame $\{X_1, X_2, N\}$ is the Gaussian frame. The Gaussian frame is in general not an orthogonal frame. Every tangential vector field y along the surface $X : U \to E^3$ can be represented in the form:

$$y(u,w) = \Delta u(u,w) \cdot X_1(u,w) + \Delta w(u,w) \cdot X_2(u,w) \qquad (A.2)$$

The bilinear form on $T_P X$ induced by the inner product of E^3 by restriction is called the first fundamental form of the surface. The matrix representation of the first fundamental form I_P with respect to the basis $\{X_1, X_2\}$ of $T_P X$ is given by:

$$\begin{vmatrix} g_{11} & g_{12} \\ g_{21} & g_{22} \end{vmatrix} = \begin{vmatrix} < X_1, X_1 > & < X_1, X_2 > \\ < X_2, X_1 > & < X_2, X_2 > \end{vmatrix} \qquad (A.3)$$

The first fundamental form I_P is symmetric, positive definite and geometric invariant. Geometrically the first fundamental form allows measurements on the surface (length of curves, angles of tangent vectors, areas of regions) without referring back to the space E^3, in which the surface lies.

The linear map $L : T_P X \longrightarrow T_P X$ defined by $L := -dN \circ dX^{-1}$ is called the Weingarten map. The bilinear form II_P defined by $II_P(A, B) := < L(A), B >$ for each $A, B \in T_P X$ is called the second fundamental form of the surface. The matrix representation of II_P with respect to the basis $\{X_1, X_2\}$ of $T_P X$ is given by:

$$h_{ij} = < -N_i, X_j > = < N, X_{ij} > \quad i,j = 1,2 \qquad (A.4)$$

The Weingarten map L is self-adjoint, the eigenvalues k_1, k_2 are therefore real and the corresponding eigenvectors are orthogonal. The eigenvalues k_1, k_2 are the principal curvatures of the surface.

$$K := k_1 \cdot k_2 = \det(L) = \frac{\det(II)}{\det(I)} \qquad (A.5)$$

is called the Gaussian curvature and

$$H := \mathrm{trace}(L) = \frac{1}{2}(k_1 + k_2) \qquad (A.6)$$

is called the mean curvature. For curves on surfaces the geometric interpretations of the second fundamental form follows. Let $A := \Delta u \cdot X_1 + \Delta w \cdot X_2$ be a tangent vector with $\|A\| = 1$. If we intersect the surface with the plane given by N and A, we get an intersection curve y with the following properties:

$$\dot{y}(s) = A \quad \text{and} \quad e_2 = \pm N \qquad (A.7)$$

where e_2 is the principal normal vector of the space curve y. The implicit function theorem implies the existence of this normal section curve. To calculate the extreme values of the curvature of a normal section curve (the normal section curvature) we can use the method of Lagrange multipliers because we are looking for the extreme values of the normal section curvature k_N with the condition $\|\dot{y}(s)\| = 1$. As the result of these considerations we obtain:

Unless the normal section curvature is the same for all directions there are two perpendicular directions A_1 *and* A_2 *in which* k_N *attains its absolute maximum and its absolute minimum values. These directions are the principle directions with the corresponding normal section curvatures* k_1 *and* k_2.

For $A = A_1 \cos\varphi + A_2 \sin\varphi$ we get Euler's formula:

$$k_N = k_1 \cos\varphi + k_2 \sin\varphi \tag{A.8}$$

If the principle directions are taken as coordinate axes, Euler's formula implies the so-called Dupin indicatrix:

$$k_1(u)^2 + k_2(u)^2 = \pm 1 \tag{A.9}$$

We use the Dupin indicatrices as a tool to visualize curvature situations on surfaces. The Dupin indicatrices at elliptic points ($k > 0$) are ellipses, at hyperbolic points ($k < 0$) pairs of hyperbolas, and at parabolic points ($k = 0$) pairs of parallel lines. Planar points ($k_1 = k_2 = 0$) are degenerated parabolic cases.

Sorting for Polyhedron Compositing

Nelson L. Max

There are basically two ways to visualize a scalar function in a volume: (1) draw contour surfaces, or (2) integrate a continuous volume density along viewing rays. The polyhedron compositing scheme of Max, Hanrahan, and Crawfis [9] combines both of these techniques by subdividing the volume cells at contour surfaces, and compositing the resulting polyhedral pieces and surface polygons in back-to-front order. The application of this algorithm to two specific situations is described, each using a special back-to-front sorting method. One application uses hierarchical data resulting from adaptive mesh refinement, and the other uses cloud data from climate simulations.

1 Polyhedron Compositing

Assume the volume to be visualized is divided up into a number of polyhedral cells, with a data value for the scalar function defined at each vertex. The polyhedron compositing method calculates the effects of the cells on the raster image one by one, in back-to-front order. Each pixel corresponds to a ray from the viewpoint. Along the segment of this ray within a single cell, a certain fraction A of the light coming from deeper cells is absorbed or scattered out, and a fraction $T = 1 - A$ is transmitted. Also an intensity C is emitted by the density within the cell. Thus, if raster B represents the light intensity coming from the background and from deeper cells, the updated raster F, including the effect of the new cell, is $F = T \cdot B + C$.

This update must be performed for each pixel in the projection of the cell. To do this, one needs a model of how the volume density emits and absorbs light. A particularly simple one is the density emitter model of Sabella [11]. Assume that a density r of particles varies linearly along the segment of the ray inside the cell. The particles have absorption and emission constants t and e, so that along a length dl of the ray, a fraction $t \cdot r \cdot dl$ of the light from behind is attenuated, and an additional intensity $e \cdot r \cdot dl$ is emitted. When attenuation between the particles and the front surface of the cell is taken into account, one can show (see [9] for details of a more general case) that $T = \exp(-t \cdot L \cdot d)$, $A = 1 - T$, and $C = (e/t) \cdot A$, where L is the length of the segment, and d is the average density along the segment.

The values for L and d are computed by scan converting the faces of the cell into two pairs of depth and density buffers, one pair for the front-facing polygons, and one pair for the back-facing ones. Standard bilinear interpolation is used to find the depth and density values at a pixel from the values at the polygon vertices. Then d is the average of the front and back densities, and L is the difference between the back and front depths. For perspective projection, a perspective depth $1/z$ is used instead of the true depth, since bilinear interpolation of $1/z$ gives the correct interpolated value for planar polygons. (See Newman and Sproull [10].) The length L can then be easily found, taking the slant of the ray into account. Note that each pixel in the projection of the cell must lie in the projections of exactly one front-facing and one back-facing polygon, in order to guarantee that the corresponding ray will intersect the cell in a single segment. This will be the case whenever the cell is convex.

The scan conversion takes advantage of the area coherence of the projections of the volume cells and their face polygons, and the analytic integration uses the coherence in the depth direction. So the combination gives "volume coherence". Thus the algorithm will be significantly more efficient per unit volume if larger cells are used.

It is often useful to have different ranges of the scalar function correspond to different colors, emissivities, and opacities, for example to distinguish air, bone, and soft tissue in medical imaging. To do this, one can divide the volume cells into subcells by contour surfaces separating the different ranges. If any contours are to be rendered as semi-transparent surfaces with highlights, these can be composited between the subcells they separate.

2 A General Sorting Algorithm

A simple topological sorting algorithm from Knuth [7] can be used to produce a consistent back-to-front sort of all the volume cells. A directed graph is constructed, with a node for each cell, and an edge for each face F separating two adjacent cells A and B. The edge is directed from A to B if the viewpoint V is on the same side of the plane of the face F as A, so that B should be drawn before A. This direction can be determined using the dot product of the normal vector to the face F, with a vector from a vertex on F to the viewpoint V. The cells can be sorted by removing cells with no incoming edges from the graph, and entering them at the head of the sorted list. (This means that the last cell added will end up first in the list.)

An initial count is made of the number of incoming edges for all cells. Any cells with count zero are entered on a "ready" list of cells ready to be removed. Whenever a cell from the ready list is removed from the graph, its outgoing edges must also be removed. This decrements the incoming count of the cell at the other end of the edge, and if the count becomes zero, that cell is entered on the ready list. If the ready list becomes empty before the graph does, then the graph contains cycles, and the algorithm reports that the sort is impossible.

Note that the topological sorting algorithm has one operation per face to determine the direction of the corresponding graph edge, one per face to establish the initial incoming count, one more per face to decrement it, one per cell to establish the initial ready list, one per remaining cell to add it later to the ready list, and one more to move each cell to the final sorted list. Thus it operates in time linear in the number of cells and faces in the volume subdivision.

The sort produced will be correct for the polyhedron compositing scheme if every ray through the data volume intersects it in a single sequence of adjacent cells. This will be the case whenever the total volume is convex. Williams and Shirley [12] deal with sorting on more general data volumes, which may contain cavities where the data is not defined. For scattered data points with no preferred subdivision into cells, the Delaunay subdivision of the convex hull into tetrahedra can always be sorted. (See [9] or Edelsbrunner [5].)

In special cases, an application-specific sort may be faster and easier than the general one just described. One such case is a rectilinear grid, which has a simple sort based on the grid indices. The adaptive mesh refinement application to be discussed next uses this simple sort recursively. In the second application, to clouds in climate modelling, the volume to be visualized is the earth's atmosphere, which is a hollow shell and is not convex. It uses the topological sort within columns of air, and sorts the columns using their latitude/longitude indices. Thus both applications combine several simpler sorts to make a complete sort.

3 Adaptive Mesh Refinement

The adaptive mesh refinement(AMR) method for solving partial differential equations subdivides the grid in regions of high variation, where finer grids are necessary to produce accurate solutions, while keeping coarser grids where they are adequate. It produces a hierarchy of finer and finer grids of data cells. In order to visualize the results of such simulations and still include the finest details, it had been necessary to subdivide the whole volume into cells of the smallest size, to produce a uniform grid suitable for a standard volume renderer. Visualization techniques which take advantage of volume coherence can efficiently render the larger grid cells, and concentrate computation time on the finely subdivided regions containing the details of interest. There is also a savings in data storage compared with the uniform fine grid.

Laur and Hanrahan [8] also use a hierarchical subdivision to speed volume rendering, in their case an octree structure fitted to data already available on a uniform fine grid. They use a semi-transparent polygon approximation to speed up the polyhedron compositing, and do not construct contour surfaces.

The AMR method of Berger and Collela [2] has recently been extended to 3-D simulations (Bell et al. [1]). It covers the simulation volume by a collection of regular grids. Each grid is a rectangular solid of cubic cells, aligned with the coordinate axes. The grids at the first coarsest level cover the volume, and have equal and aligned cubes. For example, they may be subsets of an integer lattice. The cells at a finer level are a factor of $1/n$ smaller in each dimension, for some constant even integer n, and the time step used to integrate the partial differential equation is also smaller by a factor of $1/n$. The boundaries of the finer grids are aligned with the next coarser grids, so that if a coarser cell contains any cell of a finer grid, it contains all n^3. The finer grids may overlap each other, but the solution data will agree on the overlap. Each fine grid is surrounded on all sides by cells of the next coarser level, except perhaps along the boundary of the problem domain. However, a fine grid may overlap several coarser grids, and not be entirely contained in any single one.

The integration in time uses a finite difference scheme on data values defined at the grid centers, and processes each grid independently. The time steps are interleaved, so that before a grid is advanced past time t, all finer grids have been computed at time t. At this point, any grid cell covered by a finer grid has its data values replaced by the average of the n^3 finer grid values inside, to maintain consistency in the solution. Special flux adjustments are also made at the boundaries of the finer grids. (See [2].)

At specified time intervals, the grid placements are revised, by computing an estimate of the error in the finite difference solution for each cell of the grids at all levels, and enclosing those cells where the error is too large in grids of the next finer level. (See [2] and the references therein for more details.)

4 Sorting for the AMR Method

Suppose the volume to be rendered is the rectangular solid defined by $i_1 \le x \le i_2$, $j_1 \le y \le j_2$, $k_1 \le z \le k_2$, where $i_1, \ldots, k_2$ are in the integer lattice. Suppose the viewpoint V is (x_0, y_0, z_0), with real coordinates. Let the lattice cell containing V have indices i_0, j_0, and k_0, so that $i_0 \le x_0 < i_0+1$, and so forth. The compositing order in X will be from i_1 to $i_0 - 1$ in increasing order, and then from $i_2 - 1$ to i_0 in decreasing order. If the viewpoint V is outside the X extent of the solid, one of these loops will be empty. For each fixed index in X, the rendering order in Y and Z proceeds similarly, in inner loops. If a cell is covered by a finer grid, the same sort is applied recursively to its n^3 subcells.

If a cell contains no contour surfaces, the volume compositor treats the cell as a whole. Otherwise, it is divided up into five tetrahedra. (See [9].) These five tetrahedra are easily sorted from back to front by a special case of the general topological sort. On a tetrahedron, trilinear interpolation reduces to a linear function. So contour planes for a linearly interpolated function within a single tetrahedron are all parallel, since they have the same normal, given by the coefficients of the linear function. Thus a version of the X-sort described above can sort the contour surface polygons and the pieces into which they slice the tetrahedron.

5 Interpolation and Contour Surfaces

The data values in the adaptive mesh refinement algorithm are defined at the centers of the grid cells, but the trilinear interpolation necessary for smooth volume rendering requires data at the vertices. One solution to this problem is to use dual subdivision, which has vertices at the cell centers. However the dual subdivision has a rather complex structure at the transitions between coarse and fine grids. It is simpler to interpolate the data from the cell centers to the grid vertices, because the adaptive mesh refinement proceeds until this is a good approximation.

The data at an interior vertex of a grid may be defined as the average of the values at the eight surrounding cells. However complications arise at the boundary of the grid, where some necessary cells are missing. It is not trivial to tell if these missing cells are contained in other grids of the same level, because of the complex way in which the grids are permitted to overlap. If no other grids at the same level contain the missing cells, the data for the vertex must be interpolated from the next coarsest level, in a way which will make any contour surfaces continuous.

As discussed above, cubes containing contour surfaces are divided up into five tetrahedra. Differently oriented subdivisions on the odd and even cubes (see [9]) causes cube faces to be divided by a consistent diagonal, and assures that the linear interpolations, and thus the contour surfaces, will be continuous. To maintain this consistency across a face separating a cell which has been subdivided from one which has not, the data at a fine grid vertex W on the face must be linearly interpolated from the three coarse vertices which form the triangle containing W, among the two created by the consistent face diagonal.

The contour surfaces are rendered with Phong shading, by interpolating vertex normals bilinearly across polygons. The polygon vertex normals are interpolated along grid edges from gradient estimates at the grid vertices. Since the function values are defined at the cell centers, the gradient at a vertex can be estimated using the function values at the eight cells which meet there. For example the X component of the gradient at a vertex is estimated as the average of the values at the four cells to its right, minus the average of the values at the four cells to its left. In order to keep the shading continuous across the transition between coarse and fine contours, the normals must be correctly interpolated for fine grid vertices on transition faces, in the same way as the data values.

6 Sorting for Cloud Visualization

Clouds are important in climate prediction, since they reflect sunlight coming towards the earth. They also absorb long-wave radiation leaving the earth, contributing to the greenhouse effect. These two phenomena have opposite effects on global temperature, and it is important to quantify them in order to understand global warming. (See Cess and Potter [3], Cess et al. [4].) Therefore we have developed algorithms to visualize clouds in the climate models.

In our simulations, the grid cells are bounded by equally spaced meridians of longitude, unequally spaced parallels of latitude, and surfaces of constant "geopotential", which vary with latitude and longitude according to the terrain.

The lowest geopotential contour is at the surface of the terrain and the highest is at a constant altitude. The cloud data represents the percent cloud cover between two geopotential surfaces at each latitude/longitude position. We reinterpreted the data values as lying on the geopotential surfaces, in order to give data at the vertices of a cellular subdivision. We then visualized the percent cloudiness as a semi-transparent volume density.

The curving geopotential surfaces presented three special problems. First, they could introduce non-planar quadrilaterals into the surfaces of the volume cells. Next, the varying slant of these surfaces makes the sorting more complicated. Finally, their slopes affect the gradient computations necessary for Phong shading on contour surfaces.

We needed sorting for two kinds of images: ones where the latitude and longitude were unwrapped onto a rectangle as in Fig. 4, and ones where they were wrapped onto the globe as in Fig. 5. In each case, the sorting problem can be reduced to an easy one of sorting vertical columns of air in latitude and longitude, together with a more difficult sort within a vertical column, which involves fewer cells at a time.

Let us first consider the sort within the vertical columns. These are bounded by four planes, two of constant latitude and two of constant longitude. In the unwrapped rectilinear case these form two pairs of parallel planes, but in the spherical case, the four planes are radial, and meet at the center of the sphere. Such a column, with a quadrilateral as base, can be divided by a diagonal plane into two smaller "stacks", each with a triangle as a base. (Figure 1 shows a column of a single level divided in this way.) Each stack will be the union of convex prism-like polyhedra, bounded by three quadrilaterals and two triangles, which will be called "prisms" for short. The general topological sort can be applied to each stack separately. The order of the two whole stacks can be determined by a single dot product, using the normal to the diagonal plane between them.

Note that this method discards some potential volume coherence, since sometimes two prisms on opposite sides of the diagonal plane will fit together to form a convex polyhedron. In fact, for the compositing to work properly, it is not even necessary for each cell to be convex, as long as each ray from the viewpoint intersects it in at most one connected segment. This latter condition can be guaranteed if dot products of normals to the geopotential contour triangles with the viewing rays have certain patterns of signs. Thus, the column is subdivided into two stacks of prisms, but certain pairs of prisms are rejoined into larger cells, having four quadrilateral and four triangular faces, like the one shown in Fig 1. The resulting cellular subdivision is then sorted by the general sorting algorithm. (If a cell lies entirely behind the sea level profile of the globe, it need not be copied to the output list.)

If the polygons at the bottom and top of the column are to be shaded, for example to show altitude or temperature on the terrain, or outgoing long-wave radiation at the top of the atmosphere, these polygons are entered into the

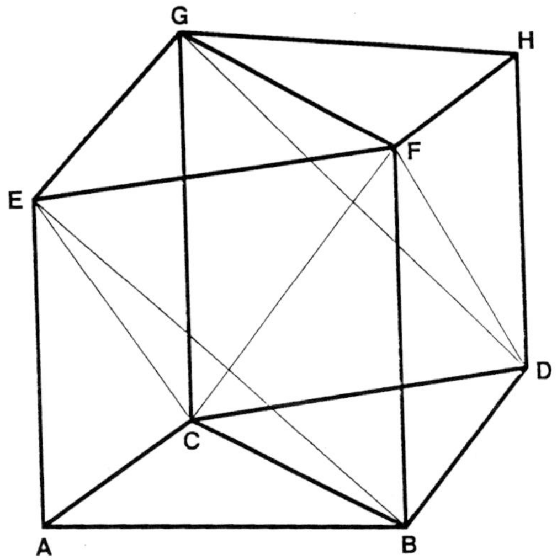

Fig. 1. The cell ABCDEFGH in a column is bounded by the four quadrilaterals ABFE, BDHF, DCGH, and CAEG, and the four triangles EFG, GFH, BAC, and BCD. It is divided bt the diagonal plane CBFG into the two prisms ABCEFG and CBDGFH. The prism ABCEFG is divided into the three tetrahedra ABCE, BCEF, and CEFG, and the other prism is divided similarly, using the same two triangular faces CFG and CBF on the diagonal plane.

directed graph as if they were cells just outside the column. They will then be rendered and composited at the appropriate stage.

If one of the prisms contains a contour surface for the scalar function, it is split into three tetrahedra, as shown in Fig. 1. In this case, the prism must not be joined with its opposite into a larger cell. The order of the three tetrahedra within a prism is easily determined by a special case of the general sort. Figure 1 shows how to make the subdivisions of two opposite prisms consistent across the diagonal plane. It is clear that it is also consistent between adjacent columns. If six tetrahedra in this pattern are used for each contour-containing cell in a cubical array, the hierarchical subdivision discussed in the previous application will also give consistent contour surfaces.

The sorting of the columns is simpler. In the unrolled case, they are sorted in latitude and longitude in a 2-D version of the rectilinear sort. In the spherical case, let i_0 be the longitude wedge containing the radial projection of the viewpoint V. Then the wedges of constant longitude are composited in order from $i_0 + pi$ down to $i_0 + 1$, and then from $i_0 - pi + 1$ up to i_0, where pi is one half the number n of longitude divisions, and the longitude index is interpreted modulo n.

Within a wedge of fixed longitude index, the columns of cells can similarly be sorted in latitude. Let Q be the projection of the viewpoint V onto the plane bisecting the wedge, and let j_0 be the latitude interval containing Q. Then the

Fig. 2. One quarter of a four-fold symmetric AMR simulation of a planar shock wave impinging on an ellipsoid of dense freon of gas, rendered as a semi-transparent density cloud.

columns are composited from the south pole up to $j_0 - 1$, and then from the north pole down to j_0. Note that the index j_0 varies with the longitude of the wedge.

Recall that the normals of the contour surfaces of the scalar function needed for the Phong shading are estimated from the gradient of the function. In this application, the data values are given at the vertices, so the gradient components can be estimated from symmetric differences in latitude, longitude, and geopotential level indices. However, even in the case of the unrolled rectangular mapping, the actual altitude is a non-linear function of the three indices, since the contour surfaces of constant geopotential are curved. The Jacobian matrix of partial derivatives for this mapping gives the first order behavior near a particular point. According to Foley et al. [6], Appendix A.5, a surface normal is transformed by the inverse transpose of this matrix. The Jacobian matrix itself can be estimated by symmetric differences of the mapping from geopotential to altitude. If the map is wrapped around the sphere, an additional Jacobian matrix for the spherical wrapping is involved.

7 Results

Figure 2 shows one quarter of a four-fold symmetric AMR simulation of a planar shock wave impinging on an ellipsoid of dense freon gas.

Fig. 3. The same density function as in Fig. 2, with two contour surfaces, one transparent and one opaque.

The two coarsest grids covered a range of $80 \times 16 \times 16$. Of these 20 480 cells, 650 were covered by finer grids which, for $n = 4$, resulted in 41 600 finer cells, giving a total of 61 430 cells. This is considerably less than the $20\ 480 \times 4^3 = 1\ 310\ 720$ cells required for a uniform subdivision at the finer resolution. The image took 491 seconds to render with a vectorized polyhedron compositor on one thread of a Stellar GS2000, at 500×500 resolution. Figure 3 shows the same data with two contour surfaces. Of the finer cells, 4 075 were crossed by contour surfaces, and thus divided into tetrahedra, which were in turn sliced by the contours. This resulted in 94 172 volume pieces and 16 372 surface pieces, and Fig. 3 took 610 seconds. Figure A of the color illustrations of this chapter at the end of the book shows a rectangular map with the clouds over Indonesia colored by an altitude scale, and upward long-wave radiation indicated by colors on a plane above the top of the atmosphere. Figure B shows a cloud image with a different assignment of cloud colors to altitude, on a round globe, at 1920 by 1035 HDTV pixel resolution.

Acknowledgements

This work was performed under the auspices of the U.S. Department of Energy by Lawrence Livermore National Laboratory, under contract number W-7405-Eng-48. Michael Welcome provided the data for the AMR figures and the input routines to read it, and explained the multi-grid data structure to me. Roger Crawfis provided the contouring routines and the command interface to the polyhedron compositing system, and lead a team effort for climate model visualization. Both Michael and Roger commented on the manuscript, and helped me with debugging. The climate model was provided by the European Centre for Medium-range Weather Forcasts.

References

[1] J. Bell, M. Berger, J. Saltzman, M. Welcome . *Three Dimensional Adaptive Mesh Refinement for Hyperbolic Conservation Laws.* In *submitted to SIAM Journal of Scientific Statistical Computing.*

[2] M.J. Berger, P. Colella (1989). *Local Adaptive Mesh Refinement for Shock Hydrodynamics. Journal of Computational Physics*, 82:64–84.

[3] R. Cess, G. Potter (1987). *Exploratory Studies of Cloud Radiative Forcing with a General Circulation Model. Tellus*, 39A:460–473.

[4] R. Cess, G. Potter, J.-P. Blanchet, G. Boer, S. Ghan, J. Keihl, H. Le Treut, Z. Li, X.-Z. Liang, J. Mitchell, J.-J. Morcette, D. Randall, M. Riches, E. Roekner, U. Schlesse, A. Slingo, K. Taylor, W. Washington, R. Wetherald, I. Yagi (1989). *Interpretation of Cloud-Climate Feedback as produced by 14 Atmospheric General Circulation Models. Science*, 245:513–516.

[5] H. Edelsbrunner (1989). *An Acyclicity Theorem in Cell Complexes in d Dimensions.* In *Proceedings of the ACM Symposium on Computational Geometry*, pp. 145–151.

[6] J. Foley, A. van Dam, S. Feiner, J. Hughes (1990). *Computer Graphics: Principles and Practice (second edition).* Addison-Wesley, Reading, MA.

[7] D. Knuth (1973). *The Art of Computer Programming, Volume 1; Fundamental Algorithms.* Addison Wesley, Reading MA.

[8] D. Laur, P. Hanrahan (1989). *Hierarchical Splatting: A Progressive Refinement Algorithm for Volume Rendering. Computer Graphics*, 25(4):284–288.

[9] N. Max, P. Hanrahan, R. Crawfis (1990). *Area and Volume Coherence for Efficient Visualization of Scalar Functions. Computer Graphics*, 24(5):27–33 and 109.

[10] W. Newmann, R. Sproull (1979). *Principles of Interactive Computer Graphics (second edition).* McGraw Hill, New York.

[11] P. Sabella (1988). *A Rendering Algorithm for Visualizing 3D Scalar Fields. Computer Graphics*, 22(4):51–55.

[12] P. Williams, P. Shirley . *A Prori Algorithms for Polyhedral Depth Sorting and Point Location. to appear in ACM Transactions on Graphics.*

Joining Volume with Surface Rendering

Martin Frühauf

Different existing methods for the rendering of different classes of data, i.e. surface and volume data, in one image are described and discussed. Two fundamentally different strategies for combining the rendering of volume data with the rendering of surface or other geometric primitives have been developed. The first strategy is to convert either volume data into polygonal data or to convert polygonal data into volume data. Applying one of these conversions results in one class of data, which can be rendered with one appropriate rendering method. The second strategy is to use different rendering methods for the different classes of data and to combine the results in a final display. The rendering modules are either independent from each other or more or less tightly coupled. This includes merging the results of different rendering processes in the z-buffer of a workstation.

1 Introduction

Many applications in scientific visualization require the rendering of sampled volume data as well as the rendering of geometrically defined objects in one single image. Examples are: the visualization of prostheses in combination with the skeleton, the visualization of clouds over terrain data or the visualization of material defects recorded using ultrasound in the context of the geometry of a workpiece obtained from a CAD system. Volume rendering, i.e. producing images from volume data, is one aspect of scientific visualization. Volume rendering is defined as volume visualization directly from volume primitives and not via surface primitives. This provides the advantage of minimal precomputation or user interaction for finding and computing object surfaces in volume data. Volume rendering is the only method that preserves the entire data set and thus, allows a detailed exploration of the volume. On the other hand, geometrically defined objects, which are to be rendered, are mostly represented using line or surface patch primitives, i. e. polylines, polygons or polygon meshes. The rendering of those primitives is well known and is highly supported by the graphics engines of workstations.

Volume data in scientific visualization is provided as a grid of data in 3D space. Normally, there is no interconnection between grid points and there is no explicit surface description in the data grid. Surface data of geometric objects is provided as a set of polygons connected or not or as a mesh of polygons in 3D space. Normals for shading can easily be computed from polygon data. Both

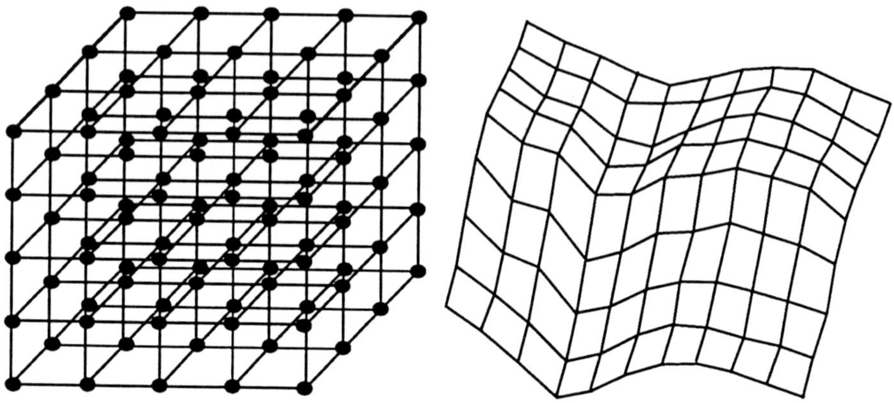

Fig. 1. Volume data versus surface primitives.

classes of primitives – volume and surface – have attributes such as color and opacity.

Many techniques for volume rendering have been developed [1, 3, 16, 21, 22]. They are very specialized in the rendering of sampled scalar volume data sets. They are more or less different from the techniques usually used in rendering of geometrically defined objects. The major difference, of course, is the calculation of surface inclination for shading. Some of these algorithms are suitable to be combined with rendering of surface primitives and some of them do not. The requirements on volume rendering for the combination with surface rendering are described exemplary in Sect. 3.4. The details of these algorithms are beyond the scope of this section. Nevertheless, the basics of the strategies for the visualization of scalar volume data are a precondition for the understanding of the following hybrid rendering methods. An almost complete collection of published volume rendering algorithms has been edited by Kaufman [12].

Highly specialized methods for either the rendering of scalar volume data or the rendering of surface primitives have been developed. Therefore, the first strategy (see Sect. 2) is to convert either volume data into polygonal data or to convert polygonal data into volume data. Applying one of these conversions results in one class of data, which can be rendered with one appropriate rendering method. The second strategy (see Sect. 3) is to use different rendering methods for the different classes of and to combine the – perhaps intermediate – results in a final display. The rendering modules are either independent from each other or more or less tightly coupled. This includes merging the results of different rendering processes in the z-buffer of a workstation.

The different approaches to combine geometrical and volume data discussed here are

- transforming polygonal data to volume data (voxelization)[10, 11],
- transforming volume data to polygonal data [18] and

Method	surface	volume rendering
Levoy 1990	ray tracing	ray casting
Goodsell 1989	scan conversion	ray casting
van Walsum 1991	scan conversion	ray casting
Frühauf 1991	optional	optional
Kaufman 1990	hardware	optional

Fig. 2. Methods.

Quality	surface	volume rendering
Levoy 1990	opaque/semi-transp.	opaque/semi-transparent
Goodsell 1989	opaque	opaque/semi-transparent
van Walsum 1991	opaque	semi-transparent
Frühauf 1991	opaque/semi-transp.	opaque/semi-transparent
Kaufman 1990	opaque	opaque

Fig. 3. Qualities achieved.

- merging the results of different sequential or parallel rendering processes [4, 6, 12, 17, 23].

The approaches of the last group have been developed in order to apply the most appropriate rendering method directly to each kind of data. They differ concerning the level of independence of the rendering algorithms applied to the different classes of primitives and concerning the kind of results, i.e. the contents and kind of information, used for merging. Ray casting algorithms for volume rendering are very close related to ray tracing algorithms for surface rendering, thus these volume rendering algorithms can be easily combined with surface rendering using ray tracing [17]. Intermediate results from the surface rendering algorithm can be used to optimize the volume rendering procedure [6, 17, 23]. Other approaches merge the results of arbitrary independent rendering algorithms into a final image [4, 12]. Figure 2 shows the rendering algorithms used by different approaches.

Figure 2 and 3 give an overview of the capabilities and the functionality of hybrid rendering methods for volume and surface data. Anyway there is no "best" method for combining the rendering algorithms or intermixing the results. The best choice depends mainly on the requirements of the application, the capabilities and the performance of the hardware and the size of the source data.

2 Converting Data

In Sect. 2 methods are discussed that convert regular grids of volumetric data into meshes of polygons or sets of dots. There are other methods converting stacks of contours, which have been computed from slices of pixel data, into polygonal representations ("triangulation"); and there are methods converting irregular grids or scattered data into polygonal representations. These methods are not discussed here.

2.1 Converting Volume Data into Polygonal Data

The "Marching Cubes" algorithm was presented by Lorensen and Cline in 1987 [18]. It constructs an iso-surface represented as a set of triangles by classifying each element in a regular grid of scalar volume data using a threshold T in the domain of the data. It is a conversion of a regular grid of scalar data (densities) into set of triangles representing an iso-surface at density T in the grid. If the algorithm is applied more than once using different thresholds and assigning different transparencies to the different surfaces, semi-transparent visualizations can be computed. This results in a representation similar to the shells of an onion. Nevertheless, main parts of the interior of the volume data set may be skipped.

The "Marching Cubes" is a high-resolution surface construction algorithm. Unfortunately, its results are a huge set of triangles. It requires powerful surface rendering engines to realize interactivity in manipulating the presentation.

The algorithm locates the iso-surface in a cube created from eight volume elements, i.e. four from each two adjacent slices of the volume data set. A volume element, i.e. a vertex of the cube, is named "outside" if its scalar value is lower than the threshold. Otherwise it is named "inside". The iso-surface intersects those cube edges where one vertex is "inside" and the other is "outside". Because of symmetry properties 14 patterns of triangles are sufficient to represent the 256 cases the iso-surface can intersect a cube [18]. The pattern or its permutation to be used for construction of the iso-surface is determined by the classification ("inside" or "outside") of the eight cube's vertices. If a specific pattern has been determined, the exact location of the triangle vertices is computed by linear interpolation from the cube's vertices depending on the threshold. Finally, a vertex normal is computed for each triangle vertex using a grey-level gradient [8].

The original paper introducing the Marching Cubes algorithm does not solve the ambiguous cases using the patterns 3, 6, 7, 12, 13. If this ambiguity is neglected, holes in the computed iso-surface may arise. In [7] seven additional patterns are defined to solve this problem. Figure 4 shows these patterns. The alternative patterns are chosen depending on the patterns already used in the adjacent cubes, i.e. evaluating a greater neighborhood to determine the iso-surface orientation. Other corrections were suggested by Nielson, Hamann [19].

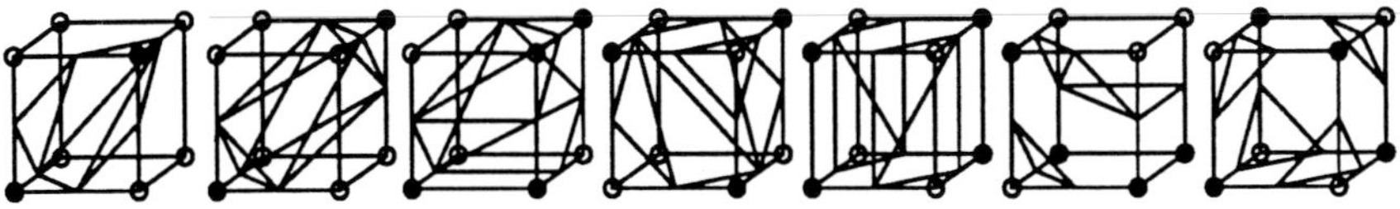

Fig. 4. Additional patterns to solve ambiguous cases.

In [24] the quality of the iso-surface approximation of the original and a modified Marching Cubes algorithm is investigated.

2.2 Converting Polygonal Data into Volumetric Data

The approach of converting polygonal data into volumetric data, which is the reverse of the method in Sect. 2.1, is called *voxelization*. Geometric data representing a surface in 3D space is converted into a set of connected discrete volume elements (voxels) in the 3D space that represent that object. Volume rendering methods are applied to the resulting volume data set for visualization purposes. Algorithms for the three-dimensional scan-conversion of 3D continuous geometric representations into the discrete voxel space have been published by Kaufman in 1987. This includes "3D scan-conversion of polygons" [10] and "3D scan-conversion of parametric curves, surfaces and volumes" [11]. The 3D scan-conversion of planar polygons serves as an example to describe the fundamentals of the conversion of continuous geometric data into volumetric data. A set of planar polygons may form the boundary surface of an object in 3D contiguous space:

- The discrete points, for which the region contained by them is entirely inside the continuous object, are in the converted object.
- The discrete points, for which the region contained by them is entirely outside the continuous object, are not in the converted discrete object [10].

All the discrete points of the two above cases are determined by sorting the polygons according to one dimension and applying a fill area algorithm. The discrete points, which are not covered by either of these two cases, are determined by converting the surface polygons into a set of connected discrete points. The converted surface polygons complete the voxelized object. Figure 5 shows a polygon represented as a set of voxels.

The generated representation of a surface in 3D voxel space has to have a minimal "thickness" in order not to allow the rays used in the volume rendering process to penetrate it. Different categories of holes and tunnels which may appear in a voxelized surface characterize the quality of the surface. Adjacent voxels in the surface representation are either connected by faces, only by edges or only by vertices. There are six neighbors of a voxel connected by faces; there are twelve neighbors connected only by edges; and there are eight neighbors

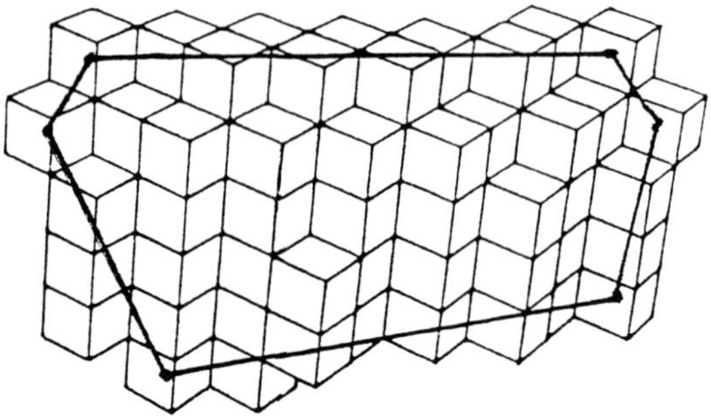

Fig. 5. A polygon represented as a set of voxels.

connected only by vertices. A voxelized surface representation consists either of voxels each connected by faces, or of voxels connected by faces or by edges, or it consists of voxels connected by any of the three cases. This depends on the algorithm for the 3D scan-conversion.

For the 3D scan-conversion of planar polygons the edges of a polygon are scan-converted by a 3D Bresenham algorithm. The interior of a surface polygon is filled as in a common 2D scan-line algorithm with a computation of the third coordinate for each voxel filled in using the plane equation of the planar polygon.

The algorithm published by Kaufman scan-converts objects with a complexity that is linear to the number of voxels generated. The complexity of the conversion as well as the complexity of volume rendering using ray casting is independent from the number of objects. This is an interesting feature because the complexity of most "traditional" rendering algorithms depends on the number of objects in a scene. Anyway, voxelized lines of linewidth one, e.g. paths of voxels connected by vertices, cannot be rendered correctly using ray casting for volume rendering because they will not be hit reliable by the rays in the discrete pixel space.

3 Combining Rendering Methods

The systems, algorithms and methods in this section apply specialized rendering algorithms to the different classes of data and combine the – perhaps intermediate – results in a final display. The rendering modules are either independent from each other or more or less tightly coupled. This includes merging the results of different rendering processes in the z-buffer of a workstation. If independent modules are used for rendering, obviously, a certain set of parameters has to be used by all rendering modules in order to guarantee the consistency of the whole scene. Such a set includes viewing parameters; thus, viewing transformations have to be applied to all rendering modules consistently. This is described extensively in Sect. 3.4, whereas it is applicable to other methods too, but it has not been published in detail.

3.1 Hybrid Ray Tracer

The hybrid ray tracer for rendering polygon and volume data of Levoy [17] was the first published method using two algorithms, one for rendering polygon data and one for rendering volume data, running in parallel. Both are ray tracing algorithms and intermediate results from the polygon rendering process are used to improve the quality of the volume rendering algorithm (Fig. 6). Parallel rays – one per pixel – are traced into the volume data from an observer position. For each ray a vector of colors and opacities is computed by resampling the volume data at evenly spaced locations along the ray and trilinearly interpolating from the colors and opacities in the eight voxels surrounding each sample location. Independently, all intersections between the ray and polygons in the environment are computed and shaded, yielding a color and an opacity for each point of intersection. Finally, the resampled volume colors and opacities are composited (alpha blended) with each other and with the polygon colors and opacities in depth-sorted order to yield for each ray a pixel color [17].

The hybrid ray tracer of Levoy employs two strategies to improve the quality of the resulting visualization. This is performed using intermediate results from the volume ray tracer for rendering polygons and vice versa. Whether to supersample the image at polygon edges or not, in order to reduce aliasing effects, is decided by alpha-blending only the results from the polygon rendering process attenuated by the volume sample opacities (Fig. 6). If the difference of the volume-attenuated polygon colors of two adjacent rays is greater than a threshold, but the difference of the polygon-only colors is lower, the color difference is due two the volume data and not due to polygon edges. In that case no supersampling is necessary. The second improvement is either to adjust the volume sampling position along a ray or to supersample the volume on a decision of the type of volume-polygon intersection. For that purpose the locations of the ray-polygon intersections and the angles between the rays and the polygons are used in the volume rendering process. Both strategies reduce aliasing effects significantly but also increase the rendering costs.

3.2 Rendering Volumetric Data in Molecular Systems

The work of Goodsell, Mian and Olson [6] is the earliest one using a strategy of rendering different classes of primitives by different rendering algorithms. However, this algorithm renders polygon data in the first step and volume data afterwards. It works as follows:

1. Render surface primitives.
2. Add the results of surface rendering (color and depth of each pixel) over the background while saving the depth of each pixel in a z-buffer.
3. Render the volumetric data using a ray casting method.

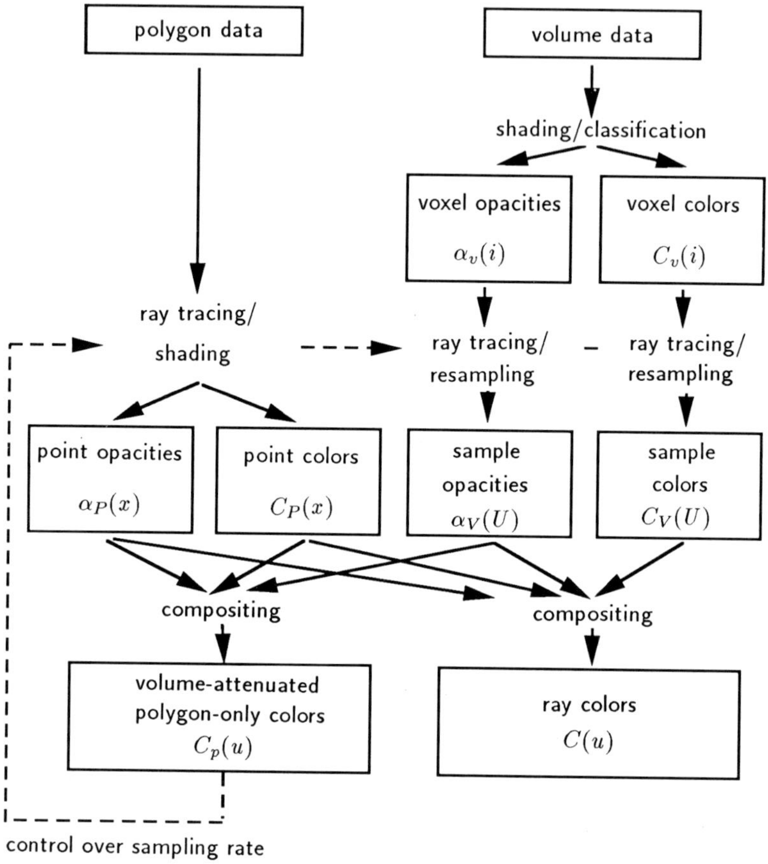

Fig. 6. The hybrid ray tracer.

4. For each ray: obscure the surface primitive or background with the summed opacity from the volume rendering process.

The criteria for terminating a ray in step 3 are:

- the ray passes through the back surface of the volume data grid,
- a surface primitive or the background is reached, or
- the summed opacity exceeds a given threshold.

The main disadvantage of this method is that only opaque surface primitives can be handled correctly. This is due to the fact that the results of rendering surface primitives are added to the background prior to the volume rendering

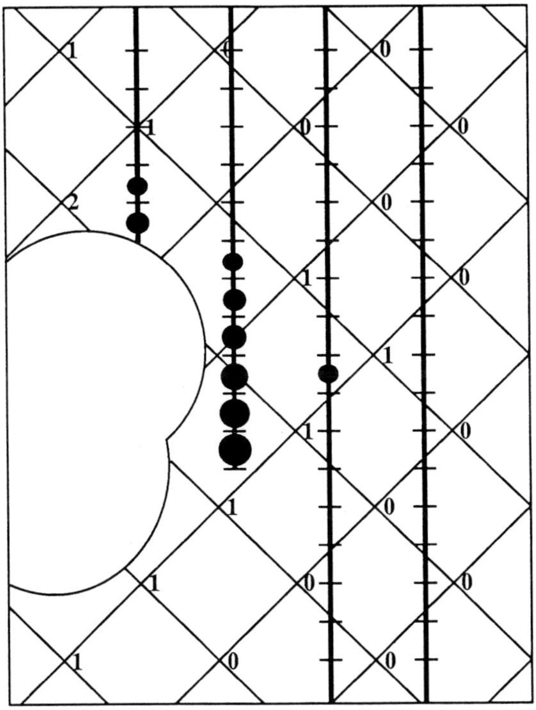

Fig. 7. Ray casting in a scene consisting of a scalar data grid and surfaces of spheres.

process; and that rays in volume rendering are terminated if its depth exceeds the value in the background z-buffer (Fig. 7).

Two different volume rendering methods are used to obscure the surface primitives or the background:

- volume rendering by incremental summation, and
- volume rendering by thresholding.

The incremental summation volume rendering algorithm uses alpha blending of colors at sample positions along the ray with different transfer functions (Fig. 8). The different transfer functions of the volume cloud allow to create a wide range of different images. The opacity and the color of an isovalue volume rendered surface is blended over the surface primitives with an empasize on highlights and with fog for distance attenuation in the second case. This is performed to support the perception of volume rendered surface features and the impression of depth in the final image.

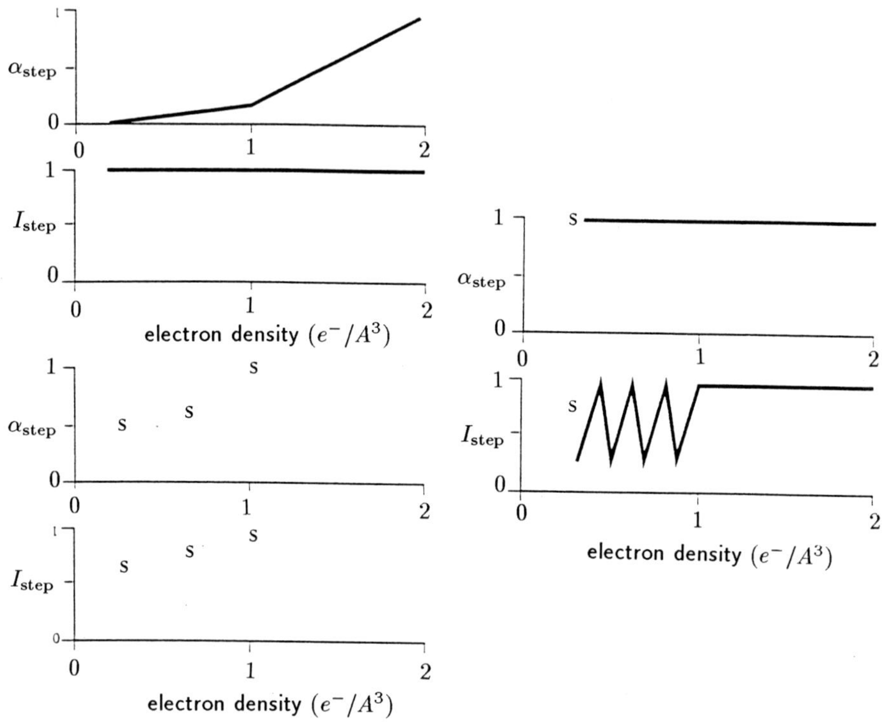

Fig. 8. Transfer functions for the opacity (α) and the intensity (I).

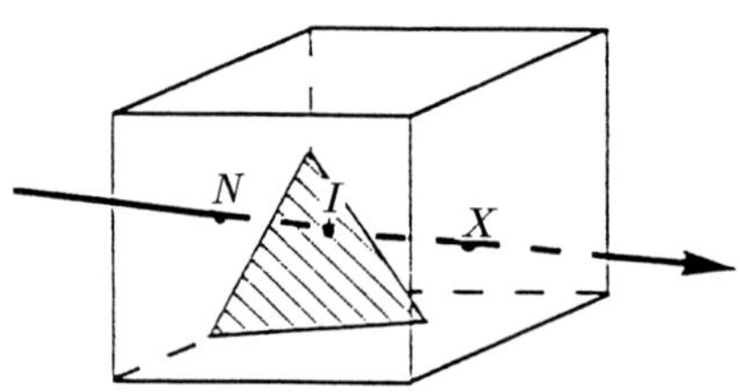

Fig. 9. Interval for sampling volume data is either NI or NX.

3.3 Hybrid Rendering of Volume Data and Polygons

The method of van Walsum and his colleagues [23] follows a similar strategy as the previous method, i.e. to render the polygonal surface data prior to the volume data. In addition information from the surface rendering process is used to speed up the volume rendering process. A polygon scan conversion algorithm with depth-buffering is used. During this step, a reference into the polygon set is stored for each pixel. This reference determines which polygon from the set will be first hit by the ray from the viewpoint through the current pixel. The intersection point of the ray and the polygon determines the interval in which the volume data is sampled in the following ray casting process (Fig. 9). No extra effort is necessary to find intersections of rays with polygons during the ray casting process. In this process, for each pixel the integral over the volume density is computed in the appropriate interval along the ray. The final pixel is computed from the approximated integral of density along the ray, the portion of diffuse reflection of the polygon hit and a user-defined color assignment.

The method can be sketched as follows:

1. scan convert polygons recording depth and polygon reference for each pixel
2. cast rays through volume data integrating the density in the appropriate interval
3. for each pixel: compute the final pixel from the results of steps 1 and 2.

The method and the intermediate data structures are only suitable to handle and render opaque polygons. Moreover, the volume rendering effort is decreased since only those areas of the volume data have to be rendered, that are not obscured by opaque polygons. Since the area of the volume data, which is rendered is determined by the location of the polygons, semi-transparent polygons can not be considered.

3.4 Combining Volume Rendering with Line and Surface Rendering

The approach of [4] merges the results of rendering geometrically defined objects with the results of rendering scalar volume data sets. It is capable to handle opaque as well as translucent volumes or objects. In both cases, the independence of the rendering processes is preserved. The independence of rendering processes in this approach is very important since the performance of volume rendering processes on workstations differs extremely from the performance of surface rendering. Furthermore, it provides the option to use different rendering modules, even rendering in hardware, if they match the interface definitions described here. The intention is to identify the requirements on volume rendering algorithms capable of performing the same transformations and applying the same shading methods, as most surface rendering algorithms in standardized graphics libraries do. The algorithm proposes lists of so called image space elements to be merged.

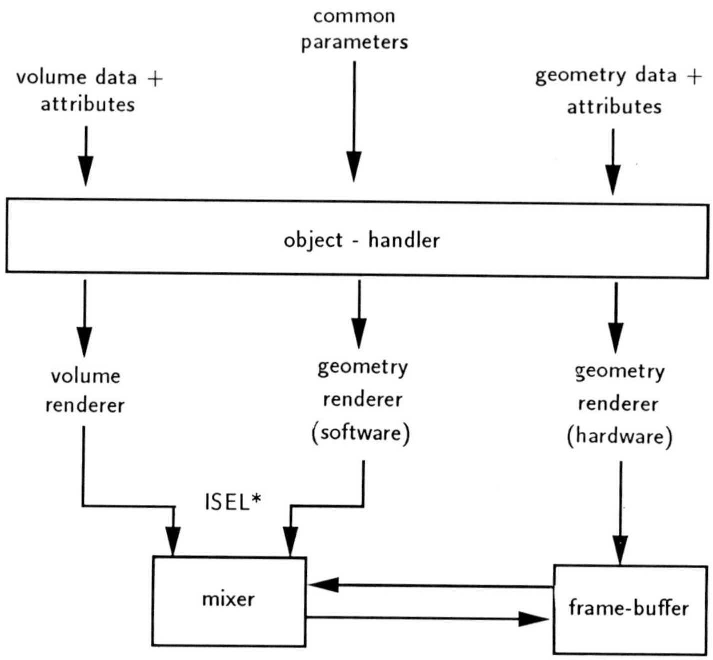

Fig. 10. Architecture – data flow view.

After introducing the architecture of the system, a set of common parameters is identified. These parameters are shared by all renderers. They preserve the consistency of the scene and the resulting image. Common parameters are lightsources in the scene and geometrical transformations.

3.4.1 Architecture of the System

The architecture of the combined rendering system couples three different types of renderers for the purpose of calculating combined visualizations of volume data sets and geometrically defined objects. Every renderer produces a list of image space elements (ISEL*) which are merged in the mixer module. Even rendering in hardware directly to the framebuffer of a workstation can be included by reading the results from the framebuffer to the mixer. Figure 10 shows the architecture of the system.

Note that the geometry rendering process is independent from the volume rendering process and vice versa. This is correct as long as no ray tracing (introducing secondary rays) is used for geometry rendering. The independence aspect permits the usage of hardware rendering capabilities for geometry rendering in the combined system. In general, every geometry rendering algorithm can be used if the implementation provides the required output to the mixer. Furthermore, a result of a volume rendering process, which is the more time consuming task, can be used several times for merging with the results of rendering different geometrically defined objects.

3.4.2 Common Parameters

In order to adapt the volume rendering algorithm to the geometry rendering algorithm and vice versa, a set of common parameters which control the algorithms has to be identified. The parameters are supplied to, and are stored in the object handler together with the volume objects, and the geometrically defined objects. The rendering modules can retrieve parameters and objects from the object handler.

Common parameters in our system are:

Group (i):

- modeling transformation parameters (rotation angles, scale factors, shift vectors)

Group (ii):

- point lightsources (direction L and intensity k_l)
- ambient light (intensity k_a)
- material reflection parameters (ambient r_a, diffuse r_d, specular r_s and shininess r_n)

Group (iii):

- viewpoint (V)
- center of interest (R)
- twist angle (δ)

Group (iv):

- projection type (orthogonal or perspective)
- locations of clipping planes (left, right, bottom, top, front, back plane)

Group (v):

- viewport or image size (width and height)

Figure 11 shows the coordinate system in which the parameters of the different groups *(i) - (v)* are defined. The used coordinate systems and transformations are described in the following chapter.

3.4.3 Transformations and Coordinate Systems

The specification of parameters of groups *(i)*, *(iii)*, *(iv)* and *(v)* demands that transformation matrices are applied to the objects. All transformations are applied in exactly the same manner to volume data as to geometrically defined objects, i.e. to voxels the same as to vertices.

The transformation pipeline is structured as follows:

1. modeling transformations
2. viewing transformation
3. projection transformation
4. window to viewport mapping.

Figure 11 shows which parameter group influences which transformation.

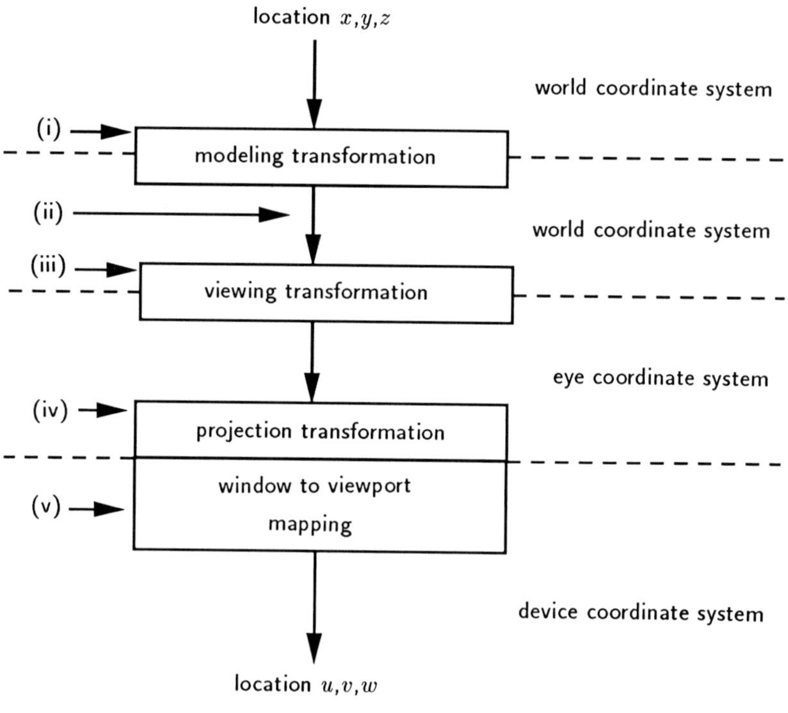

Fig. 11. Transformations and coordinate systems.

The modeling transformations are influenced by the demands of the volume rendering algorithm, but they can be applied easily to geometric data as well. It is assumed that $0 \le x, y, z < n$ for the coordinates (x, y, z) of all elements in the volume data set. The order of modeling transformations is:

1. translate $(-n/2)$
2. scale (s) with $s \leq 1$
3. rotate (x, alpha)
4. rotate (y, beta)
5. rotate (z, gamma)
6. translate $(n/2)$.

Translation and scaling are uniform in all three dimensions. For rotation, the rotation axis and the rotation angle are denoted. Steps 1, 2 and 6 are necessary in order not to exceed the bounding box of the volume data set, i. e. not to exceed the memory space allocated for the data. These modeling transformations are multiplied to one single modeling transformation matrix.

Modeling transformations, the viewing transformation and the projection transformation are multiplied to one transformation matrix TM. Thus, all transformations can be evaluated by applying only once a transformation matrix TM to the volume data set. This reduces the computational effort significantly.

The computational effort while applying a transformation matrix to a regular volume data set, is reduced making use of the regularity [5].

Figure 12 shows how the transformations (in particular the viewing transformation) are applied to regular volume data. The volume elements are rotated, scaled and translated within the volume space for the purpose that, after applying the transformation matrix TM rays can be sent through the volume data set parallel to the z-axis of the volume space. Volume data in the eye coordinate system is sampled by trilinear interpolation from the original volume data set. Window to viewport transformation in volume rendering is accomplish by sending as many rays through the volume as pixels are within the viewport. The locations of rays in the eye coordinate system are:

$$x = \frac{(right - left)}{(width - 1)} * u + left \quad \text{with } u = 0, 1, 2, \ldots, width - 1,$$

$$y = \frac{(bottom - top)}{(height - 1)} * v + top \quad \text{with } v = 0, 1, 2, \ldots, height - 1.$$

3.4.4 Requirements on the Volume Rendering Algorithm

The resolution of images computed by the volume rendering algorithm must be independent from the resolution of the volume data set. This is an essential precondition for combining volume rendering with the rendering capabilities of todays graphics libraries. If one voxel in the data set is projected to exactly one pixel in the image (like in back-to-front (BTF) projection algorithms), the size of the viewport on the screen is always restricted to the size of the volume data set.

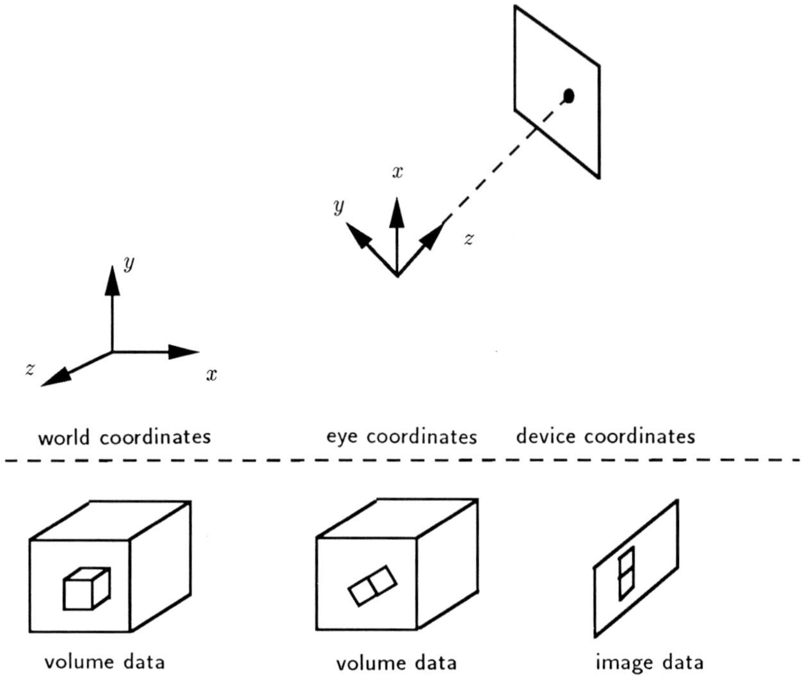

Fig. 12. Transformations applied to the volume data set.

There are two examples of the disadvantages of such a simple volume rendering algorithm. First, no zooming of details in the scene would be possible. Second: since minimum linewidth in graphics packages without anti-aliasing capabilities is one pixel, linewidth is always in the size of one voxel. Lines of that thickness often hide too much information from the volume data set in the scene.

Therefore, BTF or FTB volume rendering algorithms are not suitable to be used in combination with line and surface rendering. Versions of ray casting algorithms which are able to cast rays on arbitrary positions through the volume data set should be used. The implementation of the volume rendering algorithm must allow the application of modeling, viewing and projection transformations as specified in graphics libraries like PHIGS or GL^{TM}. Special attention has to be kept on details of the algorithm, e.g. the exact positioning of the sample locations along the ray. This is described in the following chapter.

3.4.5 The Volume Rendering Algorithm

We use a ray casting technique for volume rendering. The underlying theory of that technique is explained in [15, 21]. The idea of ray casting techniques in volume rendering is to send a virtual test particle on rays from the viewpoint through the scalar volume data set. At evenly spaced locations on the rays test samples are computed. All those test samples on one singular ray are accumulated to one singular pixel of the resulting image.

The volume rendering algorithm processes scalar data fields arranged in a regular grid of three spatial dimensions. The volume is assumed to be cubic and volume data is assumed to be located at evenly spaced integer coordinates in the world coordinate system. If data is not available in that format, it can be transformed in a preprocessing step. A singular data value in the volume data set is called a volume element or voxel.

The following volume rendering algorithm can be parameterized with the parameters described in chapter 3. Additionally, parameters which are unique for volume rendering or which are not used commonly with the geometry rendering algorithm are supplied.

Those parameters are:

- cut planes through the volume,
- the volume rendering method (opaque or semi-transparent)
- the method for gradient calculation
- the interpolation order and
- the distance of samples on the ray.

In our algorithm a voxel $V(x, y, z)$, $x, y, z = 0, 1, 2, \ldots, n-1$, is a tuple $[d, r, g, b, o, r_a, r_d, r_s, r_n]$, where

d is the scalar data value of the voxel,

r is the red component of the voxel color,

g is the green component,

b is the blue component,

o is the opacity of the voxel,

r_a is the ambient reflectance of the voxel,

r_d is the diffuse reflectance,

r_s is the specular reflectance and

r_n is the specular scattering exponent.

The element d is called data, the other elements are called attributes. Attributes can be specified individually for each voxel, constant for the whole volume data set or as a function $f(d)$, where the latter cases can be used to save memory space. Currently, functions $f(d)$ for determining attributes are implemented as look-up-tables (LUT). The length of the look-up-table is equal to the number of elements in the domain of the scalar data in the volume.

In ray casting techniques, rays are sent into the data volume according to the viewing direction (Fig. 13). In its explanation, the following terminology is used: voxel location (x, y, z), voxel data $d_v = [d]$, voxel color $c_v = [r, g, b]$, voxel opacity $o_v = [o]$, voxel shading s_v, sample location (x', y', z'), sample data d_s, sample color c_s and sample opacity o_s. Data is located at integer positions in the regular grid; those positions are referred to as voxel locations. Voxel shading s_v is computed using the voxel's gradient G and voxel color c_v in Phong lighting [2],

$$s_v = c_v(k_a r_a + k_l(r_d N * L + r_s(S * V)^{r_n})),$$

with $N = G/|G|$, $S = S(L, N)$ the direction of specular reflection, G the gray-level gradient (G_x, G_y, G_z) [8] with

$$G_x = \sum_{j=-1}^{1} \sum_{i=-1}^{1} (d(x-1, y+i, z+j) - d(x+1, y+i, z+j)),$$

$$G_y = \sum_{j=-1}^{1} \sum_{i=-1}^{1} (d(x+i, y-1, z+j) - d(x+i, y+1, z+j)),$$

$$G_z = \sum_{j=-1}^{1} \sum_{i=-1}^{1} (d(x+i, y+j, z-1) - d(x+i, y+j, z+1))$$

computed from 26 neighbors.

Sample colors c_s and sample opacities o_s are computed at evenly spaced sample locations along the rays through the volume. Sample locations are independent of integer voxel locations in the volume. The sample data d_s, sample color c_s and sample opacity o_s is computed by trilinear interpolation from the data d_v, the shading s_v or the opacity o_v, respectively, of the eight voxels which are closest to the sample location $(x', y', z') \in I\!R^3$.

If the data volume is rotated in advance, so that rays follow the lines of the volume, i.e rays are parallel to one axis of the volume space, computing the sample location is simplified, and only three layers of the data volume are required for the computation of one image line at the same time. This reduces the size of the required main memory of the workstation. The three layers are required for the calculation of the gray-level gradient.

Ray casting provides the capability of generating opaque or semi-transparent views of the data. For opaque views, visibility is mostly defined by a threshold T in the domain of the data. If visibility is defined by a threshold, the ray can be stopped when a data value d_s at a sample location (x', y', z') greater than the threshold T is computed by trilinear interpolation from the data of the eight

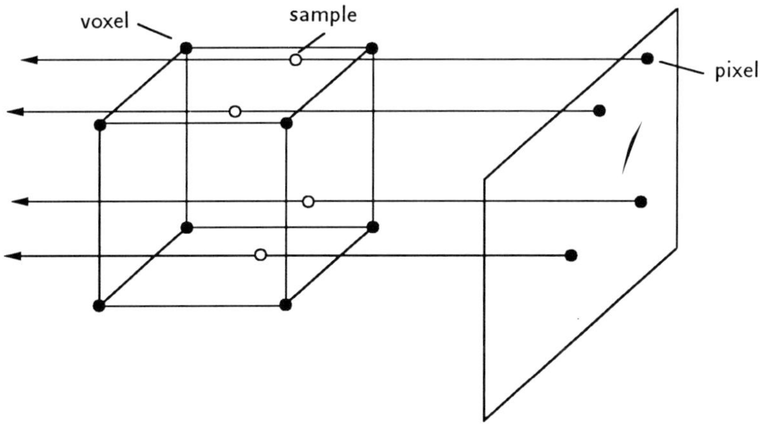

Fig. 13. Ray casting with variable resolution.

nearest neighbor voxels. For image quality, it is very important that the sample color c_s is computed at the location in the volume where trilinear interpolated data d_s is equal to the threshold T. This would increase the number of samples to be computed. To accelerate this, only one sample per voxel is computed along the ray. If a sampled data d_s greater than the threshold T is detected, a binary subdivision of the distance between the two last computed sample locations on the ray is performed in the object space. This is carried out to find a location where

$$T - e < d_s < T + e \quad \text{with } e > 0.$$

At the calculated location the sample color c_s is computed from the surrounding voxel shadings s_v by trilinear interpolation as described above.

The output of the volume rendering algorithm is a sorted list of so called image space elements (ISEL) I_i for each ray. Each $I_i = I_i(u, v)$ is a tuple $[r, g, b, \alpha, w]$, where u and v are coordinates in image space with $u = 0, 1, 2, \ldots, width - 1$ and $v = 0, 1, 2, \ldots, height - 1$. r is the red component of the color of the image space element, g is the green component, b is the blue component, α is the coverage or transparency and w is the normalized depth of the image space element.

The color tuple $[r, g, b]$ of an image space element I_i is the sample color c_s at a location x', y', z'. Let i be the element of I_i. It is calculated from the sample opacity o_s weighted with the distance between two consecutive sample locations on the ray. Let z_i' be the z-coordinate of sample location i and z_{i+1}' the z-coordinate of sample location $i + 1$ then

$$\alpha_i = 1 - (o_s * |z_{i+1}' - z_i'|).$$

The weighting is necessary to obtain the same accumulated opacity O along a ray independent from the number of samples on that ray. w is computed as

$$w = \frac{(z' - near)}{(far - near)} \quad \text{with } z' \in [near..far].$$

3.4.6 Rendering of Geometrically Defined Objects

Any algorithm for line and surface rendering, that computes a list of image space elements I_i can be used in the combined rendering system. A precondition, however, is that it can be parameterized with the parameter common set (and that it interprets the parameters in the same manner the volume rendering algorithm does). Applying transformations, lighting parameters and clipping planes to geometrical primitives is well known in opposite to applying those parameters to voxel primitives.

Material reflection parameters can be chosen differently for volumes and for geometrically defined objects. Usually, we use the same reflection parameters for all objects.

3.4.7 Geometry Rendering in Hardware

If a standardized graphics library like PHIGS+ or Silicon Graphics' GL^{TM} is used for geometry rendering, parts of the rendering process are performed in hardware and the output is directly written to the workstation's framebuffer. Only one image space element per pixel can be read back from the framebuffer into the mixer. Therefore, using rendering in hardware is restricted to special cases of scenes containing geometrically defined objects. Either, only opaque objects are allowed or transparent surfaces are not allowed to overlap each other. Otherwise, image space elements are already accumulated in the framebuffer and the mixer cannot work correctly.

3.4.8 Merging Image Space Elements

Image space elements are expected to arrive depth-sorted from each renderer. Depth-sorted means, that the list element I_0 has the lowest depth w (i.e. is nearest to the viewpoint) and w increases monotonously. Fortunately, ray tracers and ray casters compute image space elements in that order. Therefore, presorted lists from different renderers have to be merged only. This can be performed with less complexity than sorting all image space elements for each ray.

If long ISEL-lists are produced by different renderers, the number of ISEL's to be buffered at a time must be reduced. Therefore, each renderer has to process subframes in the same order like all other renderers. If the results from all renderers for a specific subframe are available, the image space elements are merged and compressed to an array of pixels and the buffer space is released. The size of the subframes depends on the size of the ISEL-lists of image space elements, the period for buffering and the size of available buffer space.

For each pixel $P(u,v) = [r,g,b]$ in the image space the lists $I(u,v)$ of image space elements from all renderers are merged to a new list $I^*(u,v)$. Let c_i^* be the color tuple $[r,g,b]$ of list element I_i^*. Let α_i^* be the transparency element $[\alpha]$ of list element I_i^*. Then pixels $P(u,v)$ are calculated by the equation for transmission of light through translucent particles or surfaces in a discrete system [21],

$$P(u,v) = \sum_{i=0}^{k} c_i \cdot (1-\alpha_i) \cdot \prod_{j=0}^{i-1} \alpha_j, \text{ with } c_k \text{ the background color and } \alpha_k = 0.$$

If a list I has been computed by rendering only opaque objects, it contains only one element I_0. Merging of two lists, each containing a single opaque ISEL is performed by a simple comparison of w (this is z-buffering).

The system is able to handle translucency and lighting of objects. In the current implementation opacity is denoted as a single scalar to reduce the size of the image space elements. Opacity can easily be expanded to an RGB-triplet, if necessary. Because of the independence of the rendering modules, inter-object reflection and shadow generation cannot be performed.

In ray casting algorithms for volume rendering image space elements are sampled at discrete locations. Each sampled image space element represents the average of the data and the attributes of a small region in the volume space. Thus, if the distance between two samples is to large, artifacts at geometry/volume intersections may appear. To minimize these artifacts in opaque volume rendering the sample location is refined as described above.

In semi-transparent volume rendering no adaptive refinement is possible without any information from the geometry renderer. If the independence of the volume renderer and the geometry renderer shall be maintained, sample distances have to be kept small along the entire ray.

3.5 Z-Buffer Merging

The method presented by Kaufman in 1990 [12] makes use of the high-speed intermixing capabilities of the z-buffer of a graphics workstation. Due to the restrictions to the information per pixel that can be stored in a common frame-buffer and a z-buffer, the method is not able to handle arbitrary combinations of semi-transparent layers of surface or volume primitives. Any algorithm for volume rendering and any algorithm for surface rendering can be used since only the results are intermixed in a post-processing step. The rendering processes for volume data and surface primitives are carried out independently from each other. Therefore, re-rendering of one class of data is only performed if it is required by a change of that specific class of data or by a change of viewing parameters. Like in all methods, that use independent rendering algorithms the complexity of the volume rendering algorithm can not be decreased using the results from the surface rendering process.

Algorithm The Z-merging algorithm [12].

Data structures.

- *volume_z* and *geometry_z* are 2D arrays of depth values.
- *volume_image* and *geometry_image* are 2D arrays of color values.
- *Screen* is a 2D array of color values representing the screen frame buffer.

```
FOREVER DO
  Clear_Buffers(volume_z, volume_image);
  Render_Volume(volume_z, volume_image);
  WHILE (no change in viewing parameters or volume data)
    Clear_Buffers(geometry_z, geometry_image);
    Render_Geometry(geometry_z, geometry_image);
    FOR EACH (i,j) in the screen DO
      if volume_z[i][j] < geometry_z[i][j]
        Screen[i][j] := volume_image[i][j];
      ELSE
        Screen[i][j] := geometry_image[i][j];
    END FOR EACH;
    wait for any change;
  END WHILE ;
END FOREVER.
```

Using the hardware z-buffer and a hardware geometry engine of a workstation the algorithm looks as follows:

Algorithm Z-buffer based intermixing [12].

```
FOREVER DO
  Clear_Buffers(volume_z, volume_image);
  Render_Volume(volume_z, volume_image);
  WHILE ( no change in viewing parameters or volume data)
    copy volume_z to Z-buffer;
    copy volume_image to Screen;
    Render_Geometry(Z-buffer, Screen);
    wait for any change;
  END WHILE ;
END FOREVER.
```

4 Applications – Two Examples

The first example is showing a study of the air flow in the human nose. For rhinosurgeons, it is very important, to be able to visualize the air flow in the context of a natural presentation of the anatomy, for the purpose of getting best insight of the behavior and the characteristics of the human nose. Only volume rendering of computer tomograms (CT) provides best representation of anatomy. Results of air flow simulation are typically presented using geometric objects, e.g. lines, polylines, stripes or polygon meshes.

The rendering of lines and stripes was carried out using Silicon Graphics' GL^{TM} graphics library on a Silicon Graphics 4D/380VGX. Several different combinations of particle paths can be visualized using the same set of image space elements from one volume rendering process. Visualization of particle paths is carried out in real-time, whereas high-resolution volume rendering takes about 100 seconds on a 4D/380. Figure A of the color illustrations of this chapter at the end of the book shows a human head using a cut plane through the left side of the head near the nasal septum for clipping. Particle paths start at the top right and are visualized using polylines.

Figure B of the color illustrations is a semi-transparent visualization of a 3D ultrasonic inspection of a metallic cube. A T-formed pipeline has been drilled into the cube and appears in magenta as well as large regions of defects left and right from the middle. The surface of a green cylinder was rendered by the graphics engine of the 4D/380VGX and inserted into the volume data to simulate the drilling of another hole.

References

[1] R. Drebin, L. Carpenter, P. Hanrahan (1988). *Volume rendering. Computer Graphics*, 22(4):65–74.

[2] J.D. Foley, A. Van Dam, S.K. Feiner, J.F. Hughes (1990). *Computer Graphics – Principles and Practice (2nd ed.).* Addison-Wesley, Reading, Massachusetts.

[3] G. Frieder, D. Gordon, R.A. Reynolds (1985). *Back-to-front display of voxel-based objects. IEEE Computer Graphics & Appl.*, 85(1):52–60.

[4] M. Frühauf (1991). *Combining volume rendering with line and surface rendering.* In F.H. Post, W. Barth, editors, *Eurographics '91*, pp. 21–32. North-Holland, Amsterdam.

[5] M. Frühauf (1991). *Volume visualization on workstations: image quality and efficiency of different techniques. Computers & Graphics*, 15(1):101–107.

[6] D. Goodsell, S. Mian, A. Olson (1989). *Rendering volumetric data in molecular systems. J. Mol. Graphics*, 7(3):35–36,41–47.

[7] W. Heiden, T. Goetze, J. Brickmann (1991). *"Marching Cube"-Algorithmen zur schnellen Generierung von Isoflächen auf der Basis dreidimensionaler Datenfelder.* In M. Frühauf, M. Göbel, editors, *Visualisierung von Volumendaten*, pp. 112–117. Springer-Verlag, Berlin.

[8] K.H. Höhne, R. Bernstein (1986). *Shading 3D-images from CT using gray-level gradients. IEEE Trans. Med. Imag.*, MI-5(1):45–47.

[9] J.T. Kajiya, B.P. Von Herzen (1984). *Ray tracing volume densities. Computer Graphics*, 18(3):165–174.

[10] A. Kaufman (1987). *An algorithm for 3D scan-conversion of polygons.* In *Proc. Eurographics '87*, pp. 197–208. North-Holland, Amsterdam.

[11] A. Kaufman (1987). *Efficient algorithms for 3D scan-conversion of parametric curves, surfaces and volumes. Computer Graphics*, 21(4):171–179.

[12] A. Kaufman, editor (1990). *Volume visualization.* IEEE Computer Society Press, Los Alamitos.

[13] A. Kaufman (1991). *State of the art in volume visualization.* In S. Coquillart, editor, *Eurographics '91, State of the art reports*, pp. 173–186. Eurographics Technical Report Series, Vol. EG91 STAR.

[14] A. Kaufman, R. Yagel, D. Cohen (1990). *Intermixing surface and volume rendering.* In K.-H. Höhne, H. Fuchs, S.M. Pizer, editors, *3D imaging in medicine*, pp. 217–225. Springer-Verlag, Berlin.

[15] W. Krüger (1990). *The application of transport theory to visualization of 3D scalar data fields.* In A. Kaufman, editor, *Proc. Visualization '90*, pp. 273–280. IEEE Computer Society Press, Washington Brussels Tokyo.

[16] M. Levoy (1988). *Display of surfaces from volume data. IEEE Computer Graphics & Appl.*, 8(5):29–37.

[17] M. Levoy (1990). *A hybrid ray trace for rendering polygon and volume data. IEEE Computer Graphics & Appl.*, 10(March):33–40.

[18] W.E. Lorensen, H.E. Cline (1987). *Marching Cubes: A high resolution 3D surface construction algorithm. Computer Graphics*, 21(4):163–169.

[19] G.M. Nielson, B. Hamann (1990). *The asymptotic decider: Resolving the ambiguity in marching cubes.* In *Proceedings of Visualization '90*, pp. 83–91. IEEE Computer Society Press, Los Alamitos, California.

[20] T. Porter, T. Duff (1984). *Compositing digital images. Computer Graphics*, 18(3):253–259.

[21] P. Sabella (1988). *Rendering algorithm for visualizing 3D scalar fields. Computer Graphics*, 22(4):51–58.

[22] C. Upson, M. Keeler (1988). *V-BUFFER: visible volume rendering. Computer Graphics*, 22(4):59–64.

[23] T. van Walsum, A. Hin, J. Versloot, F. Post (1991). *Efficient hybrid rendering of volume data and polygons.* In F. Post, editor, *Proc. 2nd Eurographics Workshop on Visualization in Scientific Computing.* Delft University of Technology.

[24] J. Wilhelms, A. van Gelder (1990). *Topological considerations in isosurface generation. Computer Graphics*, 24(5):79–86.

The Volume Priority Z-Buffer

Rolf-Hendrik van Lengen

Modern medical imaging modalities, such as computed tomography (CT) and magnetic resonance imaging (MRI), produce sequences of cross-sectional images. They represent internal spatial structures of underlying medical features. Effective strategies for the direct display of the 3D information are essential to support the diagnostic process. In this chapter, we present a new display technique which allows fast volume rendering on low cost workstations.

1 Introduction

Sequences of cross-sectional images obtained from computed tomography (CT) and magnetic resonance imaging (MRI) generally represent a three-dimensional grid of scalar values, referred to as volumetric data. By inspection of the two-dimensional cross-sections slice by slice, the observer tries to determine the internal spatial structures of the underlying medical features. In many clinical applications, such as surgery planning, it is not very suitable to assess 3D relationships from the planar views. In recent years a variety of data representations and rendering algorithms have been developed for the direct display of the 3D information contained in the volumetric data set. They make an important contribution to facilitate the diagnostic process and enhance the interpretability of the images.

At the present time, most techniques for the creation of display representations of volumetric data sets fall into the two broad categories of *surface-based rendering* and *volume-based rendering*. The classification of a method into one of these families depends basically on the chosen intermediate data representation.

Surface-based strategies originate in Computer Aided Design (CAD) and were the first methods used to display three-dimensional medical features. At the beginning, a set of planar contours, representing the boundary of the anatomical structures, are extracted from the individual slices. They can either be manually traced (see Mazziotta and Huang [15]) or automatically obtained by an edge tracking algorithm (see Vannier et al. [21]). The different points of adjacent contours are then connected by heuristic algorithms in order to optimize a specific constraint, such as the length of segments, for example. The resulting mesh of surfaces, generally triangles (see Keppel [12], Christiansen and Sederberg [2], Fuchs et al. [5], Ganapathy and Dennehy [7]) or higher order surface patches (see Sunguroff and Greenberg [18]), is a common data representation, which can easily be displayed with conventional computer graphics techniques (see Foley et al. [3]).

More recently, a new class of algorithms, referred to as *binary voxel techniques*, has been introduced by Herman and Liu. A voxel in this context is a rectangular volume element, formed by three parallel planes each being orthogonal to the other two. By setting a specific threshold, which depends on the region of interest, the *cuberille algorithm* (see Herman and Liu [9]) first produces a three-dimensional binary voxel array. Each voxel is treated as a small opaque cube having six polygonal faces. Rendering these faces with standard techniques yields an image of the object. To achieve an efficient object representation, adjacent voxels are merged together to form an octree (see Meagher [16]). Traversing the octree in back-to-front order, voxels can be directly displayed on the screen (see Frieder et al. [4]).

Another approach belonging to this class is the *Marching Cubes* algorithm presented by Lorensen and Cline [14]. An octuple of 2^3 neighboring sampled data is placed at the vertices of an voxel at a time. Depending on a binary classification, such as in the Cuberille method, up to four triangles are placed within the cube. In this way, a polygonal structure is created which estimates where the surface of the object cuts through the voxel. Finally, for each triangle of the structure a surface normal is calculated by linear interpolation of the gray level gradient (see Höhne and Bernstein [10]).

The mentioned techniques have several advantages. Geometric primitives such as planar contours or volume elements are very compact, easy to handle and therefore inexpensive in terms of the required storage capacity and system configuration power. On the other side, automatic connection of corresponding contours as well as optimal placing of triangles are serious problems, especially when processing complicated scenes with branching structures, for example. In almost every case the user is forced to intervene. Recently Boissonnat (see Boissonnat [1]) proposed a method which seems to provide an acceptable solution for the first problem. Another undesired effect is the result of the inevitable binary classification. In the presence of small or poorly defined anatomical structures binary classification leads to visual artifacts such as holey surfaces or even not existing surfaces.

Volume-based rendering was first demonstrated by Pixar in 1985. The main feature of this method is the direct display of a volumetric data set as a function of different attributes assigned to each voxel. They basically determine the behavior of the light interacting with each set of the material, which might be present in the individual voxel. Common attributes are the opacity or the gradient value of a voxel for instance (see Höhne et al. [11], Schlusselberg et al. [17], Levoy [13]).

The general idea behind these techniques is based on ray tracing (see e. g. Whitted [22], Glassner [8], Tuy and Lee [20], Fujimoto et al. [6]). From the observer's position, a ray is traced back through the center of each pixel of the image plane into the volumetric data set. At every discrete sample point along the ray, a color and a partial opacity is determined. They depend on an occupancy fraction, which is estimated only once per voxel for each set of material present there. Color and a partial opacity are blended together and yields together with the projection formed by previous slices a light intensity, which is assigned to the individual pixel.

The received image quality by volume rendering techniques is very sophisticated. Aliasing artifacts are dramatically reduced by avoiding a binary classification and carefully resampling the data during projection. Thus even small and poorly defined anatomic structures may be displayed, which is essential for the diagnostic process. Unfortunately, ray tracing volumetric data sets is very expensive in terms of computational power, because potentially all voxels contribute to a single pixel of the image plane.

In this chapter we present a *volume-based rendering* technique to visualize large volumetric data sets on low cost workstations equipped with limited memory. The main advantage of our algorithm is, that during data processing only four slices are kept in the main memory at the same time.

2 Overview of the Algorithm

Raw data obtained from medical imaging modalities are in most cases defined on a regular rectilinear grid in 3D space (see Fig. 1a). Each discrete grid node $\mathbf{v}_n$ is associated with a scalar intensity value $f_{i,j,k}$, so that the scanned data is written as

$$\begin{aligned} V = \{ & (\mathbf{v}_n = (v_i, v_j, v_k)^T, f_{i,j,k}) \mid \mathbf{v}_n \in \mathbb{R}^3, f_{i,j,k} \in \mathbb{R}, n \in \mathbb{N} \\ & i = 0, \ldots, (n_x - 1),\ j = 0, \ldots, (n_y - 1),\ k = 0, \ldots, (n_z - 1)\}. \end{aligned} \tag{2.1}$$

The expression *volumetric data* is in common use to describe data organized in this way. In general the sampled data is delimited by a domain cube D which determines the region of interest:

$$D = \{\mathbf{p} = (x, y, z)^T \mid x \in [x_{\min}, x_{\max}], y \in [y_{\min}, y_{\max}], z \in [z_{\min}, z_{\max}]\}$$

where

$$\begin{aligned} v_i &= x_{\min} + i\,\Delta x;\ i = 0, \ldots, n_x - 1;\ \Delta x = \frac{x_{\max} - x_{\min}}{n_x - 1} \\ v_j &= y_{\min} + j\,\Delta y;\ j = 0, \ldots, n_y - 1;\ \Delta y = \frac{y_{\max} - y_{\min}}{n_y - 1} \\ v_k &= z_{\min} + k\,\Delta z;\ k = 0, \ldots, n_z - 1;\ \Delta z = \frac{z_{\max} - z_{\min}}{n_z - 1} \end{aligned}$$

With reference to Eqn. (2.1) it is conceivable to decompose the volumetric data set V in n_x three-dimensional slices S_i (see Fig. 1b), so that

$$S_i = \bigcup_{\substack{j=0,\ldots,n_y-1 \\ k=0,\ldots,n_z-1}} \{(\mathbf{v}_n = (v_i, v_j, v_k)^T, f_{i,j,k}) \mid \mathbf{v}_n \in \mathbb{R}^3, f_{i,j,k} \in \mathbb{R}\}$$

The corner nodes of two neighboring slices S_i and S_{i+1} form a quadratic block B_i (see Fig. 1c), with

$$
\begin{array}{rlll}
B_i = & \{ & \mathbf{b}_i^0 = (v_i, v_0, v_0)^T, & \mathbf{b}_i^1 = (v_{i+1}, v_0, v_0)^T, \\
& & \mathbf{b}_i^2 = (v_{i+1}, v_0, v_{n_z-1})^T, & \mathbf{b}_i^3 = (v_i, v_0, v_{n_z-1})^T, \\
& & \mathbf{b}_i^4 = (v_i, v_{n_y-1}, v_0)^T, & \mathbf{b}_i^5 = (v_{i+1}, v_{n_y-1}, v_0)^T, \\
& & \mathbf{b}_i^6 = (v_{i+1}, v_{n_y-1}, v_{n_z-1})^T, & \mathbf{b}_i^7 = (v_i, v_{n_y-1}, v_{n_z-1})^T \\
& | & \mathbf{b}_i^m \in I\!R^3, \; m = 0, \ldots, 7\}. &
\end{array}
$$

Each block B_i contains $(n_y - 1)(n_z - 1)$ different voxels, which in this context are defined by eight adjacent grid nodes and their associated intensity values.

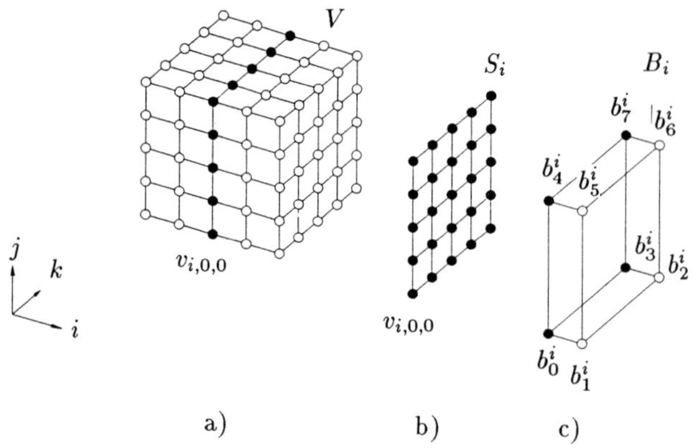

Fig. 1. a) Volumetric Data Set V, b) Slice S_i and c) Block B_i.

The following outlines the general idea behind our algorithm. The blocks B_i, $i = 0 \ldots (n_x - 2)$ are first sorted by increasing distance with respect to the observer's position. Then each block is projected according to its particular order by an enlarged viewing transformation $\varphi : B_i \rightarrow \overline{B}_i$, $(B_i \subset I\!R^3, \overline{B}_i \subset I\!R^2)$ onto the viewing plane. $\overline{B}_i$ is the two-dimensional closed polygon formed by the convex hull of the projected block B_i. For each undetermined pixel lying inside the boundary of $\overline{B}_i$ a ray from the observer's position through the pixel center is calculated and intersected with block B_i in 3D space. Following Levoy's algorithm (see Levoy [13]) a color and a partial opacity along the ray segment traversing B_i is determined and assigned to a z-Buffer entry, which corresponds to the pixel under consideration. In contrast to the conventional z-Buffer technique each entry is marked as fully determined, and therefore can be ignored in further processing. The number of entries is identical to the number of pixels obtained by scan converting the bounding box of the projected domain cube D.

3 Projection Strategy

As a first step of the algorithm, the blocks B_i are listed by increasing distance to the center of projection O. The order of enumeration depends on the relation between the position of O_{xz}, which denotes O projected onto the xz-plane, and projection D_{xz} of domain cube D onto the same plane.

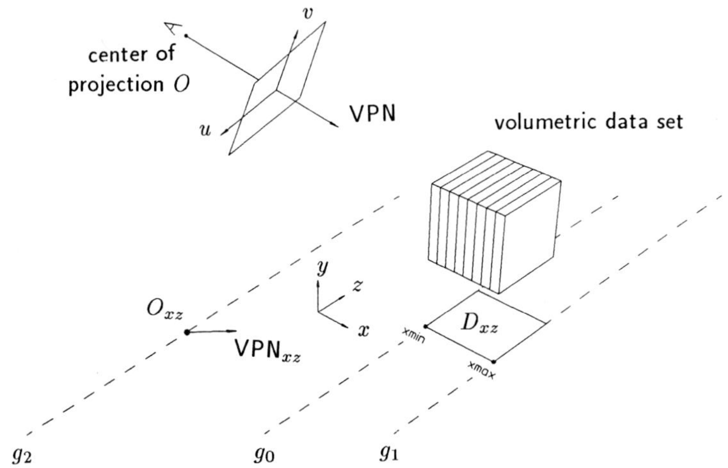

Fig. 2. Point O_{xz} located on the left side of D_{xz}.

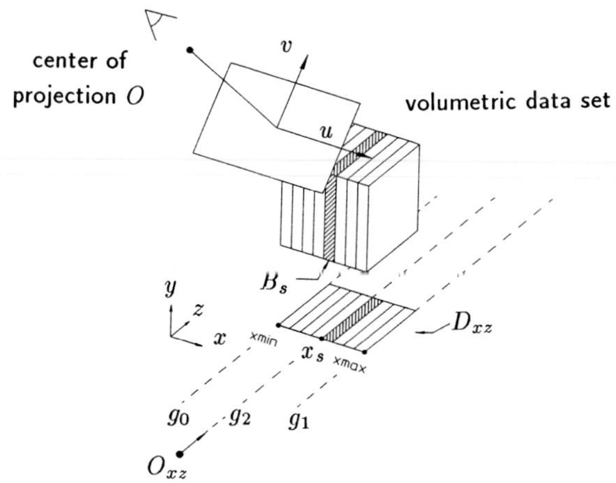

Fig. 3. Point O_{xz} located between g_0 and g_1.

The position of O_{xz} with respect to D_{xz} is categorized into three major types. In the first case point O_{xz} is located on the left side of D_{xz} defined by the straight line $g_0 = \{(x, z) \mid x = xmin;\ z \in \mathbb{R}\}$ (see Fig. 2). Sorting the blocks B_i in this case by increasing index i, yields

$$B_0, B_1, \ldots, B_i, \ldots, B_{n_x-3}, B_{n_x-2}.$$

O_{xz} on the right side of line $g_1 = \{(x, z) \mid x = x_{\max};\ z \in \mathbb{R}\}$ (see Fig. 2) leads to the second type. This time processing of the blocks B_i occurs in reverse order resulting in

$$B_{n_x-2}, B_{n_x-3}, \ldots, B_i, \ldots, B_1, B_0.$$

Finally a third type exists, if O_{xz} is positioned between g_0 and g_1. In this special case a straight line g_2 defined by O_{xz} and the z-axis is intersected with D_{xz}. The intersection point x_s determines a first sequence $B_s, B_{s-1}, \ldots, B_1, B_0$ of blocks to be processed (see Fig. 3); the remaining blocks $B_{s+1}, B_{s+2}, \ldots, B_{n_x-3}, B_{n_x-2}$ are treated in this order. As a result, all blocks B_i are projected in the following way

$$B_s, B_{s-1}, \ldots, B_1, B_0, B_{s+1}, B_{s+2}, \ldots, B_{n_x-3}, B_{n_x-2}.$$

4 Block Processing

4.1 Determination of the Convex Hull

Similar to the Cohen-Sutherland line-clipping algorithm (see Foley et al. [3]), the edges of block B_i are extended to divide the 3D space into 27 regions $R^i_{k,l,m}$, $(k, l, m = 0 \ldots 2)$. The convex hull of each projected block B_i is determined by the region $R^i_{k,l,m}$, which contains the center of projection O. Figure 4 shows an example, where O is lying above, to the left and in front of B_i and therefore in region $R^i_{0,2,0}$. In this case the corner nodes b^7_i, b^3_i, b^0_i, b^1_i, b^5_i, b^6_i are projected by the viewing transformation φ onto the viewing plane. The particular projection order is essential for the scan-conversion of the convex hull $\overline{B}_i$, described in Sect. 4.2.

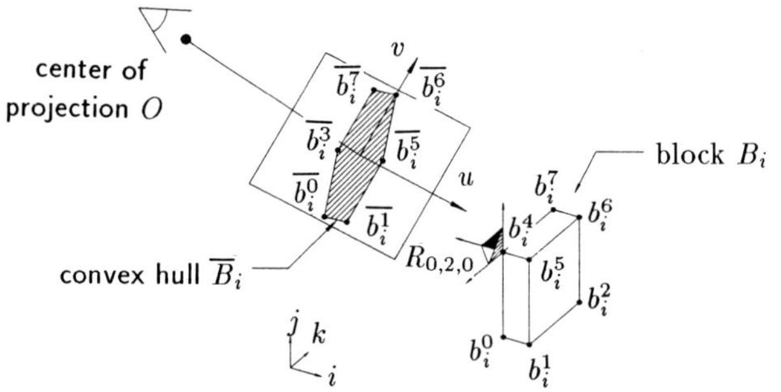

Fig. 4. Convex hull $\overline{B}_i$ regarding region $R^i_{0,2,0}$.

The convex hulls $\overline{B}_i$ of the remaining 26 regions are determined in the same manner. Figure 5 for example presents the convex hulls $\overline{B}_i$ regarding all regions $R^i_{k,l,m}$ lying in front of block B_i.

4.2 Scan-Conversion of the Convex Hull

To save rays in further processing, the convex hull $\overline{B}_i$ is first clipped against the window of the viewing plane. For this purpose we use the well known Sutherland-Hodgman polygon-clipping algorithm (see Foley et al. [3]).

Region $R^i_{k,l,m}$	Convex Hull $\overline{B}_i$
$R^i_{0,0,0}$	$b^7_i \rightarrow b^3_i \rightarrow b^2_i \rightarrow b^1_i \rightarrow b^5_i \rightarrow b^4_i$
$R^i_{1,0,0}$	$b^4_i \rightarrow b^0_i \rightarrow b^3_i \rightarrow b^2_i \rightarrow b^1_i \rightarrow b^5_i$
$R^i_{2,0,0}$	$b^4_i \rightarrow b^0_i \rightarrow b^3_i \rightarrow b^2_i \rightarrow b^6_i \rightarrow b^5_i$
$R^i_{0,1,0}$	$b^7_i \rightarrow b^3_i \rightarrow b^0_i \rightarrow b^1_i \rightarrow b^5_i \rightarrow b^4_i$
$R^i_{1,1,0}$	$b^4_i \rightarrow b^0_i \rightarrow b^1_i \rightarrow b^5_i$
$R^i_{2,1,0}$	$b^4_i \rightarrow b^0_i \rightarrow b^1_i \rightarrow b^2_i \rightarrow b^6_i \rightarrow b^5_i$
$R^i_{0,2,0}$	$b^7_i \rightarrow b^3_i \rightarrow b^0_i \rightarrow b^1_i \rightarrow b^5_i \rightarrow b^6_i$
$R^i_{1,2,0}$	$b^7_i \rightarrow b^4_i \rightarrow b^0_i \rightarrow b^1_i \rightarrow b^5_i \rightarrow b^6_i$
$R^i_{2,2,0}$	$b^4_i \rightarrow b^0_i \rightarrow b^1_i \rightarrow b^2_i \rightarrow b^6_i \rightarrow b^7_i$

Fig. 5. Convex hulls regarding regions $R^i_{k,l,m}$ lying in front of block B_i.

The succeeding scan-conversion is simplified by the consideration that the convex hull $\overline{B}_i$ is not intersected more than twice by each new scan line. We now describe a fast *scan-conversion algorithm* which takes advantage of this fact and is very similar to the technique proposed in Foley (see Foley et al. [3], pp. 96-99).

First we introduce a data structure, called the *edge table* (*ET*), which comprises a set of edges forming the convex hull $\overline{B}_i$. Each entry in the *ET* contains the y_{min} and y_{max} coordinates of the edge, the x coordinate of the bottom endpoint x, and the x increment

$$x_{inc} = \begin{cases} \dfrac{x_{max} - x_{min}}{y_{max} - y_{min}} & y_{max} - y_{min} > 0 \\ 0 & \text{otherwise} \end{cases}$$

used to step from one scan line to the other during the scan-conversion process.

The edges within the *ET* are sorted by increasing y_{min}. If there are two edges starting at the same point (y_{min}), they are kept in order of increasing bottom endpoint x. A special case occurs if two adjacent edges e_i and e_{i+1} are lying on a straight horizontal line. In this situation edge e_{i+1} is deleted in the *ET*.

Finally, we need three additional flags *left edge* (*LE*), *right edge* (*RE*) and *next edge* (*NE*). *LE* and *RE* point to edges in the *ET*, which are defining the left (respectively right) boundary of the area to be scan converted. *NE* points to an edge becoming *LE* or *RE* in the next step of the algorithm. At the beginning the flags are preset with $LE = 0$, $RE = 1$ and $NE = 2$.

Once the *ET* has been created it is processed during the scan-conversion in top to down order as follows:

1. Set scan line y to y_{min} of the first entry in the *ET*.

2. Repeat until $y = y_{max}$ of the last entry in the *ET*:

   (a) Process each pixel of scan line y (see Sect. 4.3) between suitably rounded x coordinates of edge entries indicated by *LE* and *RE*.

(b) If $y = y_{max}$ of the edge indicated by *LE* then set *LE* to *NE* and increment *NE* by 1; otherwise increment x by x_{inc} of the edge pointed out by *LE*.

(c) If $y = y_{max}$ of the edge indicated by *RE* then set *RE* to *NE* and increment *NE* by 1; otherwise increment x by x_{inc} of the edge pointed out by *RE*.

(d) Increment scan line y by 1.

4.3 Visualizing the Volumetric Data

4.3.1 Color Intensity Determination

In 1988, Levoy (see Levoy [13]) proposed an algorithm classified as a *volume rendering* technique. The fundamental idea of these methods is based on the ray tracing approach where rays are cast from the observer's position through the center of each pixel into the volumetric data set. Within the data set, each ray is subdivided into segments of equal length by discrete sample points. At every sample point a gray level gradient (see Tiede et al. [19]) and a partial opacity is determined. The opacity is in general a function of the intensity value, which is obtained by linear interpolation of the sampled data placed at the voxel vertices. To calculate the partial opacity at the sample points along the ray Levoy suggested a linear mapping function

$$\alpha(\mathbf{x_i}) = |\nabla f(\mathbf{x_i})| \begin{cases} \alpha_{v_{n+1}} \left(\frac{f(\mathbf{x_i}) - f_{v_n}}{f_{v_{n+1}} - f_{v_n}}\right) + \alpha_{v_n} \left(\frac{f_{v_{n+1}} - f(\mathbf{x_i})}{f_{v_{n+1}} - f_{v_n}}\right) & f_{v_n} \leq f(\mathbf{x_i}) \leq f_{v_{n+1}} \\ 0 & \text{otherwise} \end{cases} \tag{4.1}$$

where

$\mathbf{x_i}$	:	i-th sample location
$f(\mathbf{x_i})$	:	interpolated intensity value at i-th sample location
$\nabla f(\mathbf{x_i})$	:	local gray level gradient
f_{v_i}	:	intensity threshold of tissue type v_i
α_{v_i}	:	opacity threshold of tissue type v_i.

The magnitude of the local gray level gradient $|\nabla f(\mathbf{x_i})|$ in Eqn. (4.1) is used to enhance the boundary region between two different tissue types while putting down the opacity within the tissue at the same time.

In addition, the gray level gradient serves as an approximated surface normal and together with any lighting model defines a color intensity $c(\mathbf{x_i})$ at each sample location. Traversing the volumetric data set in front-to-back order, the partial opacities $\alpha(\mathbf{x_i})$ are accumulated and used to weight each $c(\mathbf{x_i})$. Finally, the total color intensity assigned to the ray is defined by the sum of all weighted $c(\mathbf{x_i})$

$$C = \sum_{i=0}^{k-1} c(\mathbf{x_i})\alpha(\mathbf{x_i}) \prod_{j=0}^{i-1}(1 - \alpha(\mathbf{x_j})) \tag{4.2}$$

where

k : number of sample points
$c(\mathbf{x_i})$: color at i-th sample location
$\alpha(\mathbf{x_i})$: opacity at i-th sample location.

The calculation of the color intensity in Eqn. (4.2) is terminated if the total opacity defined as

$$\alpha_{tot}(\mathbf{x_i}) = 1 - \prod_{j=0}^{i}(1 - \alpha(\mathbf{x_j})) \tag{4.3}$$

reaches unity.

4.3.2 The z-Buffer

To render the volumetric data associated with the convex hull $\overline{B}_i$, we require a z-buffer. The bounding box of the domain cube D projected onto the image plane determines the buffer size (see Fig. 6). The number of entries is identical to the number of pixels obtained by the scan-converted bounding box. Each buffer entry contains a scalar value $t(\mathbf{x_i})$ and a opacity $\alpha = 1 - \alpha_{tot}(\mathbf{x_i})$ according to Eqn. (4.3), which is initialized to unity. $t(\mathbf{x_i})$ represents the last active ray parameter and is set to zero at the beginning.

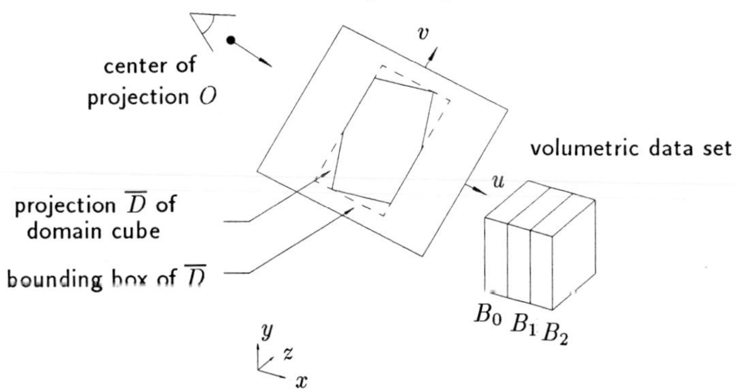

Fig. 6. Determination of the z-Buffer size.

During the scan conversion process of the convex hull $\overline{B}_i$, a ray is traced through each pixel (x, y) into the data set, if the corresponding opacity entry α in the z-Buffer is greater than zero. Otherwise, on the basis of the block processing order, this pixel is fully determined and therefore can be ignored in further processing.

If the polygon point being scan-converted at (x, y) is processed the first time ($t(\mathbf{x_i}) = 0$) the ray is intersected with block B_i. The resulting intersection values $t_{in}(\mathbf{x_i})$ and $t_{out}(\mathbf{x_i})$ define the ray segment lying in the interior of block B_i, respectively between slices S_i and S_{i+1}. The segment is subdivided by discrete sample points and a new color intensity $c(\mathbf{x_i})$ is calculated at each sample location according to Eqn. (4.1) and stored in the frame buffer. At the same time the current opacity value α in the z-Buffer is replaced by $\alpha \cdot (1 - \alpha(\mathbf{x_i}))$. If the ray leaves block B_i the last t value is stored in the z-Buffer and serves as next $t(\mathbf{x_i})$ for the processing of block B_{i+1}. Fig. 7a shows the same volumetric data set as

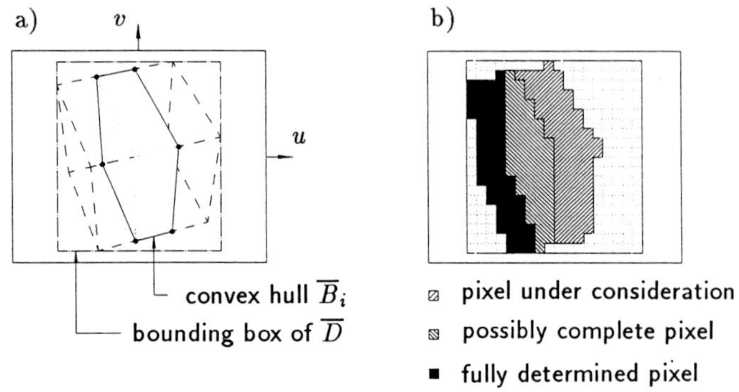

Fig. 7. Projected data and its representation in the Volume Priority z-Buffer.

in Fig. 6 projected onto the image plane. For the sake of simplicity the data set consists of only three blocks. The convex hull $\overline{B_{i-1}}$ is already scan-converted and parts of the z-Buffer (see Fig. 7b) are fully determined. Whether a pixel is marked as possibly complete depends on the opacity value α of the corresponding z-Buffer entry.

5 Conclusion

Volume-based rendering techniques have shown their clinical usefulness during the last decade. Despite their indisputable advantages, ray tracing volumetric data sets is still very expensive in terms of computational power and memory requirements. We have proposed a new technique which allows the visualization of large volumetric data sets on low cost workstations equipped with limited memory.

References

[1] J.D. Boissonnat (1988). *Shape Reconstruction from Planar Cross-Sections. Computer Graphics & Image Processing*, 44:1–29.

[2] H.N. Christiansen, T.W. Sederberg (1978). *Conversion of Complex Contour Line Definitions into Polygonal Element Mosaics. Computer Graphics (SIGGRAPH '78 Proceedings)*, 12(3):187–192.

[3] J. Foley, A. vanDam, S. Feiner, J. Hughes (1990). *Computer Graphics: Principles and Practice.* Addison Wesley, New York.

[4] G. Frieder, D. Gordon, R. Reynolds (1985). *Back-to-Front Display of Voxel-Based Objects. Computer Graphics & Applications*, 18(3):52–60.

[5] H. Fuchs, Z.M. Kedem, S.P. Uselton (1977). *Optimal Surface Reconstruction for Planar Contours. Comm. of the ACM*, 20(10):693–702.

[6] A. Fujimoto, T. Takayuki, I. Kansei (1986). *ARTS: Accelerated Ray-Tracing System. IEEE Computer Graphics & Appl.*, 6(4):16–26.

[7] S. Ganapathy, T.G. Dennehy (1982). *A New General Triangulation Method for Planar Contours. Computer Graphics*, 16(3):69–75.

[8] A.S. Glassner (1984). *Fast Ray Tracing by Space Subdivision. IEEE Computer Graphics & Appl.*, 4(10):15–22.

[9] G.T. Herman, H.K. Liu (1979). *3-D Display of Human Organs from Computed Tomograms. Computer Graphics and Image Processing*, 9(1):1–21.

[10] K.H. Höhne, R. Bernstein (1986). *Shading 3D-Images from CT Using Gray-Level Gradients. IEEE Transactions on Medical Imaging*, MI-5(1):45–47.

[11] K.H. Höhne, R.L. Delapaz, R. Bernstein, R.C. Taylor (1987). *Combined Surface Display and Reformatting for the Three-Dimensional Analysis of Tomographic Data. Investigative Radilogy*, 22(7):658–664.

[12] E. Keppel (1975). *Approximation of Complex Surfaces by Triangulation of Contour Lines. IBM Journal of Research and Development*, 19:2–11.

[13] M. Levoy (1988). *Display of Surfaces from Volume Data. IEEE Computer Graphics & Appl.*, 8(5):29–37.

[14] W.E. Lorensen, H.E. Cline (1987). *Marching Cubes: A High Resolution 3D Surface Construction Algorithm. Computer Graphics*, 21(4):163–169.

[15] J.C. Mazziotta, K.H. Huang (1976). *THREAD (Three-Dimensional Reconstruction and Display) with Biomedical Applications in Neuron Utrastructure and Display. American Federation of Information Processing Society*, 45:241–250.

[16] D.J. Meagher (1982). *Efficient Synthetic Image Generation of Arbitrary 3-D Objects. Proceedings of the IEEE Computer Society Conference on Pattern Recognition and Image Processing*, pp. 473–478.

[17] D.S. Schlusselberg, W.K. Smith, D.J. Woodward (1986). *Three-Dimensional Display of Medical Image Volumes. Proceedings of* 7th *NCGA*, III:114–123.

[18] A. Sunguroff, D. Greenberg (1978). *Computer Generated Images for Medical Applications. Computer Graphics*, 12:196–202.

[19] U. Tiede, K.H. Höhne, M. Bomans, A. Pommert, M. Riemer, G. Wiebecke (1990). *Investigation of Medical 3D-Rendering Algorithms. IEEE Computer Graphics & Appl.*, 10(3):41–53.

[20] H.K. Tuy, T.T. Lee (1984). *Direct 2-D Display of 3-D Objects. IEEE Computer Graphics & Appl.*, 4(10):29–34.

[21] M.W. Vannier, J.L. Marsh, O. James (1983). *Three Dimensional Computer Graphics for Craniofacial Surgical Planning and Evaluation. Computer Graphics*, 17(3):263–273.

[22] T. Whitted (1980). *An Improved Illumination Model for Shaded Display. Communication ACM*, 23:343–349.

A Fourier Technique for Volume Rendering

Tom Malzbender, Fred Kitson

The Fourier Projection Slice Theorem (FPST) is widely used for Medical Imaging, in particular for tomography. Here we develop a technique for using the FPST for display of volume data. Images are rendered from a frequency domain representation of the data, as opposed to the spatial representation which is more typical in computer graphics. The principal advantage of this technique lies in the increase speed at which the algorithm runs compared to standard volume rendering algorithms. In fact, one can achieve a complexity of 1-3 orders of magnitude less than either a screen space or object space volume rendering techniques. In addition, assurances of image accuracy can be made. The principal drawback of the technique is the lack of hidden surface effects which makes the images more difficult to interpret compared with more conventional approaches. We present several techniques which are useful for ensuring artifact-free imagery. Among these are resampling filter design approaches, spatial premultiplication and zero padding.

1 Introduction

Current volume rendering algorithms fall into two classes, a screen- space approach and an object space approach [8, 14, 16, 20, 21, 23]. Although capable of generating impressive imagery, these techniques are notoriously slow. This is due more to the size of the dataset than to the inherent complexity of the algorithms, making it difficult to refine these algorithms to provide real time performance for large data sets. Approaches that have met with some success at reducing the compute time capitalize on not sampling all of the data in the dataset [15]. For example, in the ray-tracing approach, one can terminate front to back rays when the ray opacity reaches some threshold value close to 1.0. Another approach is to store the data in a hierarchical data structure that keeps track of regions of empty space.

We present a alternate approach that transforms the data into a new space that has the desirable property that all the information required to compute a projection lies along a plane in this space, as opposed to the entire volume of the original data. A transform space that has this property is provided by either the Fourier or Hartley transform. We entitle this rendering technique *Fourier Volume Rendering* (*FVR*).

2 Fourier Projection-Slice Theorem

If $F(w_x, w_y)$ is the Fourier transform of $f(x, y)$ and $P_\Theta(w_r)$ is the Fourier transform of $p_\Theta(r)$, a 1D projection of $f(x, y)$ at angle Θ, then FPST states that:

$$P_\Theta(w_r) = F(w_r \cos(\Theta), w_r \sin(\Theta)).$$

Graphically this relationship is shown in Fig. 1.

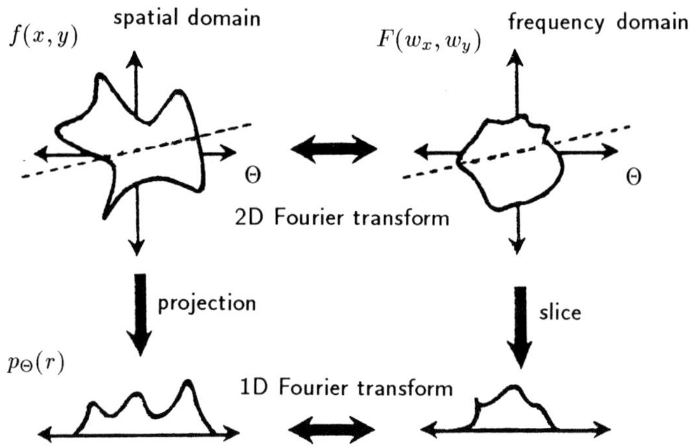

Fig. 1. Fourier Projection Slice Theorem.

The FPST states that the 1D projection of a 2D spatial function is related to a slice of the Fourier transform of that function by a forward 1D Fourier transform. This slice must pass through the origin of the frequency domain and its angle specifies the angle of the projection. This relationship is used in tomographic reconstruction. In particular, 1D projections at many angles can be measured using X-rays, and these projections are incorporated into a single spatial reconstruction of a slice of the object via the FPST [5].

3 Fourier Volume Rendering

The FPST can also be used for Volume Rendering in the following way. One can transform the 3D scalar array of data into the frequency domain via a 3D FFT. This needs to be done only once per dataset and can be considered a preprocessing operation. To generate a projection at a particular angle, one can resample in the frequency domain along a plane perpendicular to the desired projection direction, and then perform an inverse 2D FFT of the resultant array. We entitle this procedure Fourier Volume Rendering (FVR) [17]. Graphically, this approach is outlined in Fig. 2.

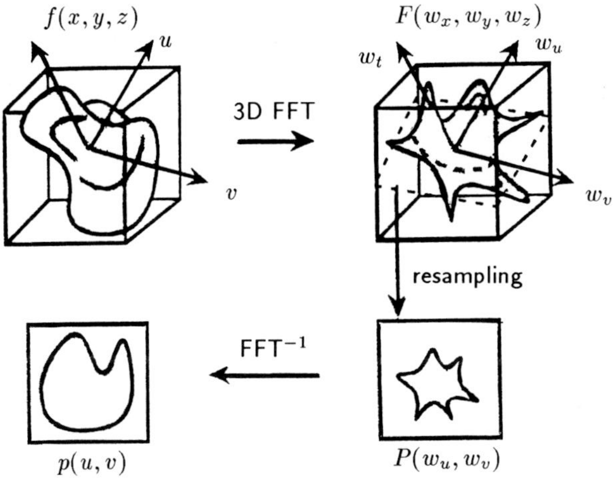

Fig. 2. Fourier Volume Rendering.

Although the use of the Fourier transform is shown above for, to, and from the transform space, in practice we use the Hartley transform for both the forward and inverse transformations. The forward 3D Hartley transform is given as

$$H(k_x, k_y, k_z) =$$

$$\frac{1}{N^3}\left(\sum_{x=0}^{N-1}\sum_{y=0}^{N-1}\sum_{z=0}^{N-1} f(x,y,z)\left[\cos\left(2\pi\frac{xk_x + \dot{y}k_y + zk_z}{N}\right) + \sin\left(2\pi\frac{xk_x + yk_y + zk_z}{N}\right)\right]\right.$$

for $f(x, y, z)$ being transformed into the (k_x, k_y, k_z) space. The inverse 2D Hartley transform that returns us into the two dimensional (u, v) space from the (k_u, k_v) space is :

$$H^{-1}(u, v) =$$

$$\sum_{k_u=0}^{N-1}\sum_{k_v=0}^{N-1} f(k_u, k_v)\left[\cos\left(2\pi\frac{uk_u + vk_v}{N}\right) + \sin\left(2\pi\frac{uk_u + vk_v}{N}\right)\right].$$

One may refer to [2, 3] for fast implementations of these. Since the Hartley transform can be seen as taking the even part of the Fourier transform and subtracting off the odd part, we can see that these two transform spaces contain the same information about the original data, in a slightly different format. For our purposes, since the input dataset will contain real data, the Hartley has an advantage over the Fourier in that a purely real transform is performed and computational speed is increased somewhat. However, due to the greater familiarity of the Fourier transform we will use it in subsequent discussions.

4 Data Shifting

Attention must be paid to the format of the Fast Fourier (or Hartley) transform data arrays since zero Hertz and zero spatial position are usually held in the first element of the respective arrays. For example, let us consider the forward 3D transformation. For a conventional projection we will want zero spatial position to correspond to the center of our dataset, so data must shifted by N/2 in each axis before the transformation is performed. After this transformation, the re-

Raw 3D Spatial Data
↓
Premultiply 3D Data (Spatial Equalization)
↓
Shift Spatial Data by (N/2, N/2, N/2)
↓
Forward 3d Fourier Transform
↓
Shift Frequency Data by (N/2, N/2, N/2)
↓
Resample Along Plane
↓
Shift Frequency Image by (N/2, N/2)
↓
Inverse 2d Fourier Transform
↓
Shift Spatial Data by (N/2, N/2)
↓
Display

Fig. 3. FVR Shifting Operations.

sulting array will have zero Hertz as the first element of the array, so if we would like to slice through the center of the array, we again need to shift by N/2 in each axis. In practice, it is possible to avoid actually shifting any data around. For example, in the first case we simply fill the array in a different order. In the second, we simply redefine our slicing plane. These various shifting operations are summarized in Fig. 3.

5 Resampling Implementation Details

As indicated in Fig. 2, after the preprocessing operation of the forward 3D transform there are two primary steps to image generation. The first involves resampling and the second is the inverse transform. Not only is the former more computationally intensive for typical datasets, but it must be performed carefully to avoid introducing artifacts in the image. We have developed several techniques to avoid any such artifacts.

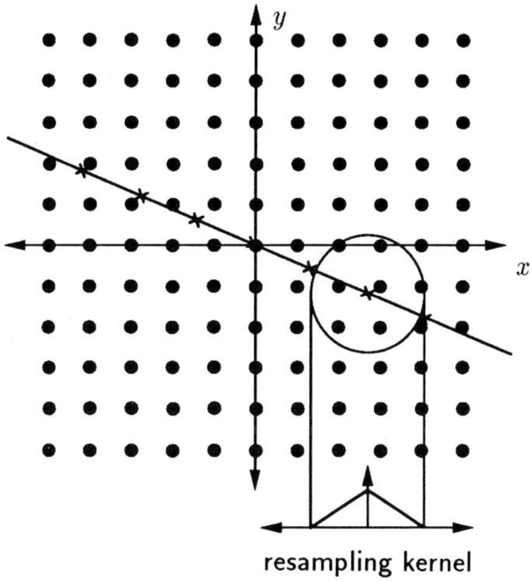

Fig. 4. Resampling a 2D array along a line.

Resampling typically involves the use of a resampling kernel, either explicitly or implicitly. For example, bilinear interpolation, as shown in Fig. 4, can be thought of as using a 2D triangle function as a resampling kernel. Typically one can think of resampling as centering this kernel at the location one would like to evaluate and then taking a sum of the samples falling underneath the kernel times the kernel weights at their respective locations. This procedure is used for both spatial domain resampling and frequency domain resampling, where we use it in FVR. This operation can also be considered as evaluating the convolution of the the discrete array with the continuous resampling kernel, at a discrete set of points.

A convolution kernel such as the triangle function shown above, corresponding to bilinear interpolation, is a poor resampling filter. As is well known, the ideal function to use for 3D resampling would be a sinc() function, given by

$$\mathrm{sinc}(w_x, w_y, w_z) = \frac{\sin(\pi w_x)}{\pi w_x}\frac{\sin(\pi w_y)}{\pi w_y}\frac{\sin(\pi w_z)}{\pi w_z}.$$

The reason for this is intuitively quite clear. Any continuous function that is sampled is made infinitely periodic in its transform domain by that sampling. For example, when we are working with a discrete set of 3D samples in the frequency domain that represents our spatial data, these samples really represent an infinitely periodic continuous function. Sampling has caused our continuous spatial data to become infinitely repeated in each axis. We can imagine recovering the original continuous non-periodic function by multiplying in the spatial domain by a three dimensional rectangle function. This rectangle function must have a value of unity for one cycle of the spatial function, and be zero outside of this cycle. Multiplying by the rectangle function in effect suppresses all the periodicities in the spatial domain except one. We know that multiplying in the spatial domain by one function is equivalent to convolving in the frequency domain by the transform of this function. We also know that resampling amounts to convolving with a resampling kernel. Therefore, if we set our resampling function to be the transform of the rectangle function, we will *exactly* reconstruct our original function. The transform of this rectangle function is just the sinc() given above.

Unfortunately, in practice this is not possible since the sinc() function has infinite support. We need to use a resampling kernel that is some finite approximation to the sinc function. We have pursued two approaches to the design of such resampling filters. The first involves multiplying the desired sinc() function by a windowing function to limit its extent. A simple windowing function that can be used is a rectangle function which passes values only within some region. A better choice of windowing function is the Hamming window, which has a smoother transition to zero than the rectangle function. It is given by

$$W(w_x, w_y, w_z) =$$

$$[\alpha + (1-\alpha)\cos(2\pi \frac{w_x}{F_1})][\alpha + (1-\alpha)\cos(2\pi \frac{w_y}{F_1})][\alpha + (1-\alpha)\cos(2\pi \frac{w_z}{F_1})],$$

where F_1 is the width of the window and $\alpha = 0.54348$.

An approach we have employed that yields better results is the use of an iterative filter design technique entitled Projection on Convex Sets (POCS) [19, 7]. We can understand the POCS approach by noticing that our filter design problem really involves two constraints. First there is the constraint that the filter should have limited extent in the frequency domain. This is simply due to the fact that we must compute some convolution in this domain and this can only be done for a finite convolution kernel. Recall that this convolution and resampling is being performed in the frequency domain. The second constraint is that the filter in the spatial domain should also have limited extent. This is due to the fact that we would like the spatial function to be an approximation to the rectangle function, which is limited in the spatial domain. Unfortunately, the Fourier uncertainty relationship [2] prohibits any solution that satisfies both constraints. If a function is limited in one domain, it must have infinite extent in the other domain. However we can use POCS to find solutions which minimize violations to both constraints.

As shown in Fig. 5, we start with some function, such as the triangle, in the frequency domain. Taking an inverse transform we wind up with an infinite support function. Here we apply the spatial constraint by truncating the tails of the function corresponding to a single cycle of the original dataset. Taking a forward transform, we wind up with a different frequency domain function that also has unlimited support. Here we apply the frequency domain constraint by eliminating the tails corresponding to the width of the convolution kernel we would like to use. This procedure is iterated until there is minimal change from one iteration to the next. We terminate with the truncation in the frequency domain to ensure that we really have a finite convolution kernel, which is required.

6 Zero Padding and Premultiplication

Either filter design technique yields a filter with finite support in the frequency domain, but of unlimited support spatially. Figure 6 shows the spatial function associated with linear interpolation in the frequency domain. Although our design techniques allow much better filter characteristics than this, it is a useful example to come to an understanding of any practical filter's shortcomings. Bear in mind that our filter is three dimensional, whereas we show here its one dimensional relative.

As an approximation to a rectangle function, there are two concerns we have with the filter's spatial characteristics. First, it is non-zero outside the desired region. Since we are convolving in the frequency domain, this is equivalent to multiplying the spatial filter function by the infinitely periodic dataset. This non-zero region will allow copies of the periodic dataset to remain, albeit at weakened intensity, and the projection of these will overlap each other. We have found that adding a small percentage of zero valued data surrounding our original spatial dataset (zero-padding) solves this problem. This is because the filter function magnitude has been allowed to drop to a negligible level before it encounters any non-zero data.

The second drawback to this filter is the fact that it does not have constant magnitude within the region corresponding to the dataset (interior of the rectangle). Ideally, this should be unity valued throughout the first cycle of the dataset. We can see that in general it is not, and this will cause an attenuation of the perimeter of the dataset relative to its center. This problem can be precisely corrected by a technique entitled premultiplication. Here we simply multiply the original dataset by the inverse of the spatial response of the resampling function, to cause an effective flat response to appear. A version of this idea, using 2D filters and functions, is illustrated in Fig. 7.

The first plot shows the spatial response of the reconstruction filter. Note that the higher amplitude in the center will cause this region to be incorrectly emphasized. The second plot shows a dataset of value 1.0 that has been multiplied by the inverse of the spatial resampling function response. The last plot shows the resultant function using this resampling function. Notice that although we still have the tails of the reconstruction filter to address, we have now flattened the response of the filter in the primary cycle.

7 Resampling Rate

Attention needs to be paid to the rate at which we resample the frequency domain as well as the filter we employ. If we use the same rate for sampling along the plane as the discrete data has in the frequency domain volume, we encounter aliasing problems.

The set of samples that we collect along the plane reflect an infinitely periodic projection of the spatial data, and we must be careful that these periodic copies

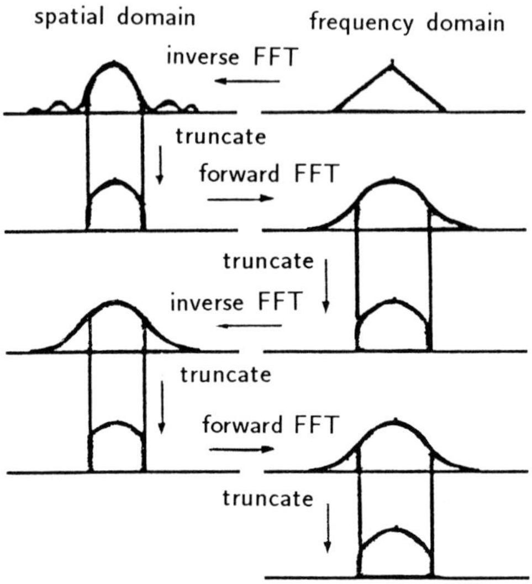

Fig. 5. Projection on Convex Sets (POCS) filter design technique.

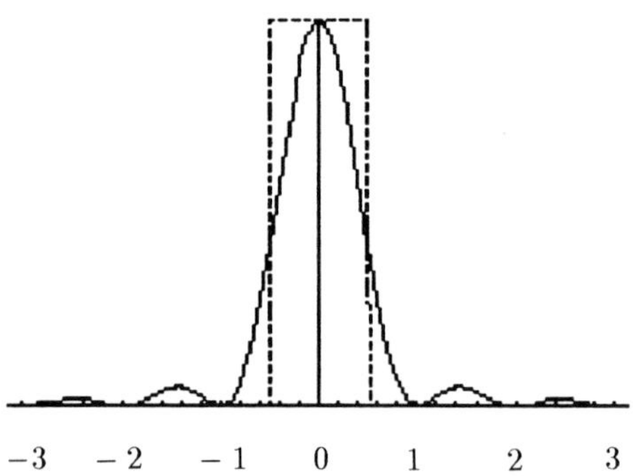

Fig. 6. Spatial response corresponding to linear interpolation.

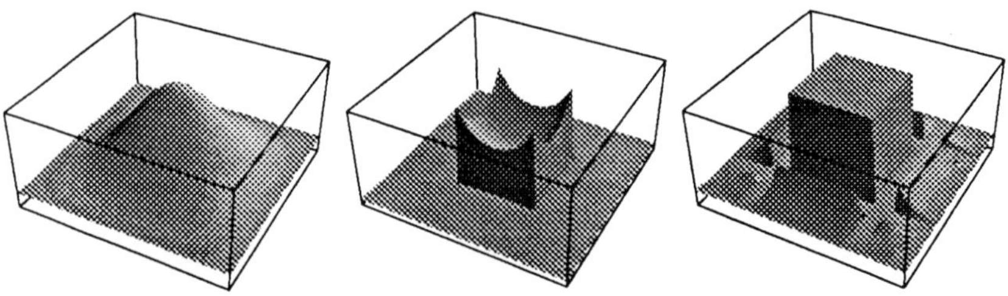

Fig. 7. Spatial Premultiplication - a) Distortion introduced by resampling filter. b) A box of uniform data premultiplied. c) Resultant function before projection.

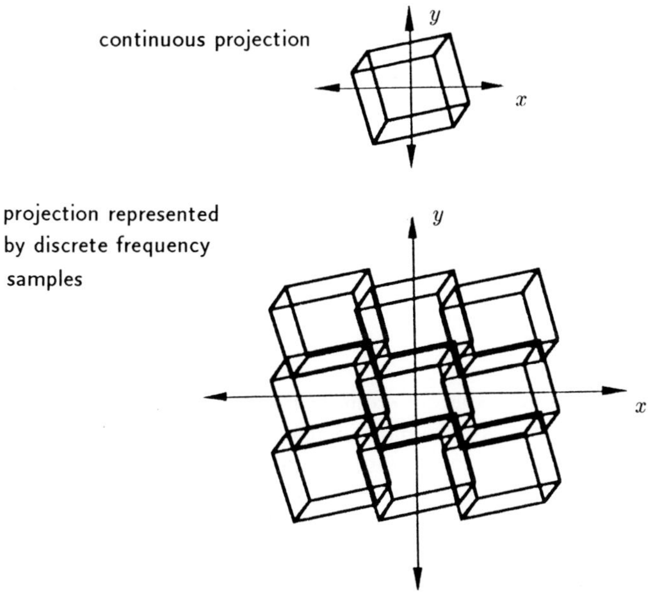

Fig. 8. 2D periodicity introduced by sampling.

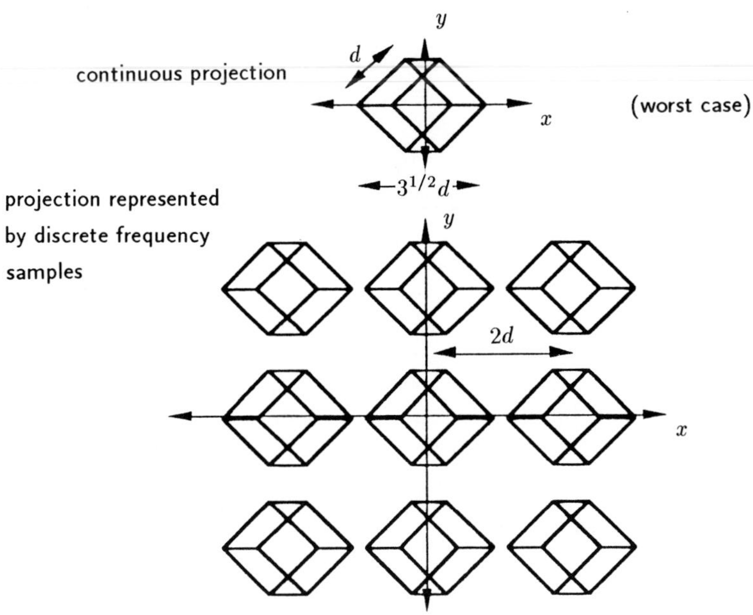

Fig. 9. Periodicity with doubled resampling rate.

do not overlap to cause aliasing. Note that the type of aliasing we are more familiar with is typically caused by sampling in the spatial or time domain at too low of a rate causing overlap in the frequency domain. Here we have the complimentary case where the overlap actually occurs in the spatial domain. Sampling with the spacing of the 3D frequency domain data will cause aliasing due to geometrical arguments shown in Fig. 8. We can avoid this overlap by sampling at a rate higher than $\sqrt{3d}$ where d is the original 3D sampling rate. This corresponds to the diagonal of a cube, which is the worst case geometry for overlap. In practice we simply double the resampling rate, as shown in figure 9, so that we can still employ radix-2 FFT or FHT routines.

8 Conclusion

We have presented a novel technique for Volume Rendering that works from a frequency domain representation of the data instead of a spatial one. The chief advantage of this approach is the lower computational complexity this method incurs when compared with other Volume Rendering approaches [17]. Shown in Fig. 10 are images computed with FVR. The left image shows a spatial repre-

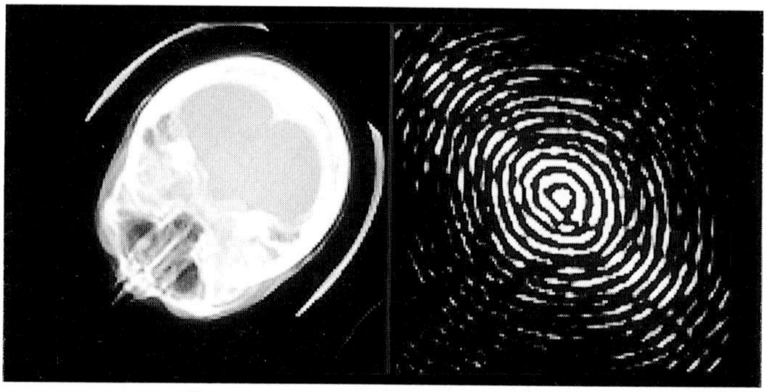

Fig. 10. Spatial and Frequency domain images of human cranial C.A.T. data.

sentation and the image on the right shows the frequency domain representation (Hartley domain). Since the images are a strictly linear projection of the dataset, they contain no hidden surface effects and appear much like "X-rays" of the dataset. The difficult problem of introducing hidden surface properties into the algorithms remains an open problem.

Acknowledgements

Ken Neighbors deserves thanks for the implementation of the Hartley transform code as well as being instrumental in helping uncover the various aliasing mechanisms present in a naive prototype. Ron Pulleyblank is responsible for suggesting spatial premultiplication and was an effective "signal processing sounding board". The need for spatial data shifting was pointed out by John Jackson, of Stanford's MRI group. FFT code was provided for early prototypes by Vasudev Bhaskaran. The medical dataset used in the figures is courtesy of I. Jackson, U. Bite and G. Forbes at the Mayo Clinic.

References

[1] P. Anuta (1970). *Spatial Registration of Multispectral and Multitemporal Digital Imagery Using Fast Fourier Transform Techniques. IEEE Transactions on Geoscience Electronics*, GE-8:353–368.

[2] R. Bracewell (1986). *The Fourier Transform and its Applications.* McGraw-Hill, N.Y. revised second edition.

[3] R. Bracewell (1990). *Assessing the Hartley Transform. IEEE Transactions on Acoustics, Speech and Signal Processing*, 38(12):2174–2176.

[4] E. Brigham (1974). *The Fast Fourier Transform.* Prentice-Hall, New Jersey.

[5] R. Brooks (1980). *Computational Principles of Transmission CT. Medical Physics of CT and Ultrasound*, pp. 37–52.

[6] M. Cartwright (1990). *Fourier Methods for Mathematicians Scientists and Engineers.* Ellis Horwood, N.Y.

[7] M. R. Civanlar, R. A. Nobakht (1988). *Optimal Pulse Shape Design Using Projections Onto Convex Sets.* In *ICASSP 88 Proceedings*, volume 3, pp. 1874–1877.

[8] R. Drebin, L. Carpenter, P. Hanrahan (1988). *Volume Rendering. Computer Graphics*, 22(4):65–74.

[9] D. Dudgeon, R. Mersereau (1984). *Multidimensional Signal Processing.* Prentice-Hall, New Jersey.

[10] S. Dunne, S. Napel, B. Rutt (1990). *Fast Reprojection of Volume Data.* In *Proceedings of the First Conference on Visualization in Biomedical Computing*, pp. 11–18.

[11] H. Fuchs, M. Levoy, S. Pizer, J. Rosenman, editors (1989). *NCGA '89 Conference Proceedings*, volume 1.

[12] P. Hanrahan (1990). *Three-Pass affine Transforms for Volume Rendering. Computer Graphics*, 24(5):71–78.

[13] A. Kaufman, editor (1991). *Volume Visualization.* IEEE Computer Society Press, Los Alamitos, CA.

[14] M. Levoy (1988). *Display of Surfaces From Volume Data. IEEE Computer Graphics and Applications*, pp. 29–37.

[15] M. Levoy (1990). *Display of Surfaces From Volume Data.* doctoral dissertation, University of North Carolina at Chapel Hill.

[16] M. Levoy (1991). *Viewing Algorithms.* In *Volume Visualization*, pp. 89–92. IEEE Computer Society Press, Los Alamitos, CA.

[17] T. Malzbender . *Fourier Volume Rendering.* to appear in ACM Transactions on Graphics.

[18] R. Mersereau, A. Oppenheim (1974). *Digital Reconstruction of Multidimensional Signals from Their Projections. Proceedings of the IEEE*, 62(10):1319–1338.

[19] A. Papoulis (1975). *A New Algorithm in Spectral Analysis and Band Limited Extrapolation. IEEE Transactions on Circuits and Systems*, CAS-22(9):735–741.

[20] P. Sabella (1988). *A Rendering Algorithm for Visualizing 3D Scalar Fields. Computer Graphics*, 22(4):51–55.

[21] C. Upson, M. Keeler (1988). *V-Buffer: Visible Volume Rendering. Computer Graphics*, 22(4):59–64.

[22] R. Webber (1990). *Ray Tracing Voxel Data via Biquadratic Local Surface Interpolation. The Visual Computer*, 6:8–15.

[23] L. Westover (1988). *Footprint Evaluation for Volume Rendering. Computer Graphics*, 24(4):367–376.

[24] T. Whitted (1980). *An Improved Illumination Model for Shaded Display. Communications of the ACM*, 23(6):343–349.

[25] D. Yang (1989). *New Fast Algorithm to Compute Two-Dimensional Discrete Hartley Transform. Electronics Letters*, 25(25):1705–1706.

An Improved Shading Algorithm for Radiosity Based Renderers

Philip Jacob

This chapter presents an improved shading method especially useful for non-interactive radiosity based renderers. This method can be used to generate smooth realistic images while reducing the number of patches needed during the rendering process. It may be used for the visualization of radiosity scenes in combination with ray tracers, painters or depth-buffer systems to reduce Mach bands and other unwanted effects of linear interpolation.

1 Introduction

Global illumination models like radiosity have become more and more popular in the last years, especially where realistic synthesis of indoor scenes is required. Global illumination models all have in common a discrete rastering of the scene using sample points. In contrast to local illumination models they consider the amount of light that reaches the sample indirectly through diffuse (or specular [8]) reflection from other patches. Much effort has been spent in the development of algorithms for the determination of form factors. Simplifications have been made to speed up this calculation [9] while preserving accuracy [1].

2 Rendering

2.1 Linear Interpolation

One of the major advantages of an illumination model based on diffuse radiosity is its view independence. Once the energy has been distributed among the patches and the form factors are known, the scene can be rendered from different viewpoints without major recalculations, as long as no objects (including the light sources) change their position. Fast rendering algorithms and hardware accelerators can be applied to patches, whose brightness and colors are known. In most cases a linear, *Gouraud* [3] like interpolation of the corner intensities is used for shading. This form of interpolation has several disadvantages which are outlined in the following two subsections.

2.1.1 Mach bands

"Unfortunately" the human eye has the ability to recognize discontinuities in the first and second order derivative of intensity changes on a surface. These effects,

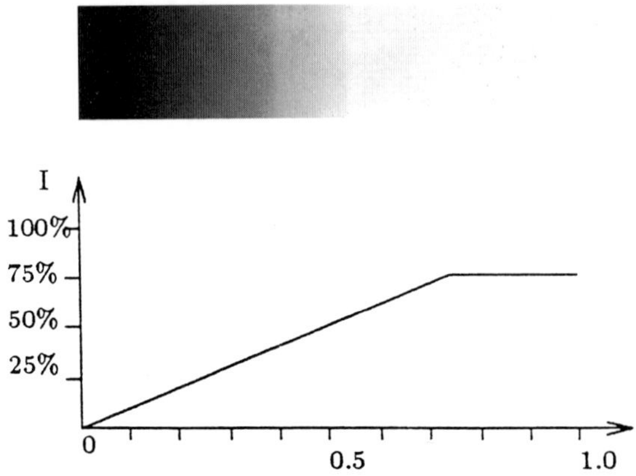

Fig. 1. Mach bands.

called *Mach bands* [1], are one of the reasons why improved shading methods like *Phong shading* [5] have been developed to display polygonal scenes.

Figure 1 shows a ramp of intensities increasing from 0 to 75%. The human eye recognizes a region that appears brighter than its environment where the slope of the intensity changes from one to zero. This effect disappears if a white sheet of paper is held over the left half of the image (covering the region from 0 to 0.75). The smaller the differences of the slopes are the harder they are to detect. *Phong shading* uses vector instead of scalar interpolation to reduce first derivative discontinuities. Thus using *Phong shading* reduces *Mach bands* but does not eliminate them. As most of the radiosity scenes where rendered using a linear interpolation of the intensities, given at the corners of each patch, the number of patches had to be increased in order to reduce the *Mach band* effects.

2.1.2 Loss of Diffuse Highlights

As only scalar intensity values are stored in the corners of each patch no information is provided on how to interpolate between them. The following example will illustrate that even on planar surfaces information is lost.

Example: Imagine a square surface patch illuminated by a single light source. If the source is positioned over the center of the surface patch the intensities calculated for each of the four sample points (located in the corners of the patch) are equal. Linear interpolation yields an uniform "grey" surface. Increasing the number of samples the scenario converges to the correct result where the center of the surface appears brighter than the edges.

[1]named after the Austrian physician Ernst Mach who recognized that: *'Whenever the light-intensity curve of an illuminated surface has a concave or convex flection with respect to the axis of the abscissa, that particular place appears brighter or darker, respectively, than its surroundings'*

The effects on curved surfaces are even worse. This is basically the same reason why highlights can be rendered much better with *Phong* than with *Gouraud shading*.

2.2 The "Perfect" Solution

Choosing the size of a patch equal to the size of a pixel projected on the viewing plane would be a method to avoid Mach bands. We used this method to generate a "reference-image" to measure the quality of our alternative approach but the computation time for determining the radiosities is immense. The large number of patches makes a real-time "walk-through" using hardware accelerators impossible.

Another method for generating a "reference-image" especially in combination with ray tracing would be to store a list of all patches that are emitting light to a certain sample in a list and sum up the light contribution of each patch when the scene is rendered. This also is only a theoretical solution, as even for a moderate sized scene, the storage of a list of all "visible" patches for each sample and the intensity computation for each surface point will be much too expensive.

2.3 Vector Form Factors

When calculating the intensity for a sample point we need to know the amount of energy that is contributed by other patches. For that reason form factors have to be derived. Different methods like the hemicube [2], single–plane [4, 6] ray tracing [9] or analytical solutions [1] to calculate form factors have been presented. In this section we will not introduce another algorithm for the calculation of form factors, but an improved shading method. Therefore the derivation of form factors has to be slightly modified.

Fig. 2. Radiosity vectors.

Our intensity calculation for a sample point due to the radiation of its environment is based on the Wallace method [9] to derive form factors using ray tracing. Assuming that each patch is disk shaped [7], the intensity I_j for a sample point $\mathcal{P}$ due to light emission from patch j is (cf. Fig. 2):

$$I_j = B_j A_j \frac{<N,\bar{\vec{v}}_j><N_j,\bar{\vec{v}}_j>}{\pi r_j^2 + A_j}, \tag{2.1}$$

with

$\vec{v}_j$	vector from the actual sample point to the center of patch j
$r_j = \|\vec{v}_j\|$	distance of the actual sample point to the center of patch j
A_j	area of patch j
B_j	radiosity of patch j
ρ_j	diffuse reflectivity of patch j
I_j	scalar intensity of patch j
$\vec{N}_j$	surface normal of patch j
$<\vec{u},\vec{v}>$	denotes the scalar product of vectors $\vec{u}$ and $\vec{v}$
$\bar{\vec{v}}$	normalized vector $\vec{v}$.

So, the total intensity for $\mathcal{P}$ is the sum ($I = \rho_{\mathcal{P}} \sum I_j$) of all patches that are visible from $\mathcal{P}$.

The basic idea of this article is to split Eqn. (2.1) into two pieces, redefining the intensity I_j via a vector,

$$\vec{I}_j = -\frac{B_j A_j <N_j,\bar{\vec{v}}_j>}{\pi r_j^2 + A_j} \bar{\vec{v}}_j. \tag{2.2}$$

Now the intensity $I_{\mathcal{P}}$ for each sample $\mathcal{P}$ can be interpreted as a sum of scalar products of surface normals $\vec{N}_{\mathcal{P}}$ and vectors $\vec{I}_j$,

$$I_{\mathcal{P}} = \rho_{\mathcal{P}} \sum_j <\vec{I}_j,\vec{N}_{\mathcal{P}}> . \tag{2.3}$$

The intensity vectors for patches that do not emit energy to sample $\mathcal{P}$ are set to zero. To reduce memory usage and computation time Eqn. (2.3) is rewritten as follows:

$$I_{\mathcal{P}} = \rho_{\mathcal{P}} <\vec{G},\vec{N}>, \text{where } \vec{G} = \sum_j \vec{I}_j. \tag{2.4}$$

In addition to the scalar intensities, the intensity vectors $\vec{G}$ are stored for each sample. After all the energy is distributed among the patches these vectors represent a vector field, where each vector is pointing in the major direction of intensity increase. Figure 3 shows a simple configuration of two parallel patches where the vectors $\vec{G}$ have been marked as red arrows. The intensities in each sample point may be retrieved by multiplying $\vec{G}$ with the surface normal.

Interpretation of the Vector Field

In case of a single point light source, energy distribution in space can be interpreted as a vector field. Depending on its position and orientation in the field a differential area receives an amount of energy that is proportional to the scalar product of surface normal and the vector field. Using area sources or multiple point light sources results in a non uniform energy distribution.

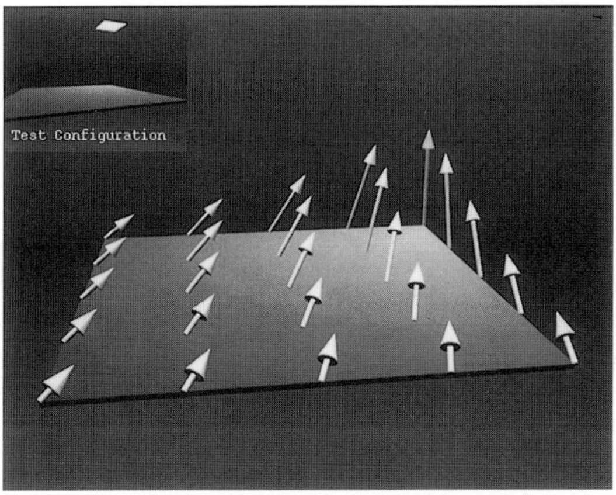

Fig. 3. Vector field.

The following interpolation algorithm exploits this additional information provided by the vectors $\vec{G}$ in each sample to achieve a smooth and more realistic shading than Gouraud shading. The brightness of the sample points remain the same, whereas the way intensity is interpolated between the samples has been modified.

2.4 An Improved Interpolation Technique

The task of a shading algorithm is to calculate the brightness of a point lying within a patch using intensity information that is available only at some specific samples on the area. Similar to the step from *Gouraud* to *Phong shading* one may have the idea to interpolate the vectors $\vec{G}$ and multiply the resulting vector with the surface normal and the diffuse reflectivity of the patch. The following example shows the reason why this idea would not work:

Example: Consider the following 2D scene (see Fig. 4) where a light emitting patch is located over a receiver. In the middle of the receiver the $\vec{G}_i$'s yield an

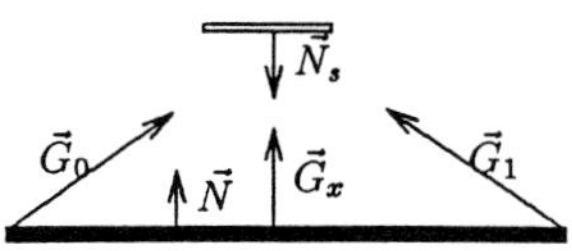

Fig. 4. Phong like shading.

interpolated $\vec{G}_x$ which has the same direction as the surface normal $\vec{N}$. If the vector $\vec{G}_0$ is $(2,1)$ and $\vec{G}_1$ is $(-2,1)$ the interpolated $\vec{G}_x$ in the middle of the receiver will be $(0,1)$. Thus the edge intensities will be $<\vec{N},\vec{G}_i> \rho = 2\rho$ (for $i = 1,2$) and the interpolated intensity is $<\vec{N},\vec{G}_i> \rho = \rho$. This is obviously not very realistic as a light emitter positioned in the center over a receiving patch does not make the edges appear brighter than the middle of the patch. The reason for this behavior is that both, distance and size information are encoded in the vectors $\vec{G}_i$ (see Eqn. (2.4)) which must be interpolated separately.

To solve this problem we propose the following (non unique) solution. As a first step *virtual light sources* for each sample point are introduced. These light emitters have a position, geometry and brightness such that exactly the scalar intensity values stored in each sample are reached. Between the sample points the attributes of the virtual sources are linearly interpolated. Finally the intensity of a single pixel is calculated by evaluating Eqn. (2.3) for a single light emitting patch, namely the virtual light source.

The position of the virtual source is calculated using the vector $\vec{G}$ and an intensity weighted sum of the distances r_v to all the patches whose light is reaching the sample point. For simplicity we assume that $\vec{G}$ and the surface normal $\vec{N}_v$ have the same orientation. The only free parameter in Eqn. (2.3) is A_v. Different values for A_v effect the smoothness of the interpolation. Setting A_v to a huge value, interpolation tends to look like a linear interpolation. In our implementation we use an average patch size, before the adaptive subdivision is done, as an estimate for A_v. Substituting all the known values in Eqn. (2.3) yields

$$I = -\frac{B_v A_v \overbrace{<\vec{N}_v,\vec{\bar{G}}>}^{-1} <\vec{N},\vec{\bar{G}}>}{\pi r_v^2 + A_v}. \tag{2.5}$$

After solving Eqn. (2.5) for B_v, we are able to determine the radiosity of the virtual source,

$$B_v = \frac{I(\pi r_v^2 + A_v)}{\rho A_v <\vec{N},\vec{\bar{G}}>}. \tag{2.6}$$

3 Comparisons

This section will compare our shading method with linear interpolation. The first example will be a relatively simple situation similar to the scene described in Fig. 4.

The second example consists of two orthogonal patches, where the upper one is a light source. In both cases the radiation of the source is 25 and the size is 3×3 (see Fig. 5). Figures 6 show the distributions of intensity over the receiving patches. The dotted lines represent the result using linear interpolation, the solid lines show the distributions using our alternative approach. Sample points are marked through the use of dashed lines. One can see that the intensity values in the sample points are exactly the same. Using more samples both methods converge to the "real distribution". Like Phong shading our method does not guarantee C^1 continuity but in most cases it helps to reduce slope differences of two neighboring patches. Figure A (left) of the color illustrations of this

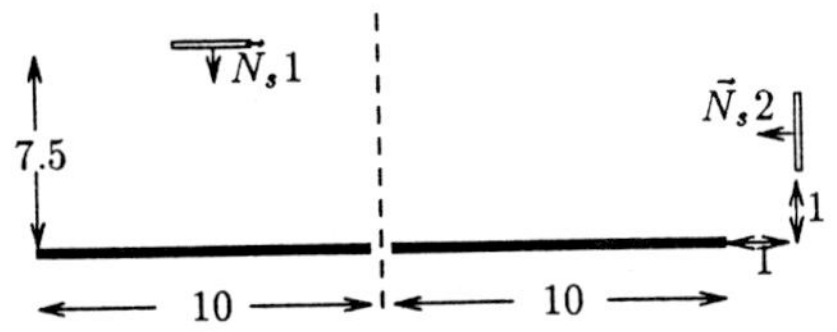

Fig. 5. Example 1 and Example 2.

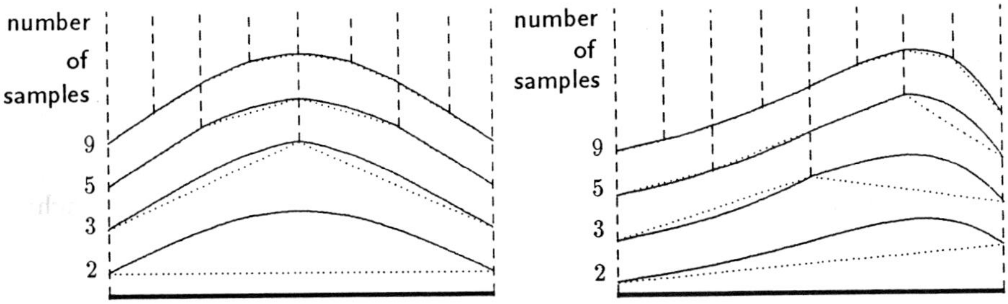

Fig. 6. Intensity distributions of Example 1 (left) and Example 2 (right).

chapter at the end of the book shows ray traced pictures of the scene described in example 1 using different sampling rates and shading methods. Figure B (right) is an image of example 2. The improved method in both cases leads to an image which is very close to the "real situation" using only 25 samples whereas linear interpolation produces Mach bands even if the number of samples is increased to 81.

4 Results

Through the use of the shading method introduced in Sect. 2.3 less patches are needed to achieve a smooth interpolation; Mach band effects are reduced. During the rendering process information on how to interpolate between the sample points is available. Thus, it is possible that a region appears brighter inside, than at the edges, which will never happen using linear interpolation. We recommend this improved shading method for producing images of high quality, necessary for photography or video recording. For previewing we use a hardware polygon renderer with built-in Gouraud shading.

5 Implementation

We implemented the above described algorithm based on a csg-ray-tracer to determine form factors using the progressive radiosity method. Adaptive substructuring of patches into elements is done to improve shadow edges. We noticed that especially for a large image, computation time and memory usage can be reduced quite a bit by using less elements in combination with our improved shading method. Figure B of the color illustrations of this chapter at the end of the book shows two images produced by our ray tracer. No adaptive subdivision has been applied to make it easier to detect the differences of the two shading methods. Using linear interpolation *Mach bands* arise at the edges of the wall and floor patches. Figure B (right) has been generated using the same subdivision as in Fig. B (left) in combination with the above described shading method.

References

[1] D. Baum, H. Rushmeier, M. Winget (1989). *Improving radiosity solutions through the use of anaylitcally determined form factors. Computer Graphics*, 23(3):323–334.

[2] M. Cohen, D. Greenberg (1985). *The hemi-cube – a radiosity solution for complex environments. Computer Graphics*, 19(3):31–40.

[3] H. Gouraud (1971). *Continuous shading of curved surfaces. IEEE Transactions on Computers*, C-20:623–629.

[4] T. Hamid . *The radiosity model.* Project Report.

[5] B. Phong (1975). *Illumination for computer generated pictures. CACM*, 18:311–317.

[6] R. Recker, D. George, D. Greenberg (1990). *Acceleration techniques for progressive refinement radiosity. Computer Graphics*, 24(2):59–66.

[7] R. Siegel, J. Howell (1972). *Thermal radiation heat transfer.* McGraw-Hill Kogakusha, Ltd.

[8] F. Sillion, C. Puech (1989). *A general two-pass method integrating specular and diffuse reflection. Computer Graphics*, 23(3):335–334.

[9] J. Wallace, A. Elmquit, E. Haines (1989). *A ray tracing algorithm for progressive radiosity. Computer Graphics*, 23(3):315–324.

Some Annotations on X-ray Tracing

Alfred Schmitt, Achim Stößer

Abstract: Some extensions of the well-known ray tracing method are proposed which are used to produce images of objects inserted into one another with enhanced visual comprehensibility of shape and relative position of the 3-D objects by simulated radio diagnostics. These extensions are based on tracing a ray called X-ray to generate non-photo-realistic images – i.e. images beyond photorealism – as well as photo-realistic images improved by making use of information that is usually redundant. The underlying geometric model is mainly based on surfaces, not on solids or voxels.

1 Introduction

In standard ray tracing each ray that hits an object may be split into two further rays: the reflected ray and the refracted ray (see Fig. 1). For a further description of ray tracing see (Glassner [1]) or (Leister, Müller, Stößer [3]) for example. As a third ray we introduce the *X-ray* that does not change its direction while penetrating the object; it is a refracting ray with refraction index 1.

X-ray tracing is done to improve the basic ray tracing algorithm to enable the human brain to gain further (especially 3-D) knowledge about the inner ("hidden") architecture of an object without losing significant information about the outer object. Two main problems have to be solved:

- A refracting outer object distorts the inner object. Therefore, transmitted X-rays are used.
- A transparent but non-refracting outer object does not "look 3-D" any more.

Possible applications include: medical imaging (for example intestines in a body, brain in a skull); computer aided design (3-D visualization of complex parts, visualizing complex assemblies of different parts, e.g. machines, cars, etc.); architecture (interior decoration in the context of a building); visualization of a large number of objects distributed in space (atoms in a molecule).

2 Underlying Principle

Foundation for the X-ray tracer is the ray tracer *Vera* (Very Efficient Raytracing Algorithm, see (Schmitt, Müller, Leister [5]) or (Leister, Müller, Stößer [3])). The

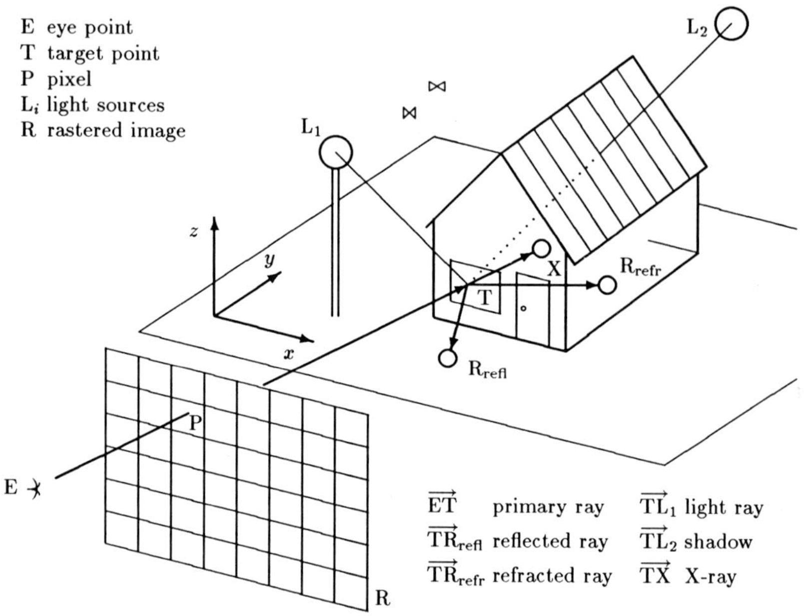

Fig. 1. The ray tracing principle with an X-ray

textual *image description* read by that program mainly consists of the *geometry* of the scene to be modeled, that is the spatial extension of the objects – by compounding simple mathematical primitives like spheres, triangles or cones – and their *properties of material* like diffuse reflection, or attenuation of reflection or refraction. Primitives may be compound to form groups, so-called *subscenes*, which may be reused in the model by applying transformations.

2.1 *Vera* Geometry

The geometry in *Vera* is described by *surfaces of primitives* – spheres, polygons, patches and bodies of revolution – which may be looked at in two ways. First, each surface (of a sphere for example) is like the infinitely thin skin of a soap bubble and has two sides, the inner side and the outside, which may have different materials. A triangle also has two sides, a front side and a back side; one of them may be defined as the inner side. Second, each surface is the border between two space areas; the surface of a sphere for example separates the interior of a water drop from the exterior – the air – with materials differing especially in their refraction index, which is 1.3 for water and approximately 1.0 for air. The same holds for six squares that build a cube of glass. This second way of consideration can append properties of solids to what in fact are only surfaces.

The interior of a sphere is obviously the side close to the center. The inner side of polygons are a matter of definition; in *Vera* an optional *inner point* is used; default is the origin. Patches in *Vera* are triangles with three additional normal vectors at the vertices; the exterior is set by the their direction. The interior of a body of revolution is determined by the relative position of the silhouette line and the rotational axis.

2.2 Color of an Object at the Target Point

To determine the total illumination $I(\vec{r})$ of a first generation target point, in *Vera* the intensities of the object color and the illumination of the recursive rays are added:

$$I(\vec{r}) = C + H + A + G + R_{\text{refl}} + R_{\text{refr}}. \tag{2.1}$$

This sum consists of the object color C, the highlight H, ambient light A, intrinsic gleam G, color of reflected ray R_{refl} and color of refracted ray R_{refr}. The vectors involved ($\vec{r}$, $\vec{n}$, $\vec{l}_i$, $\vec{r}_{\text{refl}}$ and $\vec{r}_{\text{refr}}$) are shown in Fig. 2. The terms of the sum (2.1) are defined by

$$\begin{aligned}
C &= \textstyle\sum_{i=1,\,\delta_i\neq 0}^{n} \left(\vec{l}_i^{\circ}\cdot\vec{n}^{\circ}\cdot C_{\text{d}} + \frac{1+\vec{l}_i^{\circ}\cdot\vec{r}^{\circ}}{2}\cdot C_{\text{a}}\right)\cdot L_i\cdot\ell_i\cdot\frac{1}{|\vec{l}_i|^2}\\
H &= \textstyle\sum^{n}_{\substack{i=1,\,\delta_i\neq 0,\\ \vec{l}_i^{\circ}\cdot\vec{r}_{\text{refl}}\geq h}} \Gamma_i\cdot L_i\cdot\ell_i\\
A &= (\vec{r}^{\circ}\cdot\vec{n}^{\circ}\cdot C_{\text{a}} + C_{\text{d}})\cdot\aleph\\
G &= \vec{r}^{\circ}\cdot\vec{n}^{\circ}\cdot G_{\text{a}} + G_{\text{d}}\\
R_{\text{refl}} &= I(\vec{r}_{\text{refl}})\cdot S_{\text{refr}}\\
R_{\text{refr}} &= I(\vec{r}_{\text{refr}})\cdot S_{\text{refl}}.
\end{aligned}$$

The parameters mainly depend on material and light sources; $|\vec{v}|$ is the length of $\vec{v}$; $\vec{v}\cdot\vec{w}$ is the dot product of the two vectors (with $\vec{v}\cdot\vec{w} = |\vec{v}|\cdot|\vec{w}|\cdot\cos\sphericalangle(\vec{v},\vec{w})$); the exponent $\vec{v}^{\circ}$ denotes normalization. The RGB triples (capital letters) are

- C_{d} diffuse reflection of the material;
- C_{a} angular inclined reflection of the material;
- G_{d} diffuse intrinsic gleam of the material;
- G_{a} angular inclined intrinsic gleam of the material;
- S_{refl} attenuation of specular reflection of the material;
- S_{refr} attenuation of specular refraction of the material;
- $\aleph$ ambient light of the scene;
- L_i color of light source i,

where the brightness of the highlight Γ_i is computed from the light source vector, the reflection ray and the material highlight parameter h. The other scalars used are the luminosity ℓ_i of light source i, the number of light sources n and δ_i which is 0 if there is an object between the light source i and the target point (which means shadow) and 1 otherwise.

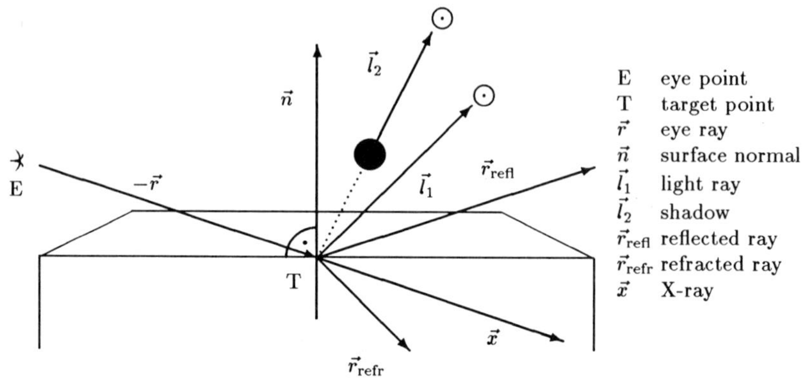

Fig. 2. Vectors at the surface of an object

The most common light source type in *Vera* is the point light source which is defined by its position, its color and its luminosity, but there are several other types that will not be discussed here.

3 X-ray Tracing Stills

Now we replace the term R_{refr} by $X = I(\vec{x}) \cdot S_{refr}$. Setting the refraction index of the transparency ray $r = 1.0$ would cause the same effect. Tracing all three rays – reflection ray, refraction ray and X-ray – is not of any use because it looks like a double exposure of two images, one of which is showing a glassy object, the other one showing an X-rayed object and, therefore, being more puzzling than helpful (see Figs. 1 and 2).

3.1 Enhanced Edges

There are two different types of edges: the border lines of objects, called *profiles* or *silhouettes*, and the lines where two faces meet, called *internal edges*. There are sophisticated methods of generating such edges (see (Saito, Takahashi [4]) for example). We will concentrate upon three *ad hoc* methods here.

First we use the *surface angle* as criterion to enhance at least some parts of a silhouette. The product of eye ray (or another incidence ray) and surface normal is 1 for rays perpendicular to the surface and 0 for parallel rays. Therefore we add $(1 - \vec{r}^\circ \cdot \vec{n}^\circ) \cdot P_1$ to increase the color the flatter the incidence angle. We may also use non-linear functions, getting

$$P = (1 - \vec{r}^\circ \cdot \vec{n}^\circ) \cdot P_1 + (1 - \vec{r}^\circ \cdot \vec{n}^\circ)^2 \cdot P_s + c(1 - \vec{r}^\circ \cdot \vec{n}^\circ) \cdot P_c \qquad (3.1)$$

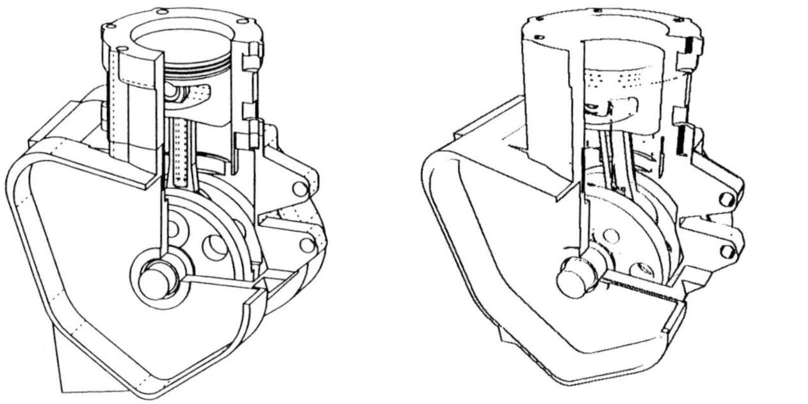

Fig. 3. The engine: CAD presentation (left), Kirsch compass operator plotting (right)

as an additional term of the sum for the total illumination $I(\vec{r})$. The function $c(t) = (1 - \cos(t \cdot \pi))/2$ is used for a smoothed blending (see Leister, Müller, Stößer [3] or Stößer [6]).

In some cases *edge detection* known from pattern recognition can be superimposed to the ray traced image. Another possibility is using the built-in *anti alias* to detect discontinuities because anti-aliasing usually has to be done at edges. Figs. 3 and 11 demonstrate this using the Kirsch compass operator (see (Young, Fu [7]) or (Horn [2])).

3.2 Texture

Another possibility to gain 3-D information on (nearly) invisible objects, e.g. pseudo wire frames, grids, geodetic lines. To generate these we use a special kind of texture. This procedural solid texture is a Boolean function to use two distinct materials on one surface. One of these materials could be transparent or even invisible. Therefore four materials are attached to an object, two on each side. Examples:

- "if $\lfloor\sqrt{x^2+y^2}\rfloor$ is odd then use $material_1$ else use $material_2$" simulates pseudo wood – a stem with equally wide annual rings. For narrower rings an additional parameter n may be used: "if $\lfloor\sqrt{x^2+y^2}\rfloor \bmod n = 0$ then $material_1$ else $material_2$" (Fig. 6);
- "if odd($\lfloor z\rfloor$) then $material_1$ else $material_2$" or "if $\lfloor z\rfloor \bmod n = 0$ then $material_1$ else $material_2$" generate something like contour lines (Fig. 7).

3.3 Miscellaneous Effects

We will only cut into several further effects here:

- depth of field;

- attenuation of medium;
- fog (add white or another color times ray length: $|\vec{r}| \cdot F \cdot f$ where F is the color of the fog and f is a factor to adapt the reduction to the total diameter of the scene);
- depth cue (allow negative RGB values or negative factor for fog, subtract white or another color times eye distance: $-|\overrightarrow{\mathrm{ET}}| \cdot D \cdot f$ where D is the depth cue color and $\overrightarrow{\mathrm{ET}} = \vec{r}$ for the eye ray, or reduce illumination (1) inclined by eye distance: $I(\vec{r}) \doteq I(\vec{r}) \cdot \frac{1}{|\overrightarrow{\mathrm{ET}}|} \cdot f$);
- darkness sources (light sources with negative luminosity: $\ell_i < 0$);
- smooth transition from invisible to visible zones;
- elimination of object parts by half plane intersection or other set operations;
- exploded assemblies via preprocessing;
- stereo images.

3.4 Animated X-ray Traced Images

Some standard animation methods to improve the 3-D impression are:

- smooth fading between two stills showing visible/invisible outer object;
- pumping animation (moving the camera forward and backward);
- moving the eye from left to right and back;
- rotation (encircling the object);
- assembling/disassembling parts.

All these animations can be done cyclically to save calculation time. To avoid jerky changes we use the function $c(t) = (1 - \cos(t \cdot \pi))/2$ as for edge enhancement (Eqn. 3.1). As a test with the engine showed even non-standard animations (piston going up and down and crank-shaft turning) are very useful to gain further 3-D information.

Acknowledgements

We like to thank Bernhard Geiger, INRIA, Sophia Antipolis, France, who reconstructed the brain and the skin from tomographical data, and Dr. Wrazidlo, Universitätsklinik Heidelberg, who provided these data. We also wish to thank Andreas Daberkow, Universität Stuttgart, for the engine data.

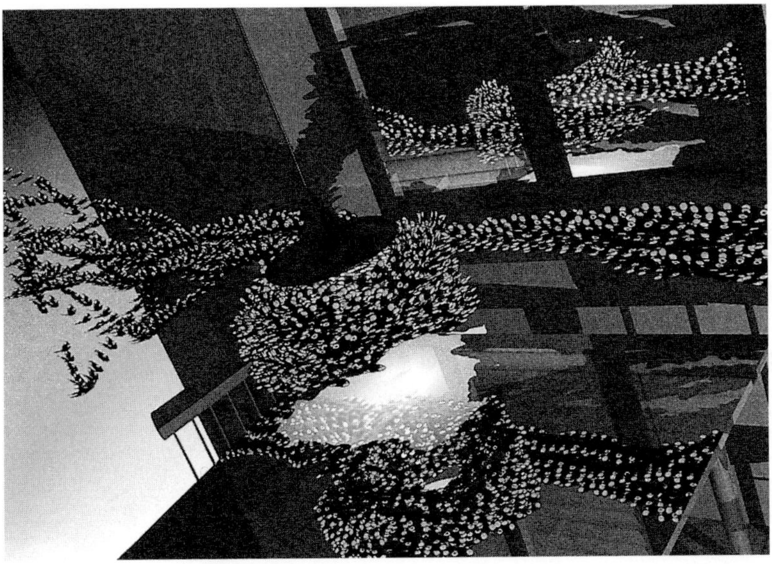

Fig. 4. Swarm of butterflies.

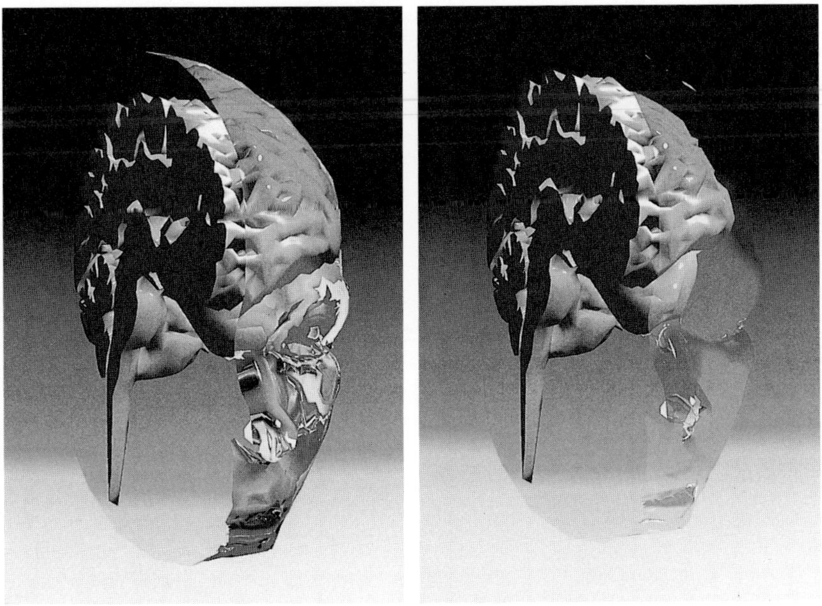

Fig. 5. Brain in a skin made of "water"; X-rayed skin.

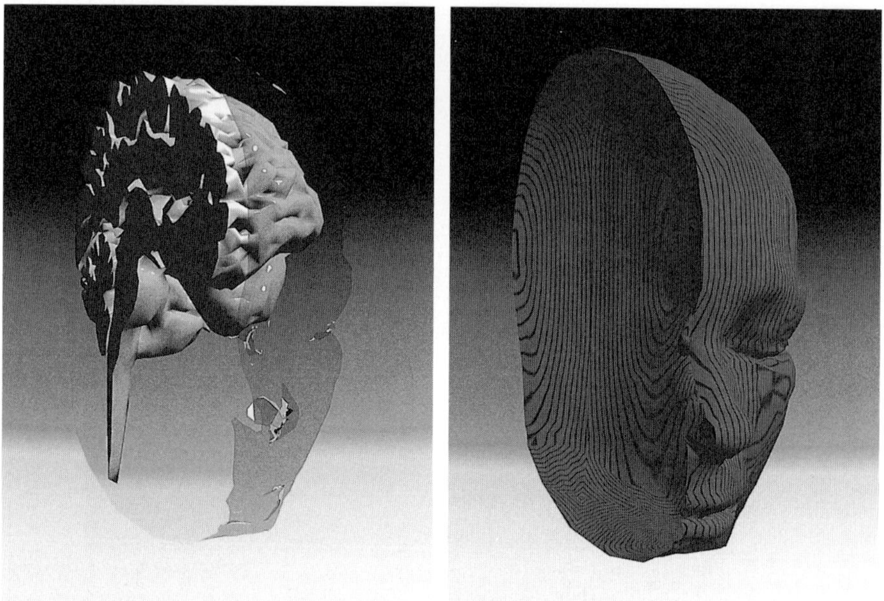

Fig. 6. X-rayed skin with some diffuse color; pseudo wood skin.

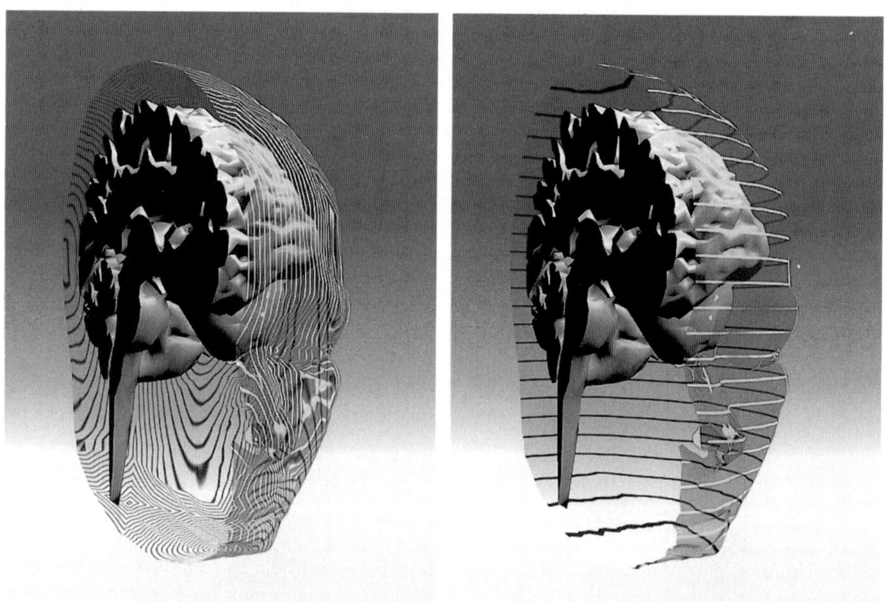

Fig. 7. X-rayed skin with pseudo wood texture; X-rayed skin with contour texture.

Fig. 8. A teapot.

Fig. 9. A teapot made of and filled with water.

Fig. 10. A spooky teapot invisible from the outside.

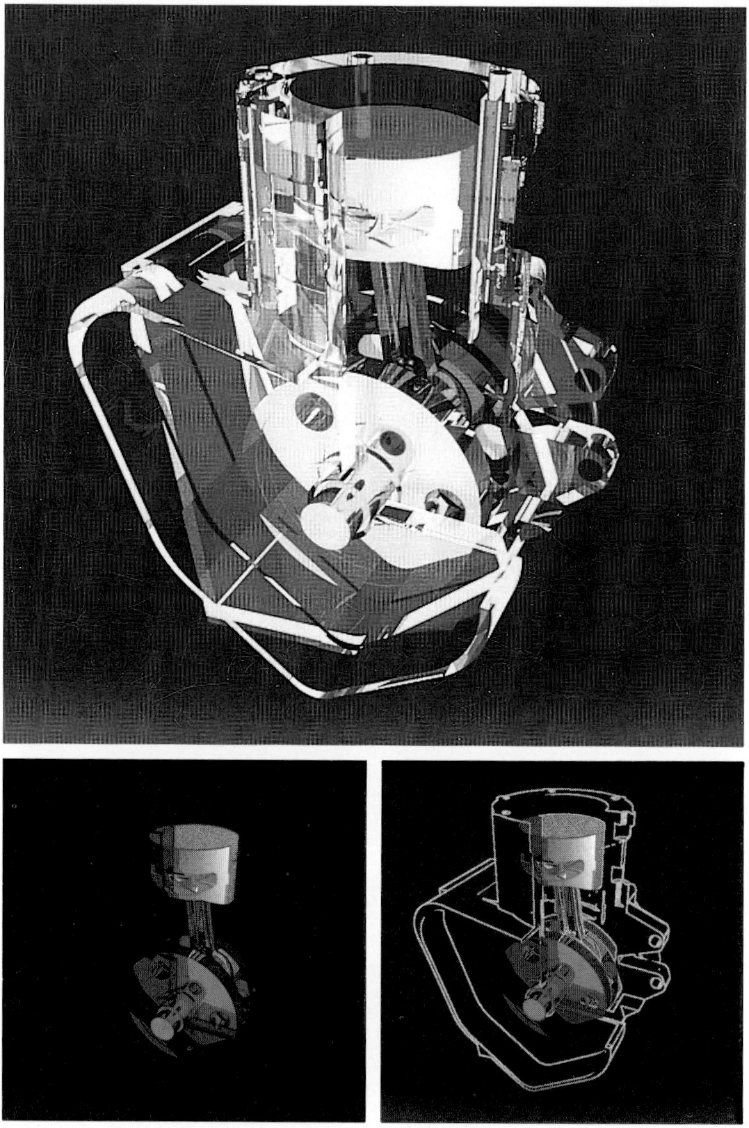

Fig. 11. Engine with glassy housing; X-ray traced engine; engine with superimposed Kirsch operator edges.

References

[1] A.S. Glassner, editor (1989). *An Introduction to Ray Tracing.* Academic Press, New York.

[2] B. Horn (1986). *Robot Vision.* The MIT Press, McGraw Hill.

[3] W. Leister, H. Müller, A. Stößer (1991). *Fotorealistische Computeranimation.* Springer-Verlag, Berlin.

[4] T. Saito, T. Takahashi (1990). *Comprehensible Rendering of 3-D Shapes. ACM Computer Graphics*, 24:197–206.

[5] A. Schmitt, H. Müller, W. Leister (1987). *Ray Tracing Algorithms – Theory and Practice.* In R.A. Earnshaw, editor, *Theoretical Fundamentals of Computer Graphics and CAD*, pp. 997–1030. Springer-Verlag, Berlin.

[6] A. Stößer (1991). *Geometrie- und Animationsmodellierung mit trigonometrischen Funktionen.* In *3. Fachtagung Computeranimation, Tagungsband, Magdeburg*, pp. 60–69.

[7] T. Young, K.S. Fu, editors (1986). *Handbook of Pattern Recognition and Image Processing.* Academic Press.

Auditory Representation of Scientific Data

Stuart Smith

The representation of data in sound is emerging as a complement to data visualization. Several pilot studies over the past decade have proved the concept of auditory data representation; however, there has been little formal research to measure the effectiveness of auditory data representation techniques or to increase our understanding of how they work. Until quite recently, appropriate computing environments for research in this area simply did not exist. This situation is now improving and an increasing number of investigators are seeking solutions to the formidable problems posed by auditory data representation. This paper surveys the present state of the field and outlines the central problems which must be solved if the auditory data representation is to become a truly useful tool for data analysis and exploration.

1 Introduction

Auditory perception provides a perceptual domain orthogonal to the visual domain. Consequently, investigators seeking alternatives to visual display or ways to raise the dimensionality of visual data representations have been attempting to develop effective methods for encoding quantitative information in sound [1, 17, 18, 20, 29, 32, 43]. The idea of using sound to represent data is not in itself new. In fact, a very effective use of sound to analyze seismographic data was reported in 1961 [35]; however, sustained interest in auditory data representation has developed only recently.

The significance of developing auditory displays for exploring scientific data is potentially large. The sheer volume of the production of scientific data warrants development of *any* technique that might help in improving our ability to adequately absorb the data. It is also likely that for some forms of data – certainly for many kinds of time series data – auditory displays will be especially well suited to human perceptual analysis and much more effective than current visual displays. Even in the area of image analysis, where we would presume vision to be the dominant mode of analysis, auditory displays may be helpful as a means of increasing the number of parameters that can be displayed together.

There has been little formal research, either to measure the effectiveness of these techniques in real-world applications or to increase our understanding of how they work and how to improve them. A small number of experiments in auditory data representation has been reported to date. Subjects in these experiments have shown modestly improved performance in a variety of data analysis tasks when using sound in addition to a visual data representation, as compared with their performance when using a visual data representation alone [1, 20, 41, 43]. While these studies have proved the concept of auditory data representation, much work remains to be done to establish the value of auditory data representation as a tool for analyzing and exploring data.

Research in auditory data representation has been severely hampered by the lack of an appropriate computing environment. In particular, work to this point has been constrained by the the limitations of the available sound generation devices, primarily "MIDI" units which were originally designed for *musical* applications. These devices provide the capability to play notes in a variety of familiar sonorities but they offer only limited control over the detailed structure of sound. MIDI devices have facilitated the development of the field to this point (mainly as inexpensive sound output devices for "toy" demonstration systems) but it is unlikely they will play a significant role in the further evolution of auditory data representation. Systematic and quantitative studies of auditory data representation require much more capability for precise manipulation of sounds and the sound displays. Fortunately, the field of computer music is developing environments suitable for work in auditory data representation.

2 Technical Considerations

Auditory data representation is fundamentally a mapping from a data variable to a property of a sonic event. Where the variable is multidimensional, the measure on each dimension of the data sample determines the value of the corresponding sound property of the event, and the values of all the measures together determine the overall character of the event. For example, if the event is a simple tone, one data dimension could be mapped to pitch, another to loudness, and a third to duration. The same approach can be used with sounds more complex than tones and with bursts of band-limited noise.

A variety of displays can be created depending on the kind of data and the kind of analysis to be conducted. Bly [1], for example, presented data as individual tones and had her subjects classify each one, while Mezrich et al. [20] presented simultaneous sequences of tones and had subjects detect correlations among the sequences. The auditory stimuli used in auditory data representation need not be discrete tones. Mansur et al. [18], for example, used a continuously varying pitch to represent the dependent variable in x-y plots of mathematical functions.

Auditory data displays are typically coupled with visual displays of the data. For example, the auditory display developed by the Institute for Visualization and Perception Research at the University of Massachusetts Lowell [12, 11, 21, 33, 34, 32] is an extension of an "iconographic" approach developed for data visualization. In the visual iconographic approach, each datum is represented by an icon whose visual features are controlled by the data. Two of the data variables, not necessarily independent of those controlling the icon features but hopefully largely independent of each other, control the position of each icon on the display surface. With sufficient density, the icons form a surface texture display, and structures in the data are revealed as streaks, gradients, or islands of contrasting texture. Icons also have auditory attributes—pitch, loudness, attack and decay rates, and depth of frequency modulation—which can be controlled by the data. The user triggers the auditory rendition of each icon by moving the mouse cursor over the icon in the graphics display. The user can select various kinds of sonorities (conventional and exotic musical instruments, natural sounds, and purely electronic sounds) and can fine-tune the auditory presentation interactively.

3 Sensory and Perceptual Considerations

While many people find the idea of auditory data representation intriguing, the use of sound in existing auditory data representation systems is not entirely satisfactory. In existing systems, one simply maps data dimensions to available sound parameters arbitrarily, allowing the resulting sounds to have whatever sonic character is determined by the given data values. In fairness to researchers in this field, it should be pointed out that current knowledge of human auditory perception provides little guidance for the design of auditory data representation systems in which several sound attributes are varied simultaneously. Most of the of the basic research on audition has focused on the perception of *single* sound attributes [30, 31] or interactions between two attributes [5, 6, 8, 23, 26, 36, 37, 38, 44]. Investigations of the perception of musical "timbre" [40, 9, 10, 22] have attempted to find a small number of auditory dimensions on which differences in the tonal qualities of sounds can be represented; however, these efforts have not produced a satisfying comprehensive theory of timbre perception. There are currently no timbre models corresponding to the various color models in vision, whose dimensions behave in psychologically simple ways and have straightforward physical definitions. In the absence of such models, auditory data representation workers have relied on intuition and practical experience to guide the design of auditory data representation systems (see, e.g., [2]).

4 Experiments in Auditory Data Representation

As was mentioned above, very little formal research has been done in auditory data representation. Six studies conducted since 1980 are widely cited and constitute the foundation of our current understanding of auditory data representation. This section summarizes these studies.

Yeung [43] explored the use of sound as an alternative to graphic presentation of data. In Yeung's experiment, test subjects were asked to classify mineral samples from four sites in California using 7-dimensional chemical data encoded in sound. Each dimension of the data was mapped to a different property of sound and the resulting tones were presented to the subjects. The subjects were able to classify the samples with 90% accuracy without training, and with greater than 98% accuracy after training.

Lunney and Morrison [16] mapped the high-frequency peaks of infrared spectra onto high-pitched tones, low-frequency peaks onto low-pitched tones, and assigned durations to the tones according to the intensities of the corresponding peaks. The resulting notes were played both sequentially as "melodies" and simultaneously as "chords." In informal tests that required subjects to match tone patterns produced by unknown substances with patterns produced by about a dozen simple organic compounds, subjects rarely failed to match identical patterns.

Bly [1] investigated the use of sound to add dimensionality to a two-dimensional graphic display. For her experiment, Bly synthesized 6-dimensional data samples and separated them into two sets according to a group of inequalities involving all six variables. Test subjects were to discriminate samples from the two sets when the data were presented visually, aurally, and in a combination of both. All six dimensions were mapped to properties of sound, but only two dimensions were mapped to the x-y scatterplot display. Trained subjects averaged 62% correct identification of samples with graphics alone, 64.5% correct with sound alone, and 69% correct with the combined graphics and sound presentation. Several months after the original experiment, subjects who had participated in the sound-only trials were given additional training. In a second experiment, these subjects correctly identified 74% of the test samples.

Mezrich, Frysinger, and Slivjanovski [20] used completely redundant visual and auditory displays to present multivariate time series data. Visually, each variable was represented by a pair of vertical lines which got longer and further apart as the value of the variable increased, shorter and closer together as it decreased. Simultaneously, each variable was represented by a tone whose pitch rose or fell according to the value of the variable. The set of lines and tones corresponding to a point in time was treated as a frame of a movie; the set of frames displayed in sequence at a suitable rate constituted the movie. In the experiment, test subjects had to discriminate correlated and uncorrelated data for four different time-series variables. The data were presented in four different formats: four separate graphs, four graphs aligned vertically, four graphs overlaid on the same pair of axes, and the movie format–four pairs of moving lines with sound accompaniment. The subjects performed this task significantly better with overlaid graphs and the movie technique than with separate graphs and vertically-aligned graphs. For sequences of ten or so frames, the subjects performed significantly better with the movie technique than with overlaid graphs.

Mansur, Blattner, and Joy [18] created "sound graphs" of data using a continuously varying pitch to represent the dependent variable and time to represent the independent variable. Subjects were tested for their ability to recognize features of the graphs such as relative slope, monotonicity, and symmetry about a given axis. After a brief training session, subjects were able to recognize the pertinent feature in between 79% and 95% of the trials.

Williams, Smith, and Pecelli [42, 41] conducted an experiment in which subjects were presented with a display consisting of a grid of patches of visual texture. The textures were composed of many small line segments having variable length and angle relative to vertical. Each patch was created from a data set clustered around a point in four-dimensional space. The center of each cluster was selected at random from one of two regions, A or B, of four-dimensional space. Two data dimensions were mapped to visual attributes (line length and angle); the other two were mapped to auditory attributes (pitch and attack rate). Test subjects were shown representative patches from region A and from region B. They were then asked to classify each patch in the display into region A or region B. They performed this task twice: once using only the visual texture of the patches and a second time using the auditory attributes as well. The subjects' performance in making the classification of patches into region A or region B was significantly better with sound than without.

5 Systems for Research in Auditory Data Representation

Research in auditory data representation requires a computing environment that integrates ordinary facilities such as a general program development environment, a graphical user interface, and a file server, as well as specialized facilities such as a real-time control system to handle user interaction, a real-time sound generation and processing system to generate precisely-specified auditory stimuli, and a set of software tools for creating and manipulating sound. In addition, systems must be provided with software tools designed specifically for used for research in auditory data representation. Among these tools are graphical user interfaces oriented towards specific data analysis tasks, appropriate sound-generating algorithms, and modules which implement the various psychometric testing procedures used to evaluate the effectiveness of auditory displays.

The development of sound-generation and processing systems is a traditional concern of the computer music community. Work on computer music system design has been under way since the late 1950's (see, e.g., [19] and [27]). The most recent efforts in this direction are the IRCAM Musical Workstation (IMW) [15, 14, 25, 24, 39], which is built around the NeXT computer system, and the KYMA/Capybara system [28, 29], which uses a Macintosh computer and a special attached sound processor that connects directly to the Macintosh bus. While neither of these systems was designed specifically with auditory data representation in mind, both are adaptable to this purpose.

The heart of each auditory data representation is a sound synthesis method.A good sound synthesis method for data representation will offer a rich set of control parameters for making perceptually relevant transformations of sound. This will assure that there are many different ways to map data to sound and that variations in data values will be perceived in the auditory representation. The development of such methods has been another central concern of the computer music community. By far the best known and most widely used method is Chowning's "FM" technique [3, 4]. This method is employed in many commercial sound generating products and it was used in the auditory data representation work reported in [32, 34, 33]. FM provides a rich palette of sounds yet is computationally efficient and conceptually simple. Another method that shares these attributes is "VOSIM" [13]. Many other synthesis methods have been reported in the literature and several manufacturers of music synthesizers have their own proprietary techniques. It is not as yet clear which sound synthesis methods are best suited to auditory data representation.

6 Evaluation of Auditory Data Representations

Most of the activity in auditory data representation to date has consisted of writing position papers and creating illustrative demonstrations of data representation techniques. There has been little formal research, either to measure the effectiveness of these techniques in real-world applications or to increase our understanding of how they work and how to improve them. Frysinger [7] has given a concise statement of what needs to be done:

> We must discover the set of truly useful auditory parameters and understand their perceptual transfer functions so that displays can be designed to take advantage of them. Likewise, we need to understand which data analysis tasks can most benefit from Auditory Data Representation, and what types of displays to apply to them. Finally, the interaction between Visual and Auditory Data Representation should be understood so that the best combination of the two can be chosen for a given analysis task. *(p. 136)*

A multidimensional data set can be represented in sound in a great many different ways, so many in fact that it is impossible to try them all out. Consequently, both quantitative and qualitative assessment of auditory data displays are needed at all stages of the development of a new display. During the initial stages of development, quantitative understanding of display requirements is essential to the development of candidate displays for meeting these requirements and for choosing the most useful directions for study. Quantitative assessment should be aimed at obtaining quick and inexpensive figures of merit for exploring among the many possibilities in this large and poorly understood domain. As understanding of the domain accumulates, quantitative assessment can and should be more focused and refined to provide a basis for systematic exploration and for the development of theory and models.

7 Summary and Conclusions

The field of auditory data representation is still in an embryonic state and very little truly substantial work has been done in this area. However, it is now clear that investigators in the field have reached rough agreement on what the significant problems are and how to attack them. Moreover, the the specialized hardware and software needed to support research in auditory data representation is finally beginning to appear. The development of auditory data displays is inherently a multidisciplinary, team-oriented undertaking. In addition to computer scientists and application domain experts, the development of auditory displays requires significant contributions from experts in sound synthesis (often musicians), auditory perception, and psychometric testing methods.

Auditory displays could become a powerful supplement to visual displays, not only of two-dimensional images but also of images in volume. For example, by moving a cursor over an image plane and listening to an auditory rendition of each pixel, or by moving a cursor through a volumetric display and listening to an auditory rendition of each voxel, one may discover whole new capabilities for detecting structure in data. It is also likely that the fields of auditory data representation, database technology, hypermedia, and virtual reality will converge on new modes of data representation that we can scarcely imagine now.

References

[1] S. A. Bly (1982). *Presenting Information in Sound.* In *Proceedings of the CHI '82 Conference on Human Factors in Computer Systems*, pp. 371–375.

[2] W. Buxton (1989). *The Use of Non-Speech Audio at the Interface.* In *Tutorial #10, CHI '89*, pp. 2.1–2.15.

[3] J. Chowning (1973). *The Synthesis of Complex Audio Spectra by Means of Frequency Modulation. Journal of the Audio Engineering Society*, 21(7):526–534.

[4] J. Chowning, D. Bristow (1986). *FM Theory and Applications.* Yamaha Music Foundation, Tokyo.

[5] J. M. Doughty, W. M. Garner (1948). *Pitch Characteristics of Short Tones II: Pitch as a Function of Duration. Journal of Experimental Psychology*, 38:478–494.

[6] H. Fletcher, W. A. Munson (1933). *Loudness, its Definition, Measurement and Calculation. Journal of the Acoustical Society of America*, 5:82–108.

[7] S. P. Frysinger (1990). *Applied Research in Auditory Data Representation.* In *Proceedings of the SPIE/SPSE Conference on Electronic Imaging, 1259*, pp. 130–139.

[8] D. M. Green, T. G. Birdsall, W. P. Tanner (1957). *Signal Detection as a Function of Intensity and Duration. Journal of the Acoustical Society of America*, 29:523–531.

[9] J. M. Grey (1975). *An Exploration of Musical Timbre.* CCRMA/Stanford Department of Music, Stanford, CA.

[10] J. M. Grey (1977). *Multidimensional Perceptual Scaling of Musical Timbres. Journal of the Acoustical Society of America*, 61:1270–1277.

[11] G. Grinstein, R. M. Pickett, M. G. Williams (1989). *EXVIS: An Exploratory Visualization Environment.* In *Graphics Interface '89, London, Ontario.*

[12] G. Grinstein, S. Smith (1990). *The Perceptualization of Scientific Data.* In *Proceedings of the SPIE/SPSE Conference on Electronic Imaging, 1259*, pp. 190–199.

[13] W. Kaegi, S. Tempelaars (1978). *VOSIM—A New Sound Synthesis System. Journal of the Audio Engineering Society*, 26(6):418–425.

[14] E. Lindemann, M. de Cecco (1991). *Animal: Graphical Data Definition and Manipulation in Real Time. Computer Music Journal*, 15(3):78–100.

[15] E. Lindemann, F. Dechelle, B. Smith, M. Starkier (1991). *The Architecture of the IRCAM Musical Workstation. Computer Music Journal*, 15(3):41–49.

[16] D. Lunney, R. C. Morrison (1981). *High Technology Laboratory Aids for Visually Handicapped Chemistry Students. Journal of Chemical Education*, 58(3):228–231.

[17] D. Lunney, R. C. Morrison (1990). *Auditory Presentation of Experimental Data.* In *Proceedings of the SPIE/SPSE Conference on Electronic Imaging, 1259*, pp. 140–146.

[18] D. L. Mansur, M. M. Blattner, K. I. Joy (1985). *Sound Graphs: A Numerical Data Analysis Method for the Blind. Journal of Medical Systems*, 9(3):163–174.

[19] M. V. Mathews, J. R. Pierce, editors (1989). *Current Directions in Computer Music Research.* Cambridge, MA: MIT Press.

[20] J. J. Mezrich, S. Frysinger, R. Slivjanovski (1984). *Dynamic Representation of Multivariate Time Series Data. Journal of the American Statistical Association*, 79(385):34–40.

[21] R. M. Pickett, G. G. Grinstein (1988). *Iconographic Displays for Visualizing Multidimensional Data.* In *Proceedings of the 1988 IEEE Conference on Systems, Man and Cybernetics, Beijing and Shenyang, People's Republic of China.*

[22] R. Plomp (1970). *Timbre as a Multidimensional Attribute of Complex Tones.* In *Frequency Analysis and Periodicity Detection in Hearing.* Sijthoff, Leiden, Netherlands.

[23] R. Plomp, M. A. Bouman (1959). *Relation Between Hearing Threshold And Duration of Pulses. Journal of the Acoustical Society of America*, 31:749–758.

[24] M. Puckette (1991). *Combining Event and Signal Processing in the MAX Graphical Programming Environment. Computer Music Journal*, 15(3):68–77.

[25] M. Puckette (1991). *FTS: a Real-Time Monitor for Multiprocessor Music Synthesis. Computer Music Journal*, 15(3):58–67.

[26] W. Reichardt, H. Niese (1970). *Choice of Sound Duration and Silent Interval for Test and Comparison Signals in the Subjective Measurement of Loudness. Journal of the Acoustical Society of America*, 47:1083–1090.

[27] C. Roads, J. Strawn, editors (1985). *Foundations of Computer Music.* Cambridge, MA: MIT Press.

[28] C. Scaletti (1989). *The Kyma/Platypus Computer Music Workstation. Computer Music Journal,* 13(2).

[29] C. Scaletti (1991). *Using Sound to Extract Meaning From Complex Data.* In *Proceedings of the SPIE/SPSE Conference on Electronic Imaging, 1459,* pp. 207–219.

[30] B. Scharf, S. Buus (1986). *Audition I: Stimulus, Physiology, Thresholds.* In K. R. Boff, L. Kaufman, J. P. Thomas, editors, *Handbook of perception and human performance 1,* pp. 14.1–14.71. New York: Wiley.

[31] B. Scharf, A. J. M. Houtsma (1986). *Audition II: Loudness, Pitch, Localization, Distortion, Pathology.* In K. R. Boff, L. Kaufman, J. P. Thomas, editors, *Handbook of perception and human performance 1,* pp. 15.1–15.60. New York: Wiley.

[32] S. Smith, R. D. Bergeron, G. Grinstein (1990). *Stereophonic and surface sound generation for exploratory data analysis.* In *Proceedings of CHI '90, Seattle, WA.*

[33] S. Smith, G. Grinstein, R. D. Bergeron (to appear). *Interactive Data Exploration with a Supercomputer.* In *Proceedings of Visualization '91, San Diego, CA, 1991.*

[34] S. Smith, G. Grinstein, R. M. Pickett (1991). *Global Geometric, Sound, and Color Controls for the Visualization of Scientific Data.* In *Proceedings of the SPIE/SPSE Conference on Electronic Imaging, 1459,* pp. 192–206.

[35] S. D. Speeth (1961). *Seismometer Sounds. Journal of the Acoustical Society of America,* 33:909–916.

[36] S. S. Stevens (1935). *The Relation of Pitch to intensity. Journal of the Acoustical Society of America,* 6:150–154.

[37] E. Terhardt (1974). *Pitch of Pure of Tones: Its Relation To Intensity.* In E. Zwicker, E. Terhardt, editors, *Facts and Models in Hearing.* Springer-Verlag, New York.

[38] J. Verschuure, A. A. van Meeteren (1975). *The Effect of Intensity on Pitch. Acustica,* 32:33–44.

[39] E. Viara (1991). *CPOS: A Real-Time Operating System for the IRCAM Musical Workstation. Computer Music Journal,* 15(3):50–57.

[40] G. von Bismarck (1974). *Timbre of Steady-State Sounds: A Factorial Investigation of Its Verbal Attributes. Acustica,* 30:146–159.

[41] M. G. Williams, S. Smith, G. Pecelli (1989). *Experimentally driven visual language design: texture perception experiments for iconographic displays.* In *Proceedings of the IEEE 1989 Visual Languages Workshop, Rome, Italy,* pp. 62–67.

[42] M. G. Williams, S. Smith, G. Pecelli (1990). *Computer-Human Interface Issues in the Design of an Intelligent Workstation for Scientific Visualization. ACM SIGCHI Bulletin,* 21(4):44–49.

[43] E. S. Yeung (1980). *Pattern Recognition by Audio Representation of Multivariate Analytical Data. Analytical Chemistry*, 52(7):1120–1123.

[44] J. J. Zwislocki (1969). *Temporal Summation of Loudness. Journal of the Acoustical Society of America*, 46:413–441.

Color Illustrations

This section contains the color illustration of the previous chapters. The references to literature refer to the bibliographies at the end of the respective chapter.

1 Fluid Flow Visualization

F.H. Post, Th. van Walsum

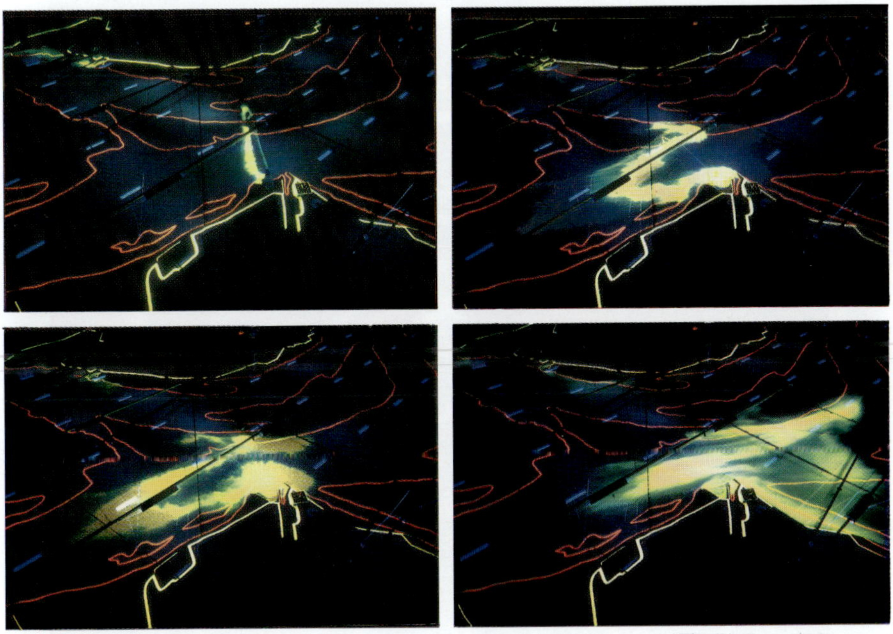

A

Fig. A: Visualization with dye to study water flow in a river model. (Courtesy Delft Hydraulics)

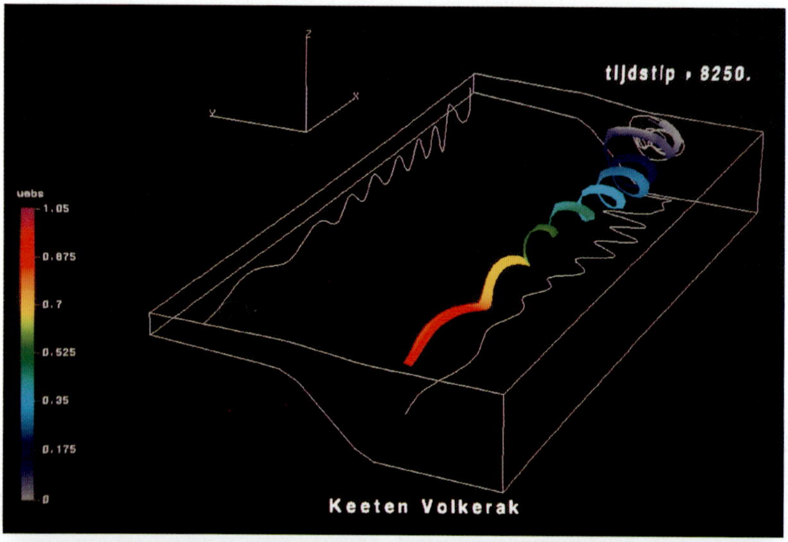

B

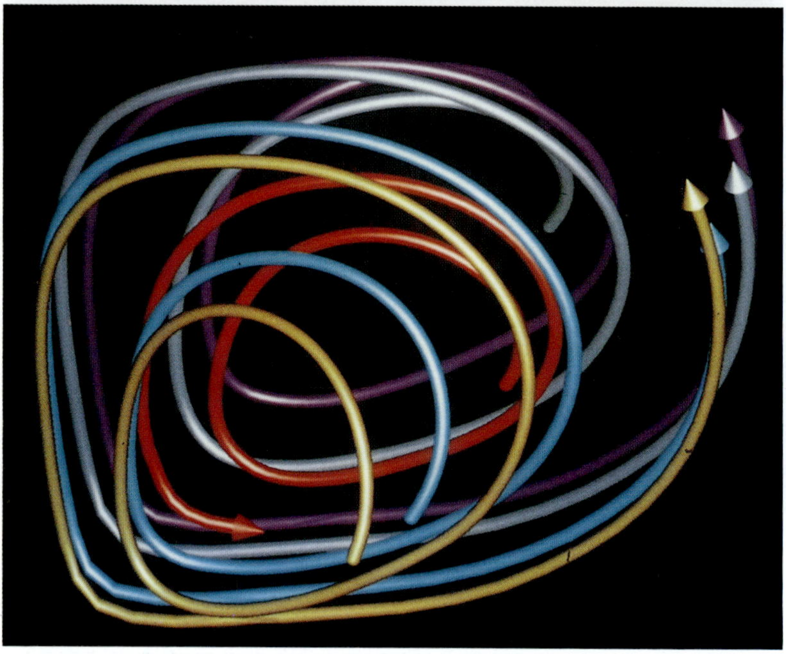

C

Fig. B: Spatial curves with projection in a coordinate plane.

Fig. C: Spatial curves rendered as pipes. (Geiben and Rumpf [17])

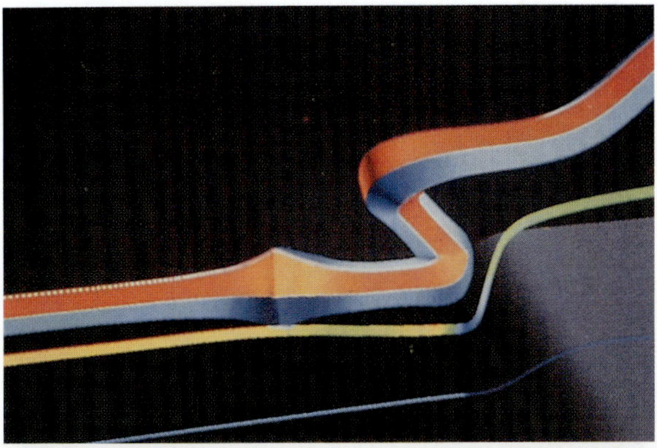

D

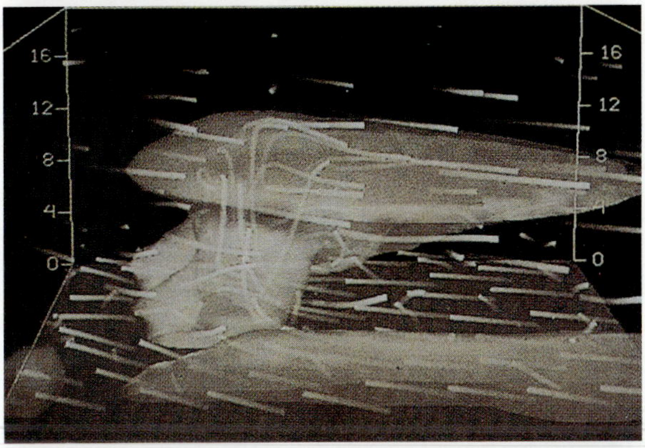

E

F

Fig. D: A stream line, ribbon, and stream tube.
(Schroeder et al. [48], ©1991, IEEE Computer Society Press. By permission)

Fig. E: Particles with tails.
(Hibbard [31], ©1989, IEEE Computer Society Press. By permission)

Fig. F: Environment geometry supports spatial orientation. (Hin et al. [32])

G

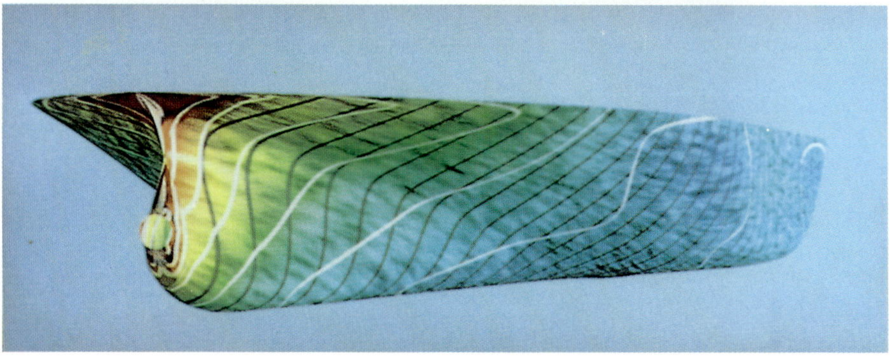

H

Fig. G: Contour lines on a curved surface. (Van Wijk [61])

Fig. H: Spot noise texture showing velocity on a ship hull. (Van Wijk [62])

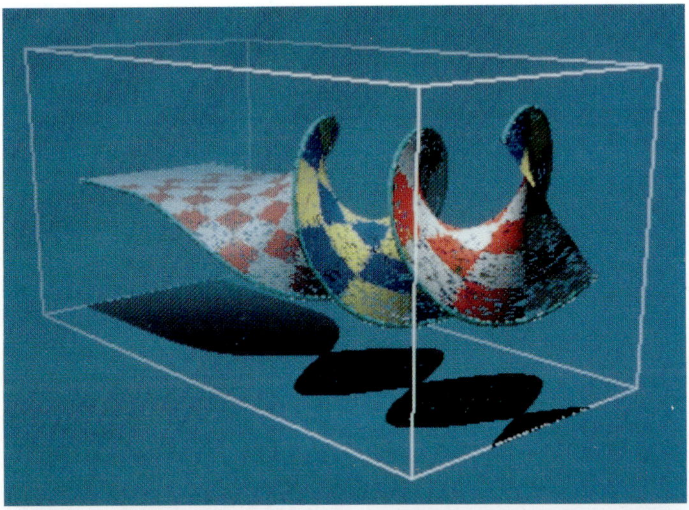

I

J

Fig. I: Surface-particles. (Stolk and Van Wijk [52])

Fig. J: Transparent contour surfaces.
(Gallagher [15], ©1989, Association for Computer Machinery, Inc. By permission)

2 Volume Visualization in Medicine: Techniques and Applications

A. Pommert, M. Bomans, M. Riemer, U. Tiede, K.H. Höhne

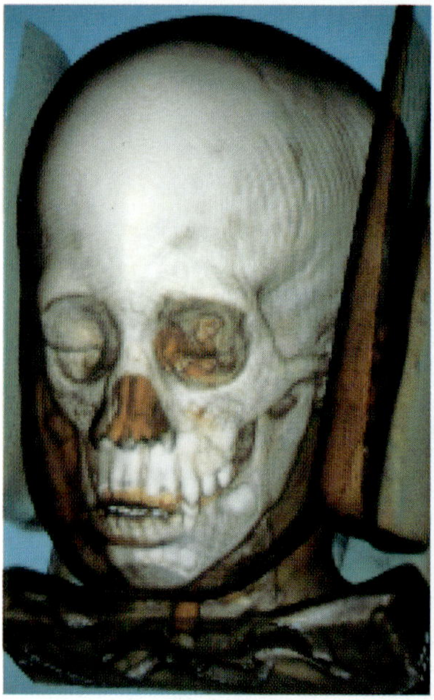

A

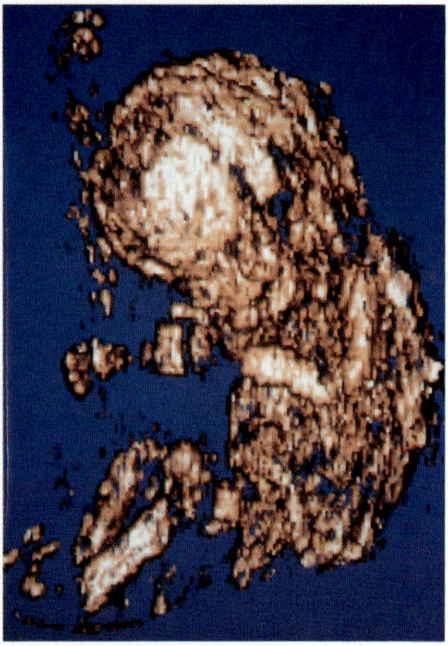

B

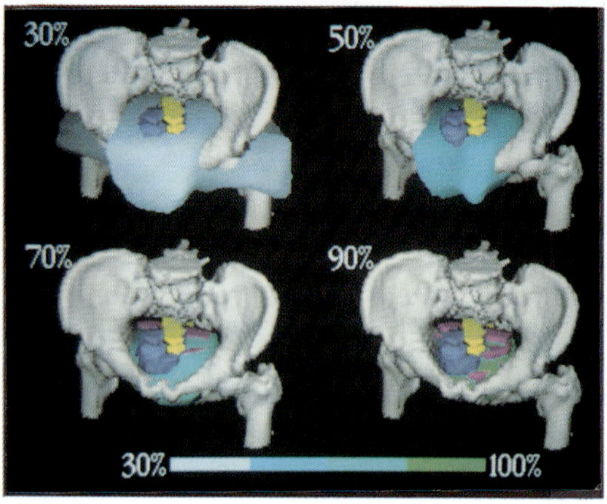

C

Fig. A: Volume rendered image of a child with a trigonocephalon from CT. Semi-transparent visualization shows a lot of different aspects, but spatial perception is quite difficult.

Fig. B: Volume rendered image of a 12 week old fetus from ultrasonography (in-vivo). The image shows some suprisingly fine details such as arms, legs and fontanelles.

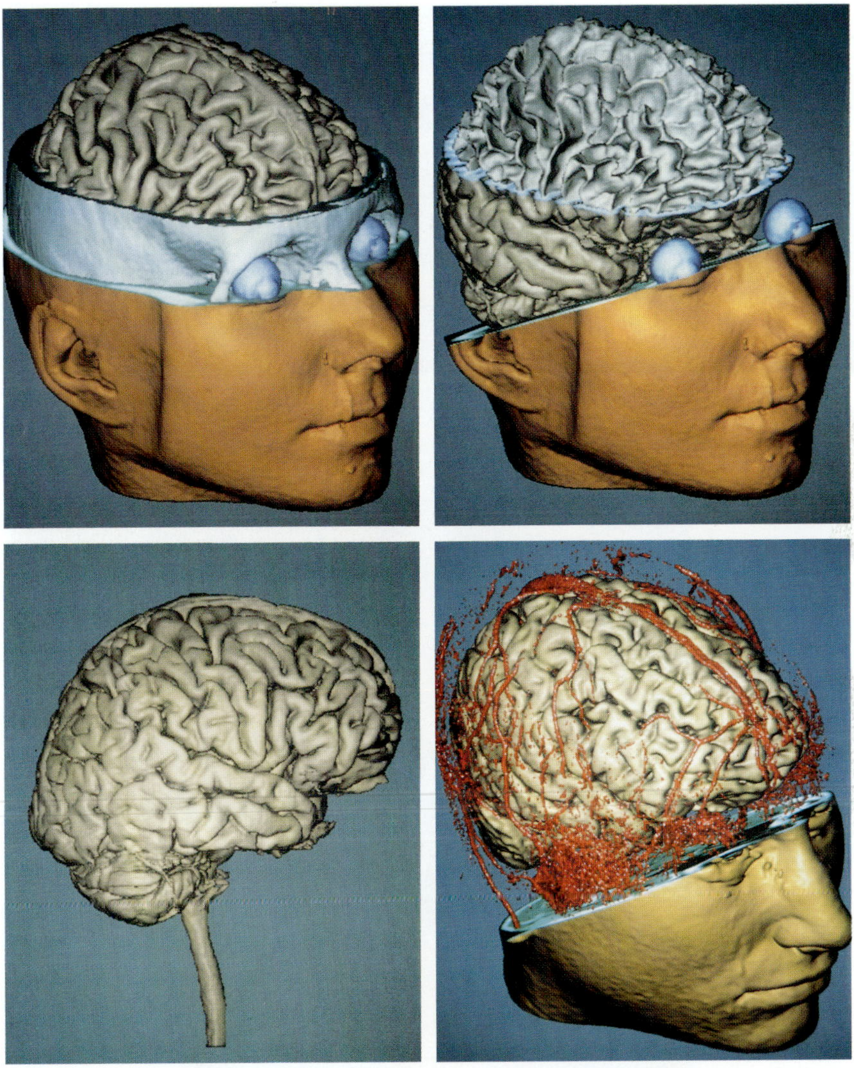

D

◁ **Fig. C:** 3D radiotherapy planning of a prostate. Pink: target volume, blue: bladder, yellow: rectum (all interactively outlined by a radiologist), green: dose volume, white: bone from CT. Numbers are indicating the visualized isodose values.

Fig. D: 3D images of a head from MRI. To right: white matter unveiled. Bottom left: brain in a lateral view. Bottom right: combined display of brain from MRI with blood vessels from MR angiography.

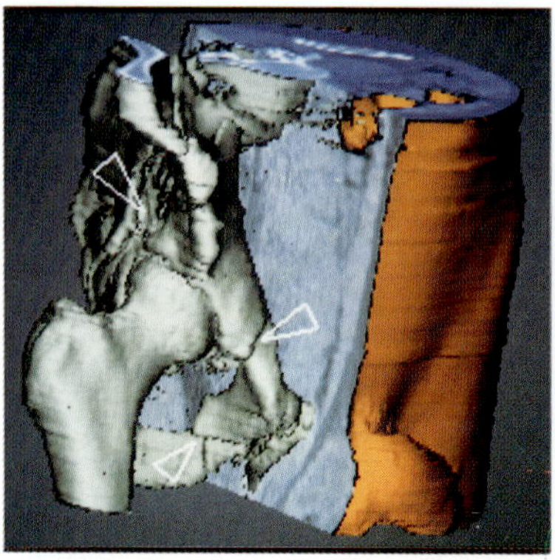

E

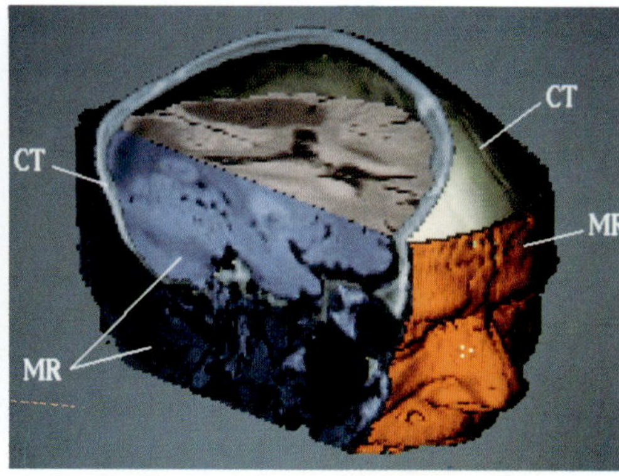

F

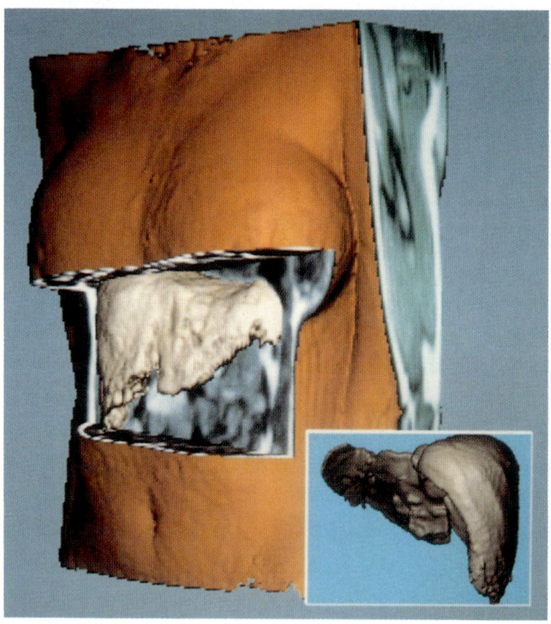

G

Fig. E: Pelvis with multiple fractures (arrows) from CT.

Fig. F: 3D views of a head with bone from CT and soft tissue from MRI. The data volumes were registered using a 3D point specifier for interactive definition of corresponding landmarks.

Fig. G: 3D image of a liver within its anatomical surroundings and a dorsal view or the excised liver from MRI. The liver was interactively segmented using morphological operations.

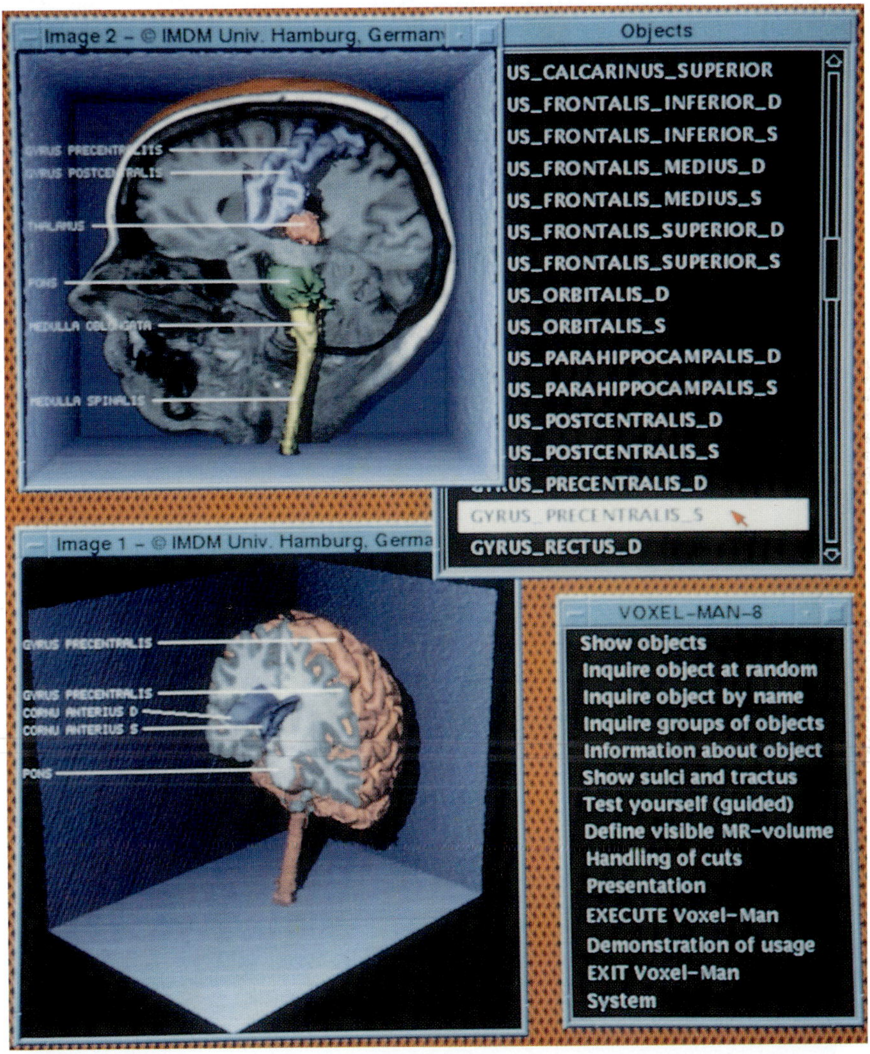

H

Fig. H: Anatomy teaching by dissection at the computer. The volume may be arbitrarily rotated and cut. By pointing to the visible surface, the selected region is highlighted and automatically annotated. The other way round, objects to be displayed can be selected from an alphabetical list.

3 Application of Visualization in Environmental Protection

R. Denzer

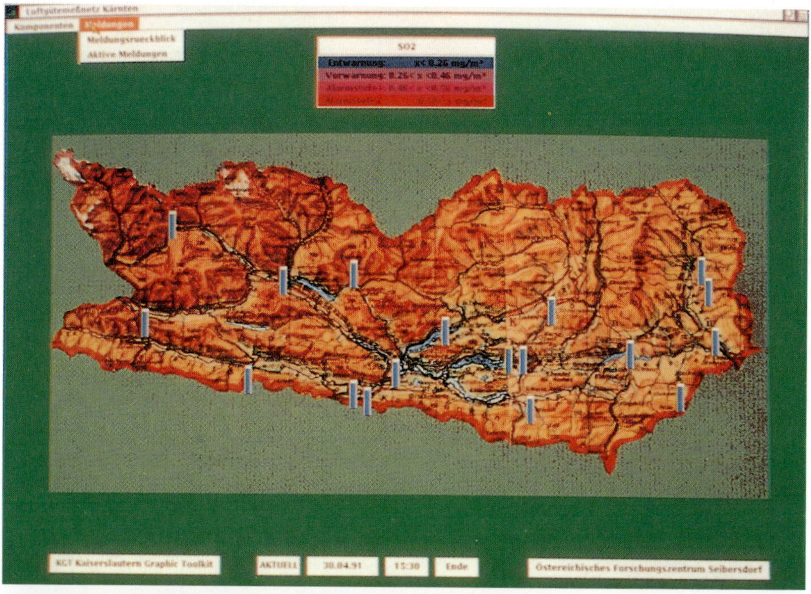

A

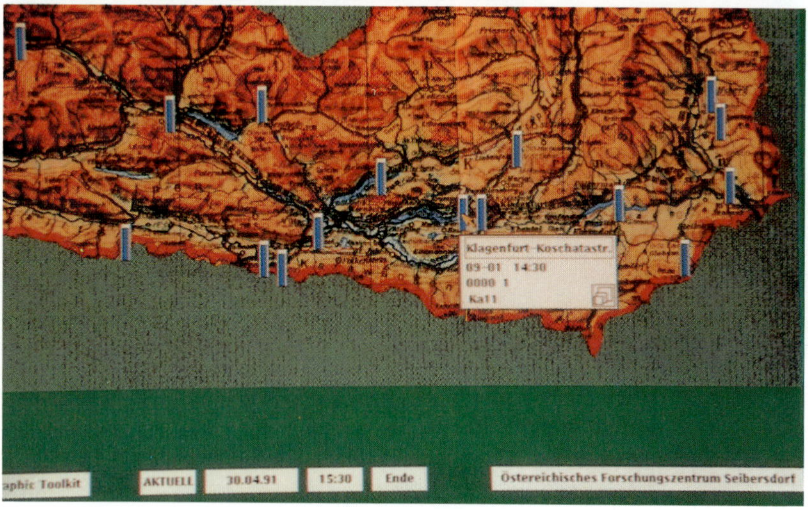

B

Fig. A: User interface for network visualization.

Fig. B: Interaction with the measurement station.

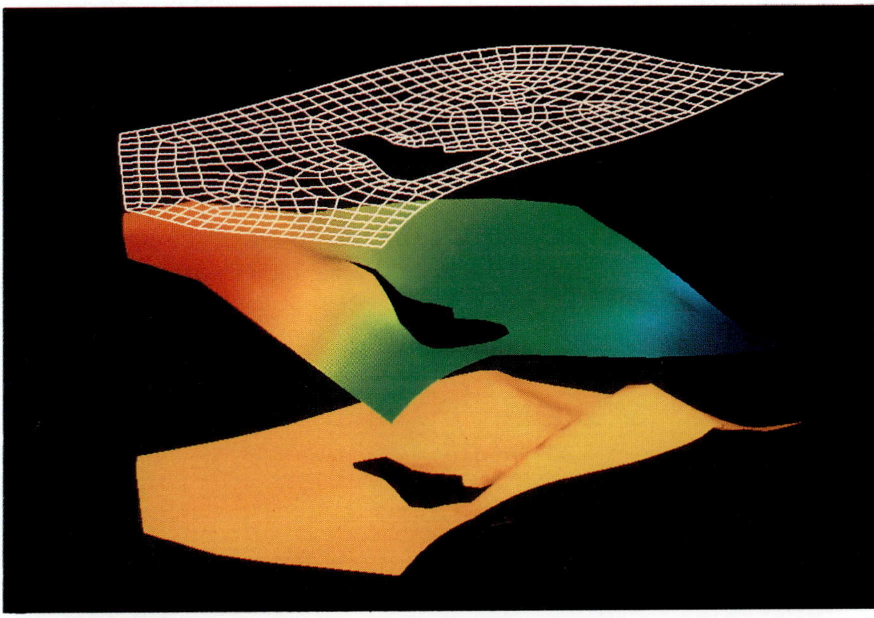

C

Fig. C: Groundwater protection using finite elements.
(Courtesy Joanneum Research, Graz, Austria)

4 Data Structures in Scientific Visualization

U. Lang

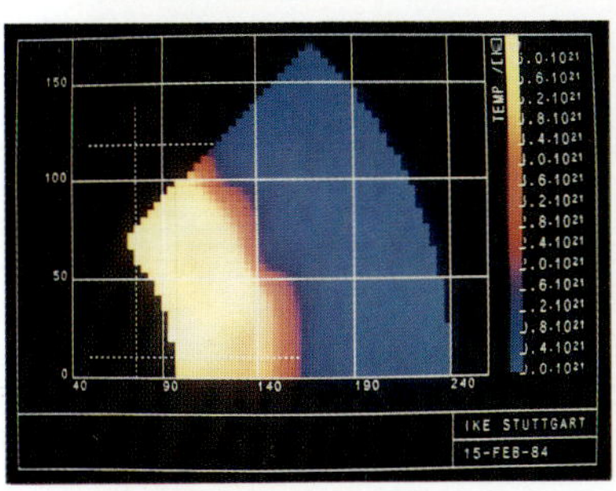

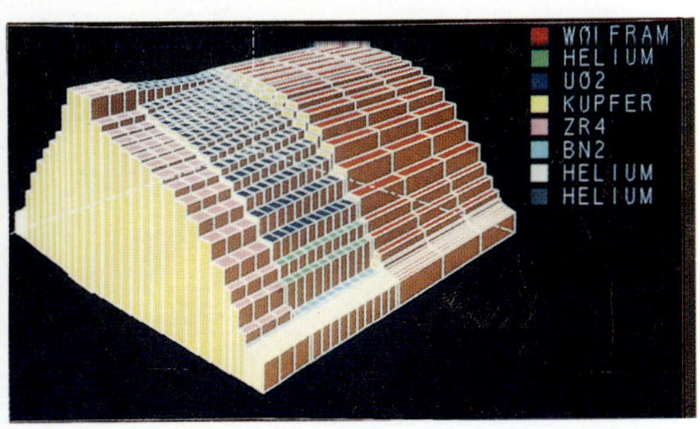

A, B

Fig. A: Display of temperatures and material zones.
Fig. B: Zone information to describe boundaries.

5 A Visualization-Based Model for a Scientific Database System

T.M. Sparr, R.D. Bergeron, L.D. Meeker

This section shows examples of iconographic data displays. All the pictures represent displays that simultaneously present satellite data containing 5 different wavelength measurements from the Eastern end of the Great Lakes region. The first three examples are all area-based icons that were generated using the Flexvis system from the University of New Hampshire (Behari [3]; Calder [6]).

A

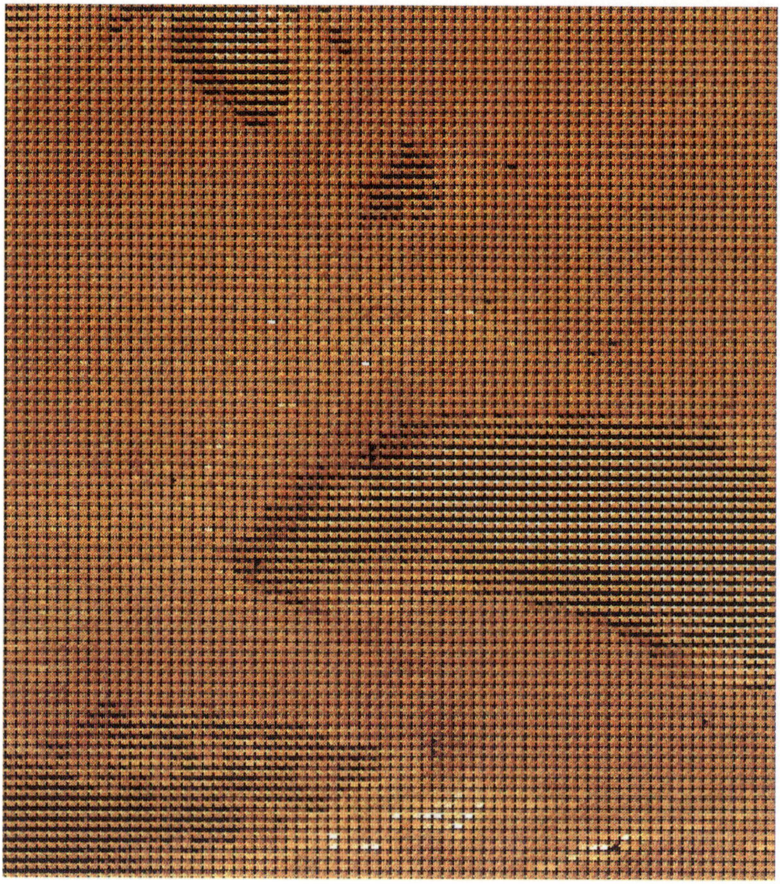

Fig. A: Example 1: Pie icon – heated object color map. Each sample point is represented by a small pie chart in which each of the 5 parameters is displayed in a wedge of the pie using color variations to represent the data value of that parameter based on the *heated object* color map (Pizer [35]).

B

Fig. B: Example 2: Pie icon – RB color map. This image shows the same data and same icon, but with a different color map. Note the distinctly different textural effects in various areas of the display.

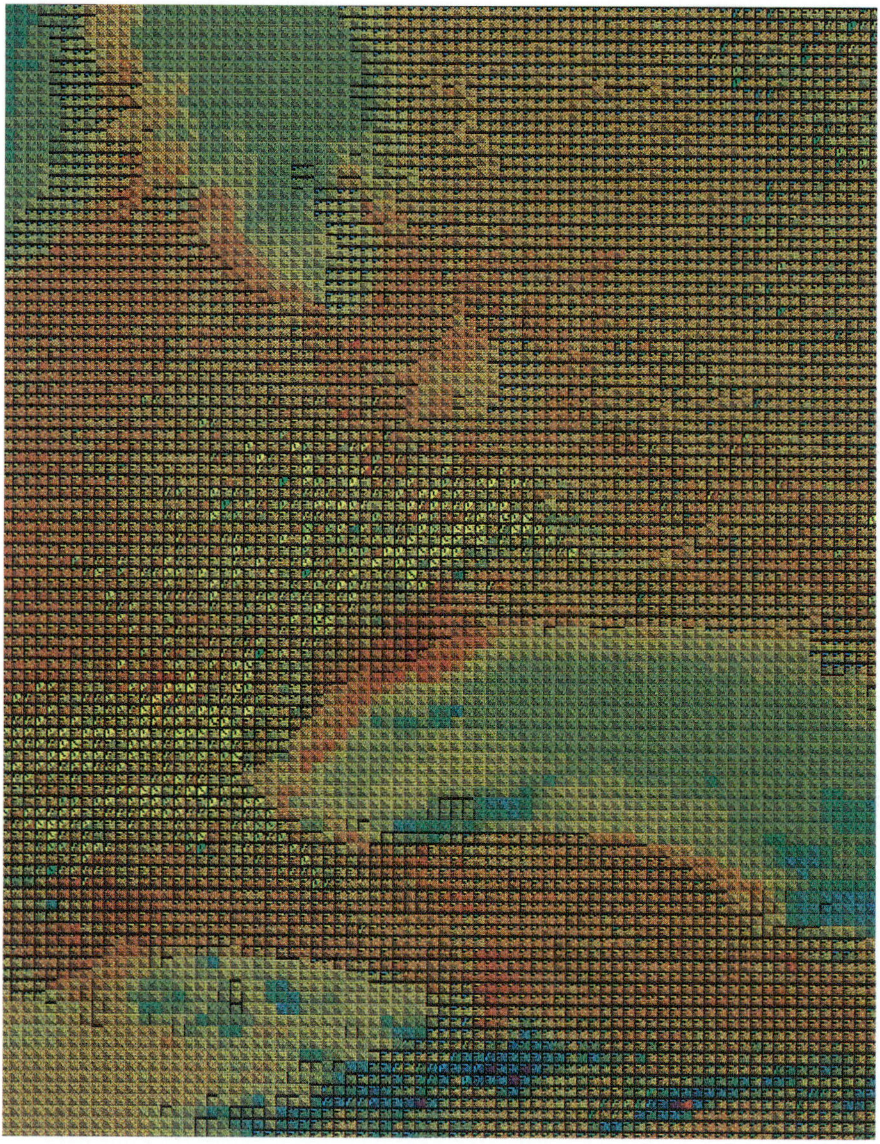

C

Fig. C: Example 3: Pyramid icon. This display uses an icon that simulates a small four-sided pyramid. Input data determines the position of the peak of the pyramid within the square area it occupies, producing different size and shape pyramid sides. The pyramid sides are colored based on the data values also.

D

Fig. D: Example 4: Stick icon. This display uses one of a family of stick figure icons developed by Pickett and Grinstein at the University of Lowell and implemented in the Exvis visualization system (Grinstein and Pickett [17]; Pickett and Grinstein [34]).

6 Modeling and Visualizing Volumetric and Surface-on-Surface Data

G.M. Nielson

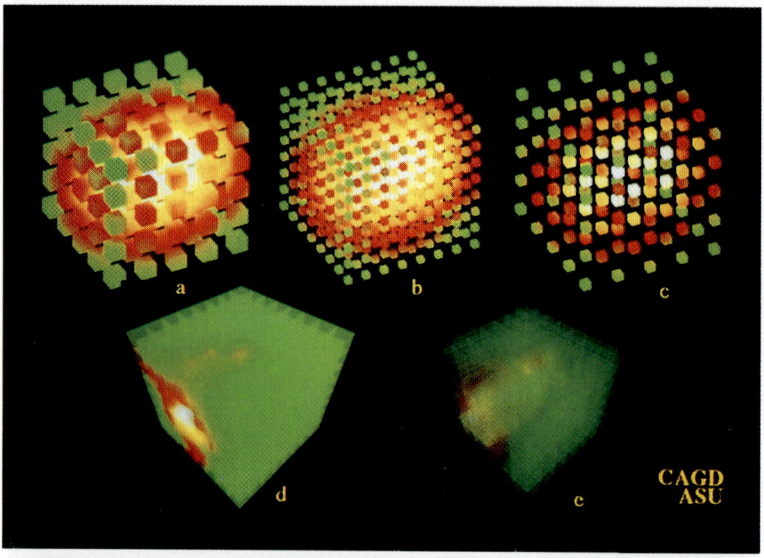

A

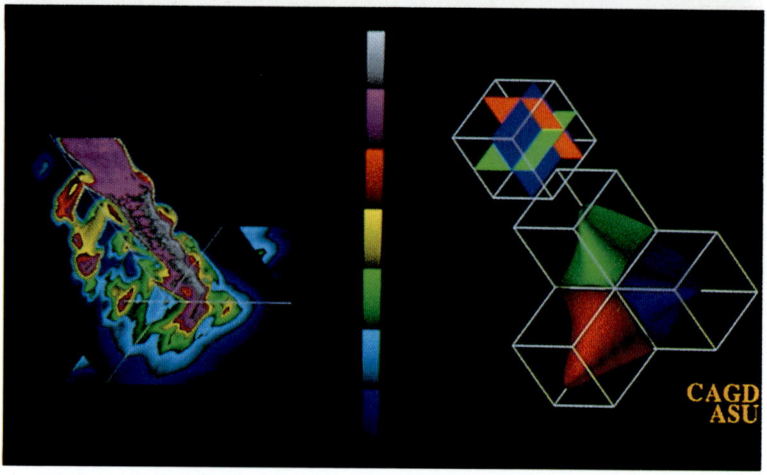

B

Fig. A: Top three images are the tiny cube method and the bottom two images are the vanishing cube method. For image b, d and e, $N_x = N_y = N_z = 8$ and for image a and c, $N_x = N_y = N_z = 5$. $M = 1, 2, 3$ for images a, b and c, respectively. $t = 0.5$ for image d and $t = 0.95$ for image e.

Fig. B: Slice methods.

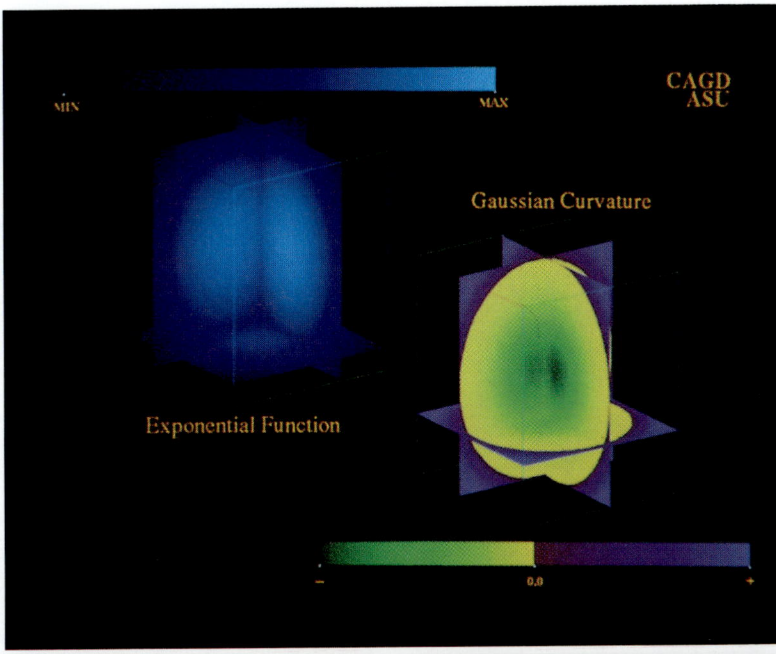

C

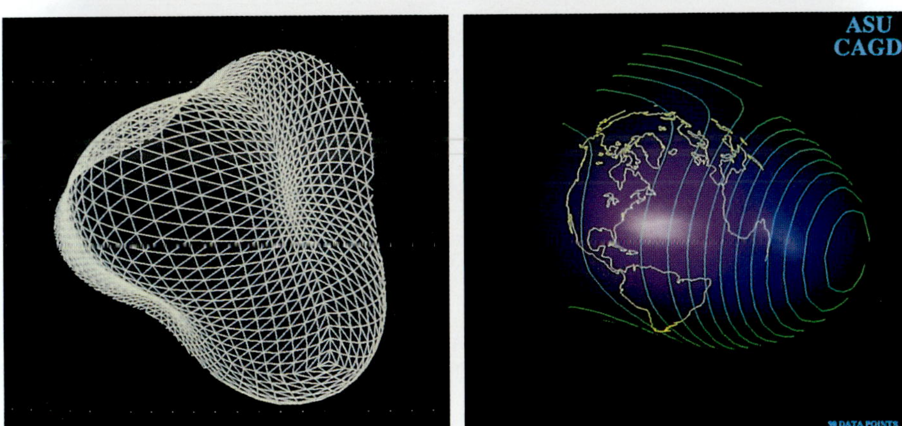

D

Fig. C: Volume interrogation tool with graph of original function on the left and Gauss-Kronecker curvature on the right.

Fig. D: Projected surface graph rendered as a wireframe surface (left) and as smooth shaded surface (right).

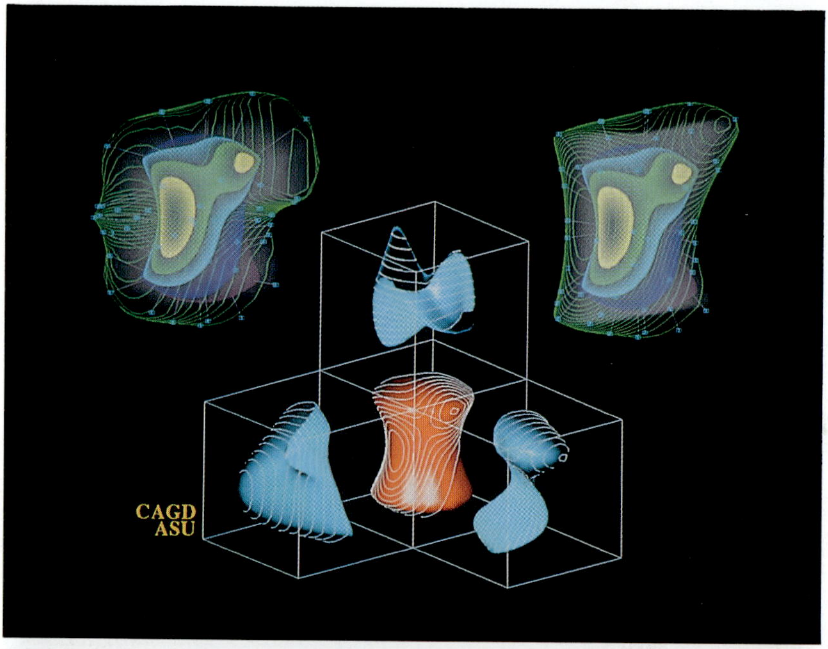

E

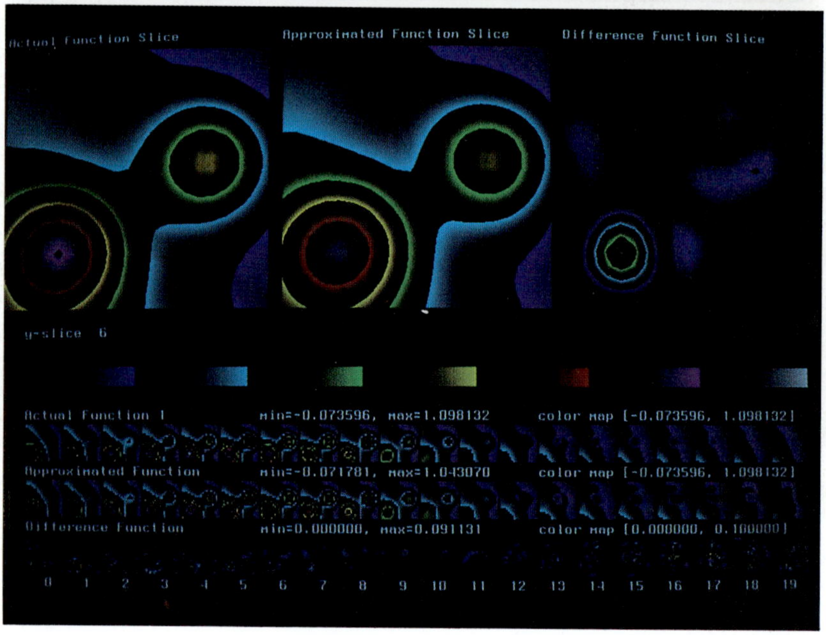

F

Fig. E: Hypersurface projection graph.

Fig. F: Distance function method for volumetric data.

7 Curve and Surface Interrogation

H. Hagen, St. Hahmann, Th. Schreiber,
E. Gschwind, B. Wördenweber, Y. Nakajima

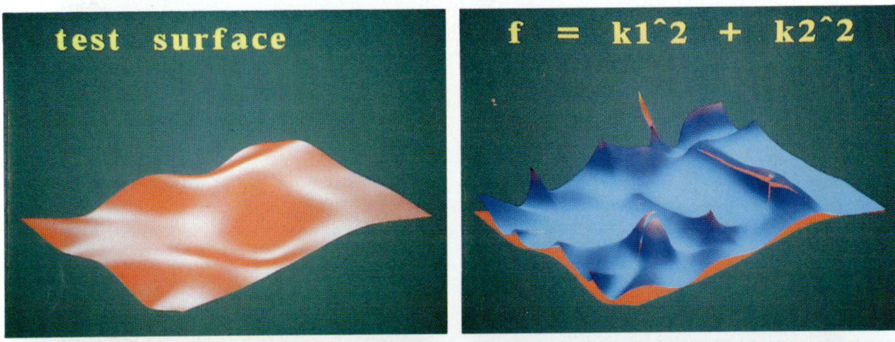

A

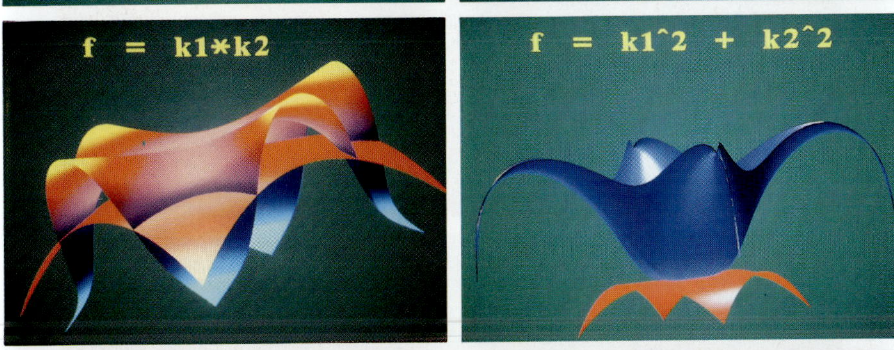

B, C

D

Fig. A: G^2-discontinuity of the test surface.

Fig. B: Convexity test of a bicubic surface.

Fig. C: Visualization of a flat point.

Fig. D: Focal analysis of smoothed surfaces.

8 Sorting for Polyhedron Compositing

N.L. Max

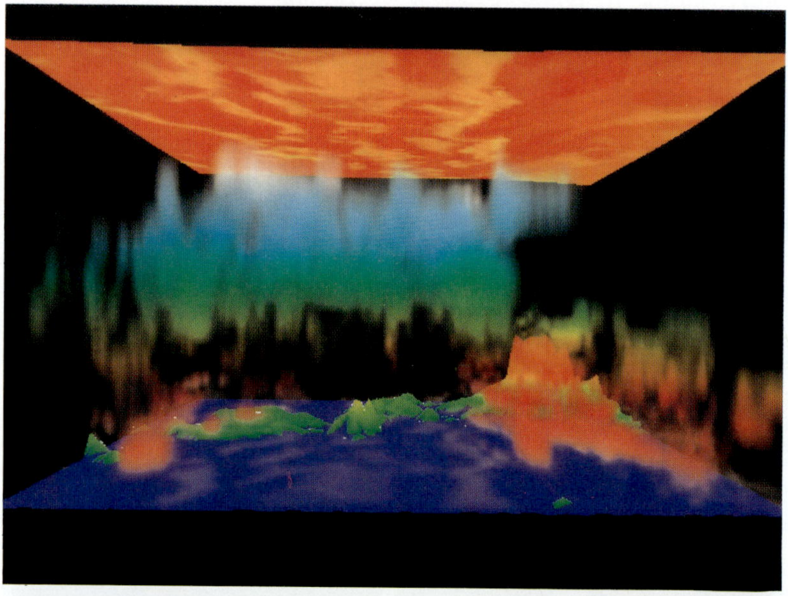

A

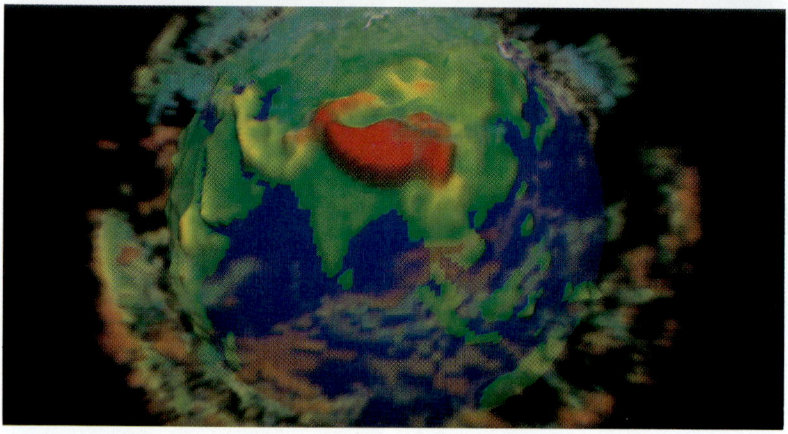

B

Fig. A: A rectangular map with the clouds over Indonesia colored by an altitude scale, and upward long-wave radiation indicated by colors on a plane above the top of the atmosphere.

Fig. B: A cloud with a different assignment of cloud colors to altitude, on a round globe, at 1920 by 1035 HDTV pixel resolution.

9 Joining Volume with Surface Rendering

M. Frühauf

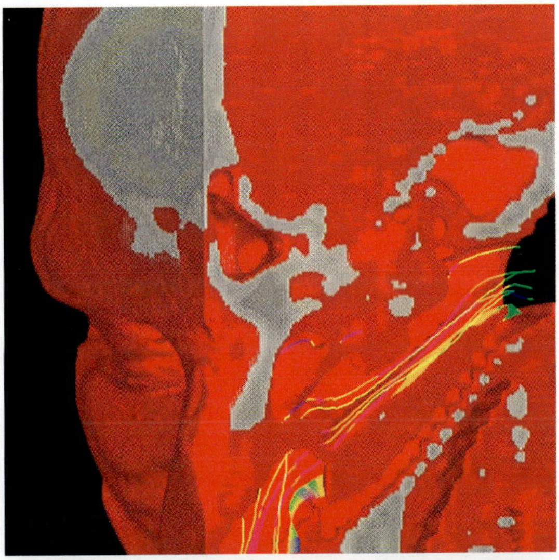

A

B

Fig. A: A human head with a cut plane through the left side near the nasal septum. Particle paths start at the top right and are visualized using polylines.

Fig. B: A semi-transparent visualization of a 3D ultrasonic inspection of a metallic cube. A T-formaed pipeline has been drilled into the cube.

10 An Improved Shading Algorithm for Radiosity Based Renderers

P. Jacob

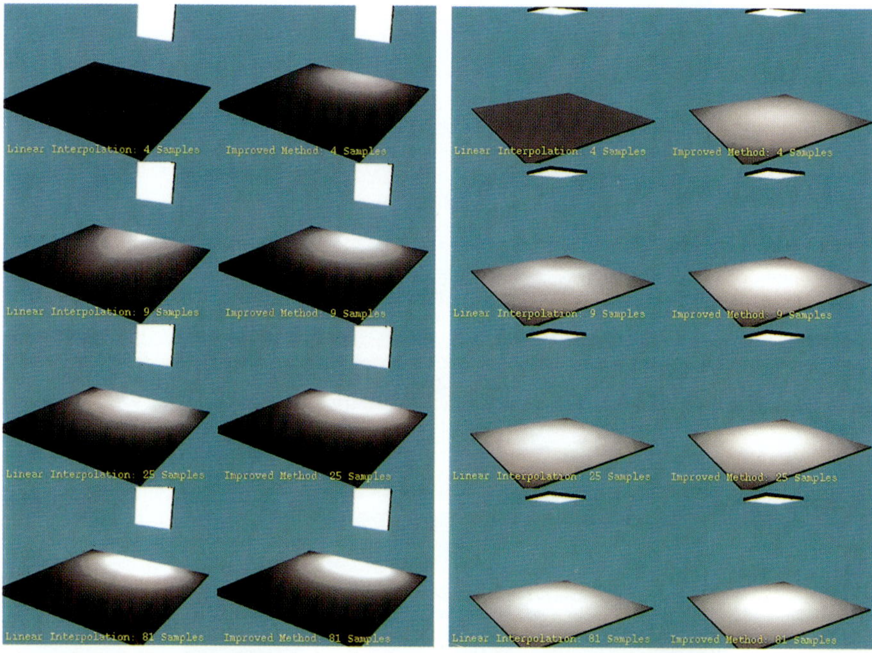

A

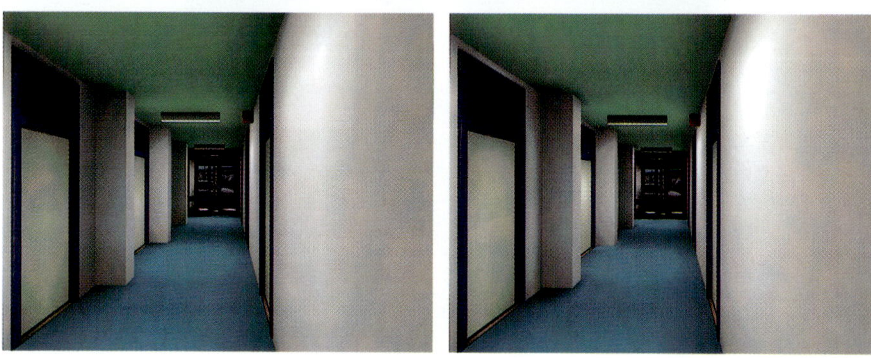

B

Fig. A: Examples 1 (left) and 2 (right).

Fig. B: Pictures generated by linear interpolation (left) and by the improved method (right).

Index